# Python
# 项目开发实战

陈　强◎编著

清华大学出版社
北　京

## 内容简介

本书通过 12 个大型项目的实现过程展示了开发 Python 项目的方法和流程。全书共 12 章，分别讲解了 AI 人机对战版五子棋游戏(AI+pygame 实现)，在线商城系统(Django+Mezzanine+Cartridge 实现)，房产价格数据可视化分析系统(网络爬虫+ MySQL+pylab 实现)，招聘信息实时数据分析系统(网络爬虫+Flask+Highcharts+MySQL 实现)，基于深度学习的 AI 人脸识别系统(Flask+OpenCV-Python+Keras+Sklearn 实现)，在线生鲜商城系统(Django+Vue+新浪微博账号登录+支付宝支付)，民宿信息可视化分析系统(网络爬虫+Django+Echarts 可视化)，实时疫情监控系统(腾讯 API 接口+Seaborn+matplotlib 实现)，个人博客系统(Flask+TinyDB 实现)，电影票房数据可视化系统(网络爬虫+MySQL+Pandas 实现)，大型 3D 枪战类冒险游戏(Panda3D 实现)，AI 人脸识别签到打卡系统(PyQt5+百度智能云+OpenCV-Python+SQLite3 实现)。

本书适合了解 Python 语言基础语法并希望进一步提高 Python 开发水平的读者阅读，可以作为大中专院校相关专业的师生用书和培训机构的专业教材。



**图书在版编目(CIP)数据**

Python 项目开发实战/陈强编著. —北京：清华大学出版社，2021.1(2022.2 重印)
ISBN 978-7-302-57286-2

Ⅰ. ①P… Ⅱ. ①陈… Ⅲ. ①软件工具—程序设计 Ⅳ. ①TP311.561

中国版本图书馆 CIP 数据核字(2021)第 005913 号

**责任编辑**：魏 莹
**装帧设计**：李 坤
**责任校对**：王明明
**责任印制**：杨 艳
**出版发行**：清华大学出版社
**网　址**：http://www.tup.com.cn, http://www.wqbook.com
**地　址**：北京清华大学学研大厦 A 座　**邮　编**：100084
**社 总 机**：010-62770175　**邮　购**：010-62786544
**投稿与读者服务**：010-62776969, c-service@tup.tsinghua.edu.cn
**质量反馈**：010-62772015, zhiliang@tup.tsinghua.edu.cn
**印 装 者**：三河市龙大印装有限公司
**经　销**：全国新华书店
**开　本**：185mm×260mm　**印　张**：21.5　**字　数**：523 千字
**版　次**：2021 年 3 月第 1 版　**印　次**：2022 年 2 月第 2 次印刷
**定　价**：79.00 元

---

产品编号：090088-01

# 前言

## Python 语言的重要性

Python 是目前国内外使用最广泛的程序设计语言之一，也是近年来最流行的编程语言之一。其清晰的语法和可读性使其成为初学者的完美编码语言。Python 具有功能丰富、表达能力强、使用方便灵活、执行效率高、可移植性好等优点，几乎可用于所有领域。

在当今科技界，大数据、AI 人工智能、数据分析是最火热的三大研究和应用领域，而 Python 是公认的开发大数据、AI 人工智能和数据分析程序的最优编程语言。

## 本书的特色

(1) 以项目为单位，每个项目都是独立的整体。

书中精心挑取了作者的经典项目案例，每个项目案例都是一个独立的整体，详细讲解了每个项目从开始到运行的完整过程。

(2) 扫二维码观看视频讲解和下载本书源代码。

我们不但提供了每个项目的完整源代码，而且还提供了每个项目具体实现过程的视频讲解，读者可通过扫描标题后面的二维码获取在线视频讲解，还可扫描下方的二维码获取全书案例源代码。

扫码下载全书案例源代码

(3) 每个实例都是精心挑选的典型代表。

书中的实例都是最典型的，涵盖了最主要、最常见的领域，并且包含了各种类型的应用。每个实例都代表了作者一个时期的成果，都极具代表性；每个实例都是作者的人生转折，发自肺腑。在讲解实例过程中，还展示了各个层次的实现技巧，为读者日后的亲身实践起到指引的作用。

(4) 结合图表，通俗易懂。

在项目讲解过程中，给出了相应的图表进行说明，方便读者领会其含义；对于复杂的程序，结合程序的具体实现流程进行讲解，方便读者理解程序的执行过程；在语言的叙述上，普遍采用短句子、易于理解的语言，避免晦涩难懂的语句。

## 本书的内容

每章一个完整的项目案例，共详细讲解了 12 个案例的具体实现过程

| 章节内容 | 技术 |
| --- | --- |
| 第 1 章介绍开发 AI 人机对战版五子棋游戏的具体流程。 | AI+pygame |
| 第 2 章介绍使用第三方框架快速开发在线商城系统的具体流程。 | Django+Mezzanine+Cartridge |
| 第 3 章介绍开发房产价格数据可视化分析系统的具体流程。 | 网络爬虫+ MySQL+pylab |
| 第 4 章介绍基于网络爬虫技术开发一个招聘信息实时数据分析系统的具体流程。 | 网络爬虫+Flask+Highcharts+MySQL |
| 第 5 章介绍基于深度学习技术开发一个 AI 人脸识别系统的具体流程。 | Flask+OpenCV-Python+Keras+Sklearn |
| 第 6 章介绍开发一个大型在线生鲜商城系统的具体流程。 | Django+Vue +新浪微博账号登录+支付宝支付 |
| 第 7 章介绍基于网络爬虫技术开发一个民宿信息可视化分析系统的具体流程。 | 网络爬虫+Django+Echarts 可视化 |
| 第 8 章介绍基于腾讯 API 接口开发一个实时疫情监控系统的具体流程。 | 腾讯 API 接口+Seaborn+matplotlib 实现 |
| 第 9 章介绍基于数据库技术开发一个个人博客系统的具体流程。 | Flask+TinyDB 实现 |
| 第 10 章介绍基于网络爬虫技术开发一个电影票房数据可视化系统的具体流程。 | 网络爬虫+MySQL+Pandas 实现 |
| 第 11 章介绍使用第三方框架开发一个大型 3D 枪战类冒险游戏的具体流程。 | 使用 Panda3D 技术实现 |
| 第 12 章介绍基于百度 AI 接口开发一个 AI 人脸识别签到打卡系统的具体流程。 | PyQt5+百度智能云+OpenCV-Python+SQLite3 |

## 致谢

本书由陈强编著，在编写本书的过程中，我们始终本着科学、严谨的态度，力求精益求精，但疏漏之处在所难免，敬请广大读者批评指正。感谢清华大学出版社各位编辑，是他们的严谨和专业才使得本书顺利出版。

最后感谢您购买本书，希望本书能成为您编程路上的领航者。

编　者

目录

# 第 1 章

# AI 人机对战版五子棋游戏

## (AI+pygame 实现)

五子棋是智力运动会竞技项目之一，是一种两人对弈的纯策略型棋类游戏。通常双方分别使用黑白两色的棋子，下在棋盘直线与横线的交叉点上，先形成五子连线者获胜。本章将详细讲解使用 AI 技术开发人机对战版五子棋游戏的过程。

## 1.1 项目介绍

经过近百年的规则演化，传统五子棋逐步复杂化、规范化，同时也成为国际上的竞技棋种之一，流行于亚欧地区。随着五子棋的发展，人们逐步发现先行方优势巨大，并得出“先行方必胜”的结论。五子棋要成为竞技类棋牌运动，必须解决传统五子棋下法中“先行方必胜”的问题。为此人们对传统五子棋的规则进行改良，先是引入禁手规则，但是禁手规则并不能完全平衡黑白棋之间的差距，黑棋依旧是必胜的。于是，人们又引入了交换行棋权等一系列新规则，禁手则作为连珠棋的特色被保留下来。传统的五子棋最终形成了现在的可以公平竞技的现代连珠棋。

在国内，五子棋是全国智力运动会竞技项目之一，是一种两人对弈的纯策略型棋类游戏。五子棋游戏非常容易上手，老少皆宜，而且趣味横生，引人入胜；它不仅能增强思维能力，提高智力，而且富含哲理，有助于修身养性。五子棋的游戏规则如下：

(1) 对局双方各执一色棋子。

(2) 空棋盘开局。

(3) 黑先、白后，交替下子，每次只能下一子。为了提高程序的灵活性，也可以规定谁先下。

(4) 棋子下在棋盘的空白点上。棋子下定后，不得向其他点移动，不得从棋盘上拿掉或拿起另落别处。

(5) 黑方的第一枚棋子可下在棋盘任意交叉点上。

(6) 轮流下子是双方的权利，但允许任何一方放弃下子权(即 PASS 权)。

## 1.2 系统架构分析

在具体编码之前，需要做好系统架构分析方面的工作。本节将详细分析五子棋的基本棋型，然后根据棋型分析划分整个系统的功能模块，并最终作为编码的依据。

### 1.2.1 五子棋的基本棋型

对于五子棋游戏来说，最常见的基本棋型有：连五、活四、冲四、活三、眠三、活二和眠二。

(1) 连五：顾名思义，五颗同色棋子连在一起，表示胜利，如图 1-1 所示。

(2) 活四：有两个连五点(即有两个点可以形成五)，图 1-2 中白点即为连五点。当活四出现的时候，如果对方单纯过来防守的话，已经无法阻止自己连五了。

(3) 冲四：有一个连五点。如图 1-3 中的 3 种情形，均为冲四棋型，图中白点为连五点。

相比活四来说，冲四的威胁性就小了很多，因为这个时候，对方只要跟着防守在那个唯一的连五点上，冲四就没法形成连五。

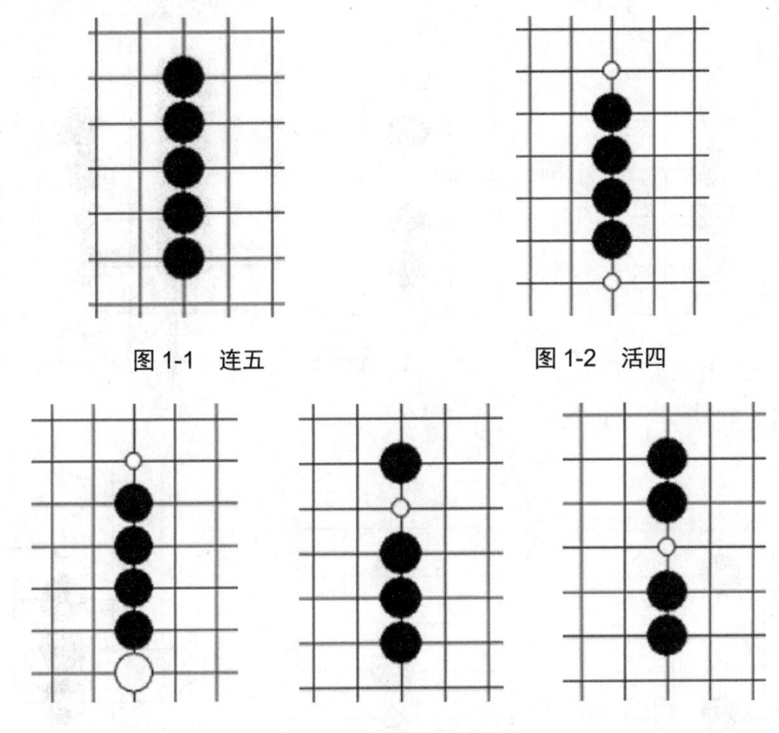

图 1-1　连五

图 1-2　活四

图 1-3　冲四的 3 种情形

(4)　活三：可以形成活四的三，如图 1-4 所示，代表两种最基本的活三棋型，图中白点为活四点。活三棋型是我们进攻中最常见的一种，因为活三之后，如果对方不予理会，将可以下一手将活三变成活四，而活四已经无法单纯防守了。所以，当我们面对活三的时候，需要非常谨慎。在自己没有更好的进攻手段的情况下，需要对其进行防守，以防止其形成可怕的活四棋型。其中在图 1-4(2)中跳着一格的活三，也可以叫做跳活三。

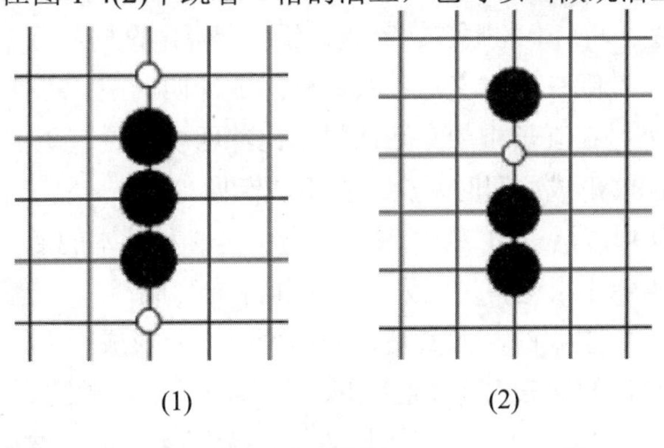

(1)　(2)

图 1-4　活三

(5)　眠三：只能形成冲四的三，如图 1-5 中的 6 种情形，分别代表最基础的 6 种眠三形状。图中白点代表冲四点。眠三的棋型与活三的棋型相比，危险系数下降不少，因为眠三棋型即使不去防守，下一手它也只能形成冲四；而对于单纯的冲四棋型，是可以防守住的。

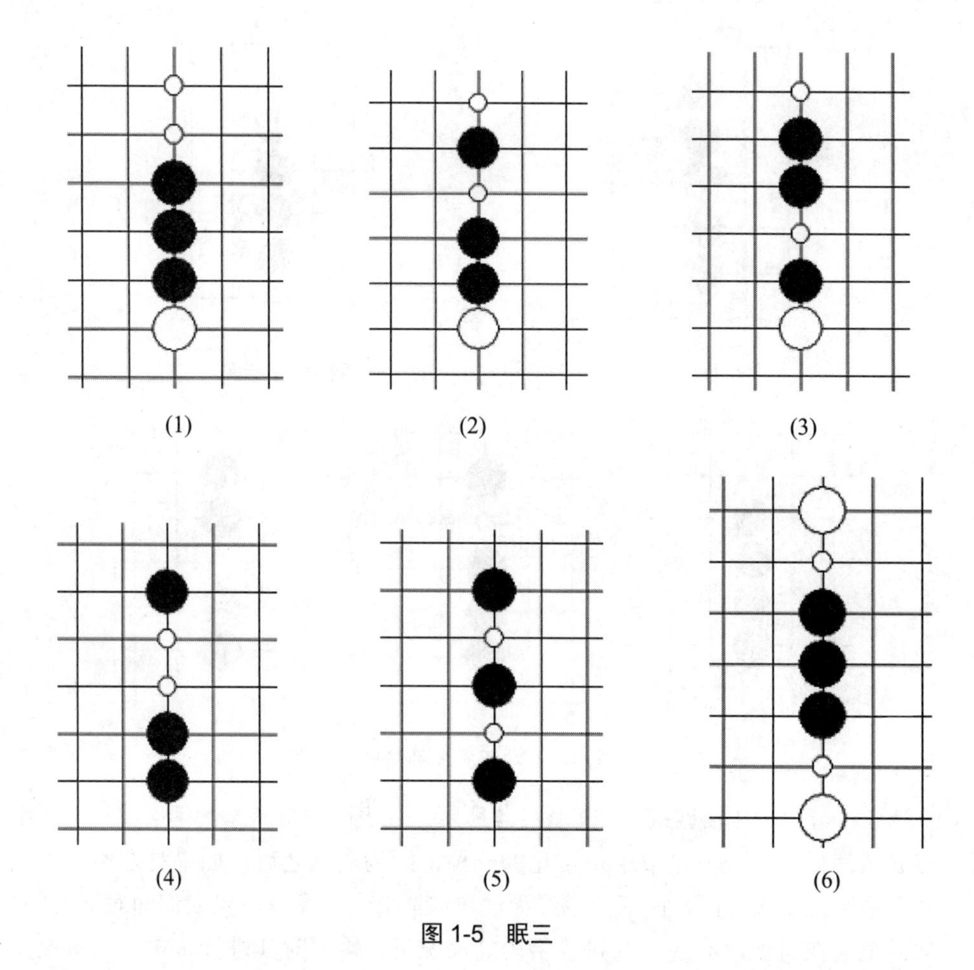


图 1-5　眠三

由此可见，眠三的形状是很丰富的。对于初学者，在下棋过程中，很容易忽略不常见的眠三形状，例如图 1-5(6)中的眠三。可能有新手会提出疑问：活三也可以形成冲四啊，那岂不是也可以叫眠三？眠三的定义是：只能够形成冲四的三。而活三可以形成眠三，但也能够形成活四。此外，在五子棋中，活四棋型比冲四棋型具有更大的优势，所以，我们在既能够形成活四又能够形成冲四时，会选择形成活四。

(6)　活二：能够形成活三的二，如图 1-6 所示是 3 种基本的活二棋型。图中白点为活三点。活二棋型看起来似乎很无害，因为下一手棋才能形成活三，等形成活三再防守也不迟。但其实活二棋型是非常重要的，尤其是在开局阶段，若形成较多活二棋型，在将活二变成活三时，才能够令自己的活三绵绵不绝微风里，让对手防不胜防。

(7)　眠二：能够形成眠三的二。在图 1-7 中展示了 4 种最为常见的眠二棋型，我们可以根据图 1-5(6)找到与 4 个基本眠二棋型都不一样的眠二。图中白点为眠三点。

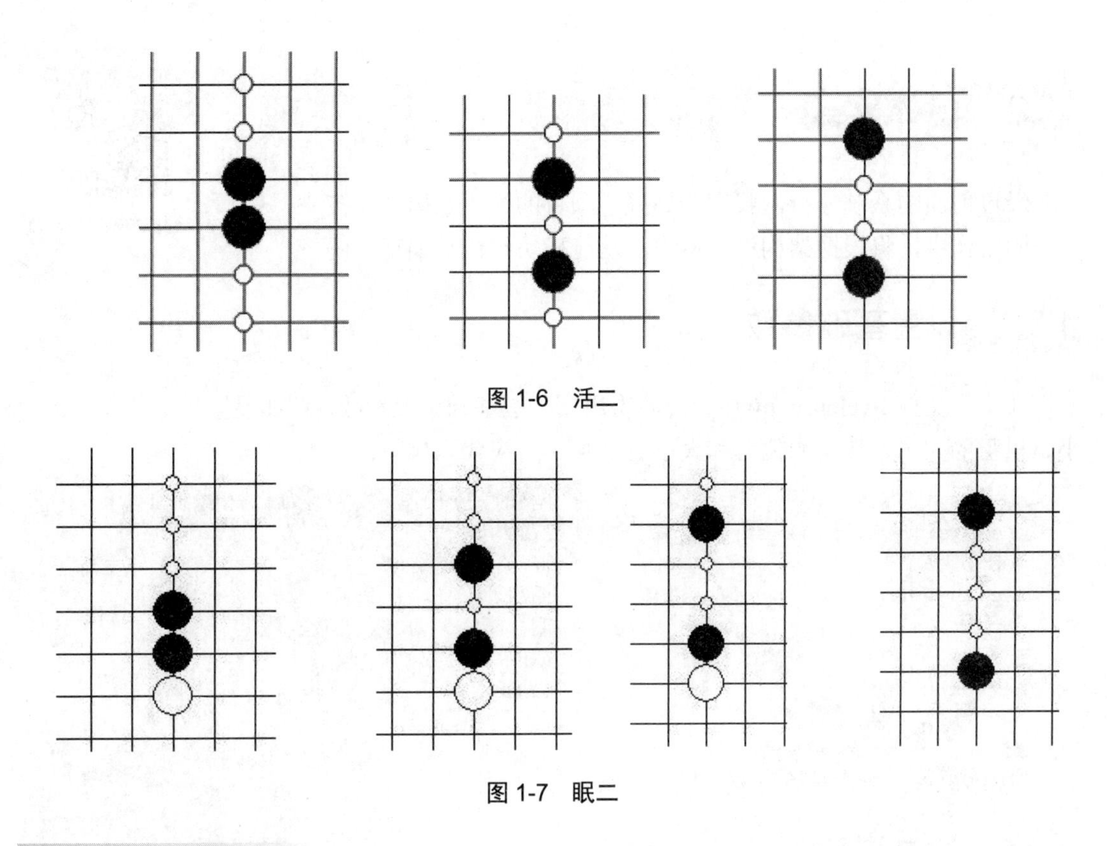

图 1-6　活二

图 1-7　眠二

## 1.2.2　功能模块

根据五子棋的游戏规则和基本棋型分析项目架构，最终得出的功能模块结构如图 1-8 所示。

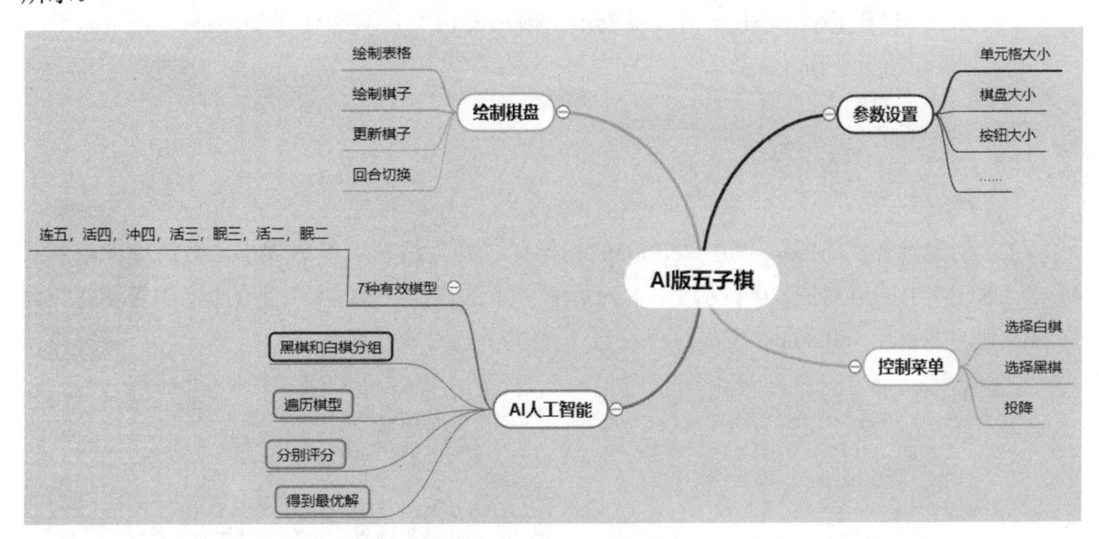


图 1-8　功能模块结构

## 1.3 具体实现

经过前面的系统分析，我们规划好了项目的模块结构。在接下来的工作中，将根据架构设计得出的功能模块结构来编写代码。

### 1.3.1 设置基础参数

在实例文件 fiveinrow.py 中，会多次用到一些基础设置参数，例如设置棋盘单元格的大小、棋盘的大小、按钮的位置和大小等信息。对应代码如下：

```
square_size = 40                       # 单格的宽度(不是格数！是为了方便绘制棋盘用的变量,
chess_size = square_size // 2 - 2     # 棋子大小
web_broad = 15                         # 棋盘格数+1(nxn)
map_w = web_broad * square_size        # 棋盘长度
map_h = web_broad * square_size        # 棋盘高度
info_w = 60                            # 按钮界面宽度
button_w = 120                         # 按钮长宽
button_h = 45
screen_w = map_w                       # 总窗口长宽
screen_h = map_h + info_w
```

### 1.3.2 绘制棋盘

在实例文件 fiveinrow.py 中，使用类 MAP_ENUM 和 Map 绘制棋盘界面。具体实现流程如下。

(1) 在类 MAP_ENUM 中使用数字表示当前格的情况，对应代码如下：

```
class MAP_ENUM(IntEnum):
    be_empty = 0,                      # 无人下
    player1 = 1,                       # 玩家一，执白
    player2 = 2,                       # 玩家二，执黑
    out_of_range = 3,                  # 出界
```

(2) 创建地图类 Map，使用 self.map 初始化二维数组表示棋盘大小，数组中的值和类 MAP_ENUM 中的值对应：0 表示空，1 为玩家一下的棋，2 为玩家二下的棋，3 表示超过允许的限制。并且用 self.steps 按顺序保存已下的棋子。对应代码如下：

```
class Map:                                  # 地图类
    def __init__(self, width, height):      # 构造函数
        self.width = width
        self.height = height
        self.map = [[0 for x in range(self.width)] for y in range(self.height)]
                                            # 存储棋盘的二维数组
        self.steps = []                     # 记录步骤先后

    def get_init(self):                     # 重置棋盘
        for y in range(self.height):
```

```
        for x in range(self.width):
            self.map[y][x] = 0
    self.steps = []
```

(3) 编写函数 intoNextTurn()，进入下一回合的比赛中，交换下棋人。对应代码如下：

```
def intoNextTurn(self, turn):
    if turn == MAP_ENUM.player1:
        return MAP_ENUM.player2
    else:
        return MAP_ENUM.player1
```

(4) 编写函数 getLocate()，功能是根据输入的下标返回棋子的具体位置，对应代码如下：

```
def getLocate(self, x, y):
    map_x = x * square_size
    map_y = y * square_size
    return (map_x, map_y, square_size, square_size)  # 返回位置信息
```

(5) 编写函数 getIndex()，功能是根据输入的具体位置返回下标。对应代码如下：

```
def getIndex(self, map_x, map_y):
    x = map_x // square_size
    y = map_y // square_size
    return (x, y)
```

(6) 编写函数 isInside()，功能是判断当前位置是否在棋盘的有效范围内。对应代码如下：

```
def isInside(self, map_x, map_y):
    if (map_x <= 0 or map_x >= map_w or
            map_y <= 0 or map_y >= map_h):
        return False
    return True
```

(7) 编写函数 isEmpty()，功能是判断在当前的格子中是否已经有棋子。对应代码如下：

```
def isEmpty(self, x, y):
    return (self.map[y][x] == 0)
```

(8) 编写函数 printChessPiece()，功能是在棋盘中绘制已下的棋子，会按照下棋的顺序加上序号，在绘制时会区分黑棋和白棋。对应代码如下：

```
def printChessPiece(self, screen):          # 绘制棋子
    player_one = (255, 245, 238)            # 象牙白
    player_two = (41, 36, 33)               # 烟灰
    player_color = [player_one, player_two]
    for i in range(len(self.steps)):
        x, y = self.steps[i]
        map_x, map_y, width, height = self.getLocate(x, y)
        pos, radius = (map_x + width // 2, map_y + height // 2), chess_size
        turn = self.map[y][x]
        pygame.draw.circle(screen, player_color[turn - 1], pos, radius)
        # 画棋子
```

(9) 编写函数 drawBoard()，功能是绘制棋盘，使用两个 for 循环分别绘制棋盘中的横

线和竖线。对应代码如下：

```
def drawBoard(self, screen):                    # 画棋盘
    color = (0, 0, 0)                           # 线色
    for y in range(self.height):
        # 画横的棋盘线
        start_pos, end_pos = (square_size // 2, square_size // 2 + square_size * y),
          ( map_w - square_size // 2, square_size // 2 + square_size * y)
        pygame.draw.line(screen, color, start_pos, end_pos, 1)
    for x in range(self.width):
        # 画竖的棋盘线
        start_pos, end_pos = (square_size //2 + square_size * x, square_size // 2),
           (square_size // 2 + square_size * x, map_h - square_size // 2)
        pygame.draw.line(screen, color, start_pos, end_pos, 1)
```

## 1.3.3 实现 AI 功能

### 1. 方法分析

经过前面的基本棋型介绍可知，在五子棋游戏中有 7 种有效的棋型(连五，活四，冲四，活三，眠三，活二，眠二)。我们可以创建黑棋和白棋两个数组，记录棋盘上黑棋和白棋分别形成的所有棋型的个数，然后按照一定的规则进行评分。

究竟如何记录棋盘上的棋型个数呢？例如，在本游戏的棋盘上设置有 15 条水平线和 15 条竖直线，不考虑长度小于 5 的斜线，有 21 条从左上到右下的斜线，21 条从左下到右上的斜线。然后在每一条线上分别对黑棋和白棋查找是否有符合的棋型。这种方法比较直观，但是实现起来不方便。本实例的方法是对整个棋盘进行遍历，对于每一个白棋或黑棋，以它为中心，记录符合的棋型个数。具体实现方式如下。

(1) 遍历棋盘上的每个点，如果是黑棋或白棋，则对这个点所在四个方向形成的四条线分别进行评估。四个方向即水平、竖直、两个斜线(\，/)，四个方向依次按照从左到右、从上到下、从左上到右下、从左下到右上来检测。

(2) 对于具体的一条线，如图 1-9 所示，以选取点为中心，取该方向上前面四个点和后面四个点，组成一个长度为 9 的数组。

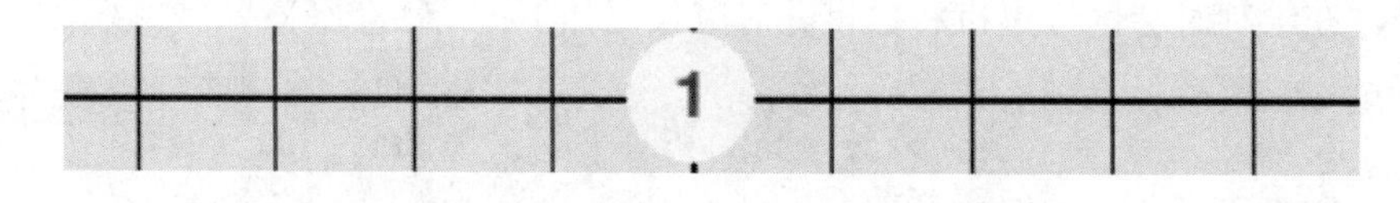


图 1-9　一条线

(3) 找出和中心点相连的同色棋子有几个，如在图 1-10 中，相连的白色棋子有 3 个，根据相连棋子的个数再分别进行判断，最后得出这行属于上面说的哪一种棋型。在此需要注意的是，在评估白棋 1 的时候，白棋 3 和白棋 5 已经被判断过，所以要标记下，下次遍历到这个方向的白棋 3 和白棋 5 时需要跳过，避免重复统计棋型。

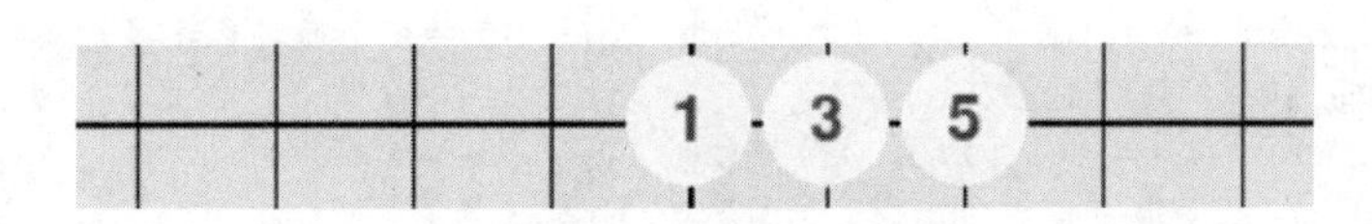


图 1-10 和中心点相连的同色棋子

(4) 根据棋盘上黑棋和白棋的棋型统计信息，按照一定的规则进行评分。假设形成该棋局的最后一步是黑棋下的，则最后的评分是(黑棋得分-白棋得分)，在相同棋型、相同个数的情况下，白棋会占优，因为下一步是白棋下。比如黑棋有个冲四，白棋有个冲四，显然白棋占优，因为下一步白棋就能形成连五。最后按照下面的规则依次匹配：

- 黑棋连五，评分为 10000。
- 白棋连五，评分为-10000。
- 黑棋两个冲四，可以当成一个活四。
- 白棋有活四，评分为-9050。
- 白棋有冲四，评分为-9040。
- 黑棋有活四，评分为 9030。
- 黑棋有冲四和活三，评分为 9020。
- 黑棋没有冲四，且白棋有活三，评分为 9010。
- 黑棋有 2 个活三，且白棋没有活三或眠三，评分为 9000。

(5) 最后针对黑棋或白棋的活三、眠三、活二、眠二的个数依次增加分数，具体评分值为(黑棋得分-白棋得分)。

### 2. 功能实现

有了上面的评估标准后，当轮到 AI 下棋时，就要针对当前的棋局，找到一个最有利的位置。AI 会尝试在每个空点下棋，每次都会形成一个新的棋局，然后用评估函数来获取这个棋局的评分，只需在最后从中选取评分最高的位置就行了。下面是 AI 获取最有利位置的逻辑：

- 首先遍历棋盘上的每一个空点，并在这个空点下棋，获取新的棋局的评分。
- 如果是更高的评分，则保存该位置。
- 然后将这个位置恢复为空点。
- 最后会获得最高评分的位置。

在实例文件 fiveinrow.py 中，通过类 MyChessAI 实现 AI 功能。具体实现流程如下。

(1) 使用构造函数实现初始化功能，在数组 record 中记录所有位置的 4 个方向是否被检测过，使用二维数组 count 记录黑棋和白棋的棋型个数统计。通过 position_isgreat 方法给棋盘上的每个位置设一个初始分数，越靠近棋盘中心，分数越高，这样在最初没有任何棋型时，AI 会优先选取靠近中心的位置。对应代码如下：

```
class MyChessAI():
    def __init__(self, chess_len):                # 构造函数
        self.len = chess_len                      # 当前棋盘大小
        # 二维数组，每一格存的是：横评分，纵评分，左斜评分，右斜评分
```

```
    self.record = [[[0, 0, 0, 0] for i in range(chess_len)] for j in range(chess_len)]
    # 存储当前格具体棋型数量
    self.count = [[0 for i in range(SITUATION_NUM)] for j in range(2)]
    # 位置分(同条件下越靠近棋盘中央分数越高)
    self.position_isgreat = [
        [(web_broad - max(abs(i - web_broad / 2 + 1), abs(j - web_broad / 2
         + 1))) for i in range(chess_len)]
        for j in range(chess_len)]

def get_init(self):                              # 初始化
    for i in range(self.len):
        for j in range(self.len):
            for k in range(4):
                self.record[i][j][k] = 0
    for i in range(len(self.count)):
        for j in range(len(self.count[0])):
            self.count[i][j] = 0
    self.save_count = 0

def isWin(self, board, turn):                    # 当前人胜利
    return self.evaluate(board, turn, True)
```

(2) 编写函数 genmove()，功能是返回所有未下棋的坐标(位置从好到坏)。也就是说，函数 genmove()能够获取棋盘上所有的空点，然后依次尝试，获得评分最高的位置并返回。对应代码如下：

```
def genmove(self, board, turn):
    moves = []
    for y in range(self.len):
        for x in range(self.len):
            if board[y][x] == 0:
                score = self.position_isgreat[y][x]
                moves.append((score, x, y))
    moves.sort(reverse=True)
    return moves
```

(3) 编写函数 search()，功能是返回当前最优解下标，此函数是上面 AI 逻辑的代码实现。先通过函数 genmove()获取棋盘上所有的空点，然后依次尝试，获得评分最高的位置并返回。对应代码如下：

```
def search(self, board, turn):
    moves = self.genmove(board, turn)
    bestmove = None
    max_score = -99999                          # 无穷小
    for score, x, y in moves:
        board[y][x] = turn.value
        score = self.evaluate(board, turn)
        board[y][x] = 0
        if score > max_score:
            max_score = score
            bestmove = (max_score, x, y)
    return bestmove
```

(4) 编写函数 findBestChess()，此函数就是 AI 的入口函数。对应代码如下：

```
def findBestChess(self, board, turn):
    # time1 = time.time()
    score, x, y = self.search(board, turn)
    # time2 = time.time()
    # print('time:%f  (%d, %d)' % ((time2 - time1), x, y))
    return (x, y)
```

(5) 编写函数 getScore()，功能是对黑棋和白棋进行评分。对应代码如下：

```
# 直接列举所有棋型
def getScore(self, mychess, yourchess):
    mscore, oscore = 0, 0
    if mychess[FIVE] > 0:
        return (10000, 0)
    if yourchess[FIVE] > 0:
        return (0, 10000)
    if mychess[S4] >= 2:
        mychess[L4] += 1
    if yourchess[L4] > 0:
        return (0, 9050)
    if yourchess[S4] > 0:
        return (0, 9040)
    if mychess[L4] > 0:
        return (9030, 0)
    if mychess[S4] > 0 and mychess[L3] > 0:
        return (9020, 0)
    if yourchess[L3] > 0 and mychess[S4] == 0:
        return (0, 9010)
    if (mychess[L3] > 1 and yourchess[L3] == 0 and yourchess[S3] == 0):
        return (9000, 0)
    if mychess[S4] > 0:
        mscore += 2000
    if mychess[L3] > 1:
        mscore += 500
    elif mychess[L3] > 0:
        mscore += 100
    if yourchess[L3] > 1:
        oscore += 2000
    elif yourchess[L3] > 0:
        oscore += 400
    if mychess[S3] > 0:
        mscore += mychess[S3] * 10
    if yourchess[S3] > 0:
        oscore += yourchess[S3] * 10
    if mychess[L2] > 0:
        mscore += mychess[L2] * 4
    if yourchess[L2] > 0:
        oscore += yourchess[L2] * 4
    if mychess[S2] > 0:
        mscore += mychess[S2] * 4
    if yourchess[S2] > 0:
        oscore += yourchess[S2] * 4
    return (mscore, oscore)  # 自我辅助效果，counter 对面效果
```

(6) 编写函数 evaluate()，功能是对上述得分进行进一步的处理。参数 turn 表示最近一

步棋是谁下的，根据 turn 决定的 me(表示自己棋的值)和 you(表示对手棋的值，下一步棋由对手下)，在对棋型评分时会用到。checkWin 用来判断是否有一方获胜。对应代码如下：

```
def evaluate(self, board, turn, checkWin=False):
    self.get_init()
    if turn == MAP_ENUM.player1:
        me = 1
        you = 2
    else:
        me = 2
        you = 1
    for y in range(self.len):
        for x in range(self.len):
            if board[y][x] == me:
                self.evaluatePoint(board, x, y, me, you)
            elif board[y][x] == you:
                self.evaluatePoint(board, x, y, you, me)
    mychess = self.count[me - 1]
    yourchess = self.count[you - 1]
    if checkWin:
        return mychess[FIVE] > 0  # 检查是否已经胜利
    else:
        mscore, oscore = self.getScore(mychess, yourchess)
        return (mscore - oscore)  # 自我辅助效果，counter 对面效果
```

(7) 编写函数 evaluatePoint()，功能是对某一个位置的 4 个方向分别进行检查。对应代码如下：

```
def evaluatePoint(self, board, x, y, me, you):
    direction = [(1, 0), (0, 1), (1, 1), (1, -1)]  # 四个方向
    for i in range(4):
        if self.record[y][x][i] == 0:
            # 检查当前方向棋型
            self.getBasicSituation(board, x, y, i, direction[i], me, you,
                self.count[me - 1])
        else:
            self.save_count += 1
```

(8) 编写函数 getLine()，功能是把当前方向棋型存储下来，方便后续使用。此函数能够根据棋子的位置和方向，获取上面说的长度为 9 的线。如果线上的位置超出了棋盘范围，就将这个位置的值设为对手的值，因为超出范围和被对手棋挡着，对棋型判断的结果是相同的。对应代码如下：

```
def getLine(self, board, x, y, direction, me, you):
    line = [0 for i in range(9)]
    # 光标移到最左端
    tmp_x = x + (-5 * direction[0])
    tmp_y = y + (-5 * direction[1])
    for i in range(9):
        tmp_x += direction[0]
        tmp_y += direction[1]
        if (tmp_x < 0 or tmp_x >= self.len or tmp_y < 0 or tmp_y >= self.len):
            line[i] = you  # 出界
```

```
        else:
            line[i] = board[tmp_y][tmp_x]
    return line
```

(9) 编写函数 getBasicSituation()，功能是把当前方向的棋型识别成具体情况，例如把 MMMMX 识别成活四冲四、活三眠三等。对应代码如下：

```
def getBasicSituation(self, board, x, y, dir_index, dir, me, you, count):
    # record赋值
    def setRecord(self, x, y, left, right, dir_index, direction):
        tmp_x = x + (-5 + left) * direction[0]
        tmp_y = y + (-5 + left) * direction[1]
        for i in range(left, right):
            tmp_x += direction[0]
            tmp_y += direction[1]
            self.record[tmp_y][tmp_x][dir_index] = 1

    empty = MAP_ENUM.be_empty.value
    left_index, right_index = 4, 4
    line = self.getLine(board, x, y, dir, me, you)
    while right_index < 8:
        if line[right_index + 1] != me:
            break
        right_index += 1
    while left_index > 0:
        if line[left_index - 1] != me:
            break
        left_index -= 1
    left_range, right_range = left_index, right_index
    while right_range < 8:
        if line[right_range + 1] == you:
            break
        right_range += 1
    while left_range > 0:
        if line[left_range - 1] == you:
            break
        left_range -= 1
    chess_range = right_range - left_range + 1
    if chess_range < 5:
        setRecord(self, x, y, left_range, right_range, dir_index, dir)
        return SITUATION.NONE
    setRecord(self, x, y, left_index, right_index, dir_index, dir)
    m_range = right_index - left_index + 1
    if m_range == 5:
        count[FIVE] += 1
    # 活四冲四
    if m_range == 4:
        left_empty = right_empty = False
        if line[left_index - 1] == empty:
            left_empty = True
        if line[right_index + 1] == empty:
            right_empty = True
        if left_empty and right_empty:
            count[L4] += 1
        elif left_empty or right_empty:
```

```
            count[S4] += 1
        # 活三眠三
        if m_range == 3:
            left_empty = right_empty = False
            left_four = right_four = False
            if line[left_index - 1] == empty:
                if line[left_index - 2] == me:  # MXMMM
                setRecord(self, x, y, left_index - 2, left_index - 1, dir_index, dir)
                    count[S4] += 1
                    left_four = True
                left_empty = True
            if line[right_index + 1] == empty:
                if line[right_index + 2] == me:  # MMMXM
                setRecord(self, x, y, right_index + 1, right_index + 2, dir_index, dir)
                    count[S4] += 1
                    right_four = True
                right_empty = True
            if left_four or right_four:
                pass
            elif left_empty and right_empty:
                if chess_range > 5:  # XMMMXX, XXMMMX
                    count[L3] += 1
                else:  # PXMMMXP
                    count[S3] += 1
            elif left_empty or right_empty:  # PMMMX, XMMMP
                count[S3] += 1
        # 活二眠二
        if m_range == 2:
            left_empty = right_empty = False
            left_three = right_three = False
            if line[left_index - 1] == empty:
                if line[left_index - 2] == me:
                    setRecord(self, x, y, left_index - 2, left_index - 1, dir_index, dir)
                    if line[left_index - 3] == empty:
                        if line[right_index + 1] == empty:  # XMXMMX
                            count[L3] += 1
                        else:  # XMXMMP
                            count[S3] += 1
                        left_three = True
                    elif line[left_index - 3] == you:  # PMXMMX
                        if line[right_index + 1] == empty:
                            count[S3] += 1
                            left_three = True
                left_empty = True
            if line[right_index + 1] == empty:
                if line[right_index + 2] == me:
                    if line[right_index + 3] == me:  # MMXMM
                     setRecord(self, x, y, right_index + 1, right_index + 2,
                     dir_index, dir)
                        count[S4] += 1
                        right_three = True
                    elif line[right_index + 3] == empty:
                        # setRecord(self, x, y, right_index+1, right_index+2,
                          dir_index, dir)
                        if left_empty:  # XMMXMX
```

```
                    count[L3] += 1
                else:  # PMMXMX
                    count[S3] += 1
                right_three = True
            elif left_empty:  # XMMXMP
                count[S3] += 1
                right_three = True
        right_empty = True
    if left_three or right_three:
        pass
    elif left_empty and right_empty:  # XMMX
        count[L2] += 1
    elif left_empty or right_empty:  # PMMX, XMMP
        count[S2] += 1
# 特殊活二眠二(有空格)
if m_range == 1:
    left_empty = right_empty = False
    if line[left_index - 1] == empty:
        if line[left_index - 2] == me:
            if line[left_index - 3] == empty:
                if line[right_index + 1] == you:  # XMXMP
                    count[S2] += 1
        left_empty = True
    if line[right_index + 1] == empty:
        if line[right_index + 2] == me:
            if line[right_index + 3] == empty:
                if left_empty:  # XMXMX
                    count[L2] += 1
                else:  # PMXMX
                    count[S2] += 1
        elif line[right_index + 2] == empty:
            if line[right_index + 3] == me and line[right_index + 4] == empty:
              # XMXXMX
                count[L2] += 1
# 以上都不是，则为 none 棋型
return SITUATION.NONE
```

## 1.3.4　实现按钮功能

在本项目的棋盘下方会显示 4 个按钮，具体说明如下。

- Pick White：选择白棋。
- Pick Black：选择黑棋。
- Surrender：投降。
- Multiple：暂时不可用。

在实例文件 fiveinrow.py 中，实现上述按钮功能的流程如下。

(1) 编写游戏的按钮类 Button。这是一个父类，通过函数 draw()根据按钮的 enable 状态填色。对应代码如下：

```
class Button:
    def __init__(self, screen, text, x, y, color, enable):  # 构造函数
        self.screen = screen
```

```
        self.width = button_w
        self.height = button_h
        self.button_color = color
        self.text_color = (255, 255, 255)  # 纯白
        self.enable = enable
        self.font = pygame.font.SysFont(None, button_h * 2 // 3)
        self.rect = pygame.Rect(0, 0, self.width, self.height)
        self.rect.topleft = (x, y)
        self.text = text
        self.init_msg()

    # 重写pygame内置函数，初始化按钮
    def init_msg(self):
        if self.enable:
            self.msg_image = self.font.render(self.text, True, self.text_color,
              self.button_color[0])
        else:
            self.msg_image = self.font.render(self.text, True, self.text_color,
              self.button_color[1])
        self.msg_image_rect = self.msg_image.get_rect()
        self.msg_image_rect.center = self.rect.center

    # 根据按钮的enable状态填色，具体颜色在后续子类控制
    def draw(self):
        if self.enable:
            self.screen.fill(self.button_color[0], self.rect)
        else:
            self.screen.fill(self.button_color[1], self.rect)
        self.screen.blit(self.msg_image, self.msg_image_rect)
```

(2) 编写类WhiteStartButton，实现选择白棋的功能，对应代码如下：

```
class WhiteStartButton(Button):  # 开始按钮(选白棋)
    def __init__(self, screen, text, x, y):  # 构造函数
        super().__init__(screen, text, x, y, [(26, 173, 25), (158, 217, 157)], True)

    def click(self, game):  # 点击，pygame内置方法
        if self.enable:  # 启动游戏并初始化，变换按钮颜色
            game.start()
            game.winner = None
            game.multiple = False
            self.msg_image = self.font.render(self.text, True, self.text_color,
              self.button_color[1])
            self.enable = False
            return True
        return False

    def unclick(self):  # 取消点击
        if not self.enable:
            self.msg_image = self.font.render(self.text, True, self.text_color,
self.button_color[0])
            self.enable = True
```

(3) 编写类BlackStartButton，实现选择黑棋的功能，对应代码如下：

```
class BlackStartButton(Button):  # 开始按钮(选黑棋)
```

```
    def __init__(self, screen, text, x, y):  # 构造函数
        super().__init__(screen, text, x, y, [(26, 173, 25), (158, 217, 157)], True)

    def click(self, game):  # 点击，pygame内置方法
        if self.enable:  # 启动游戏并初始化，变换按钮颜色，安排AI先手
            game.start()
            game.winner = None
            game.multiple = False
            game.useAI = True
            self.msg_image = self.font.render(self.text, True, self.text_color,
              self.button_color[1])
            self.enable = False
            return True
        return False

    def unclick(self):  # 取消点击
        if not self.enable:
            self.msg_image = self.font.render(self.text, True, self.text_color,
              self.button_color[0])
            self.enable = True
```

(4) 编写类GiveupButton，实现投降功能，对应代码如下：

```
class GiveupButton(Button):  # 投降按钮(任何模式都能用)
    def __init__(self, screen, text, x, y):
        super().__init__(screen, text, x, y, [(230, 67, 64), (236, 139, 137)], False)

    def click(self, game):  # 结束游戏，判断赢家
        if self.enable:
            game.is_play = False
            if game.winner is None:
                game.winner = game.map.intoNextTurn(game.player)
            self.msg_image = self.font.render(self.text, True, self.text_color,
              self.button_color[1])
            self.enable = False
            return True
        return False

    def unclick(self):  # 保持不变，填充颜色
        if not self.enable:
            self.msg_image = self.font.render(self.text, True, self.text_color,
              self.button_color[0])
            self.enable = True
```

### 1.3.5 重写功能

为了更好地在主函数中规划和控制整个游戏的代码，编写类Game，在里面调用上面的功能函数，然后分别绘制棋盘、按钮和判断获胜一方。类Game的具体实现流程如下。

(1) 通过函数__init__(self, caption)实现初始化处理，设置按钮的内容和可用性。

```
class Game:  # pygame类,以下所有功能都是根据需要重写
    def __init__(self, caption):
        pygame.init()
```

```
    self.screen = pygame.display.set_mode([screen_w, screen_h])
    pygame.display.set_caption(caption)
    self.clock = pygame.time.Clock()
    self.buttons = []
    self.buttons.append(WhiteStartButton(self.screen, 'Pick White', 20, map_h))
    self.buttons.append(BlackStartButton(self.screen, 'Pick Black', 190, map_h))
    self.buttons.append(GiveupButton(self.screen, 'Surrender', 360, map_h))
    self.buttons.append(MultiStartButton(self.screen, 'Multiple', 530, map_h))
    self.is_play = False
    self.map = Map(web_broad, web_broad)
    self.player = MAP_ENUM.player1
    self.action = None
    self.AI = MyChessAI(web_broad)
    self.useAI = False
    self.winner = None
    self.multiple = False
```

(2) 定义函数 start(self)，开始游戏，默认白棋先下。

```
def start(self):
    self.is_play = True
    self.player = MAP_ENUM.player1  # 白棋先手
    self.map.get_init()
```

(3) 定义函数 play(self)，绘制出棋盘和按钮。

```
def play(self):
    # 画底板
    self.clock.tick(60)
    wood_color = (210, 180, 140)
    pygame.draw.rect(self.screen, wood_color, pygame.Rect(0, 0, map_w, screen_h))
    pygame.draw.rect(self.screen, (255, 255, 255), pygame.Rect(map_w, 0,
      info_w, screen_h))
    # 画按钮
    for button in self.buttons:
        button.draw()
    if self.is_play and not self.isOver():
        if self.useAI and not self.multiple:
            x, y = self.AI.findBestChess(self.map.map, self.player)
            self.checkClick(x, y, True)
            self.useAI = False
        if self.action is not None:
            self.checkClick(self.action[0], self.action[1])
            self.action = None
        if not self.isOver():
            self.changeMouseShow()
    if self.isOver():
        self.showWinner()
        # self.buttons[0].enable = True
        # self.buttons[1].enable = True
        # self.buttons[2].enable = False
    self.map.drawBoard(self.screen)
    self.map.printChessPiece(self.screen)
```

(4) 定义函数 changeMouseShow(self)，在开始游戏的时候把鼠标指针切换成棋子的样子。

```
def changeMouseShow(self):
    map_x, map_y = pygame.mouse.get_pos()
    x, y = self.map.getIndex(map_x, map_y)
    if self.map.isInside(map_x, map_y) and self.map.isEmpty(x, y):
     # 在棋盘内且当前无棋子
        pygame.mouse.set_visible(False)
        smoke_blue = (176, 224, 230)
        pos, radius = (map_x, map_y), chess_size
        pygame.draw.circle(self.screen, smoke_blue, pos, radius)
    else:
        pygame.mouse.set_visible(True)

def checkClick(self, x, y, isAI=False): # 后续处理
    self.map.click(x, y, self.player)
    if self.AI.isWin(self.map.map, self.player):
        self.winner = self.player
        self.click_button(self.buttons[2])
    else:
        self.player = self.map.intoNextTurn(self.player)
        if not isAI:
            self.useAI = True
```

(5) 定义函数 mouseClick(self, map_x, map_y)，处理下棋动作，将某个棋子放在棋盘中的某个位置。

```
def mouseClick(self, map_x, map_y):
if self.is_play and self.map.isInside(map_x, map_y) and not self.isOver():
        x, y = self.map.getIndex(map_x, map_y)
        if self.map.isEmpty(x, y):
            self.action = (x, y)
```

(6) 定义函数 isOver(self)，如果一方获胜则中断游戏。

```
def isOver(self): # 中断条件
    return self.winner is not None
```

(7) 定义函数 showWinner(self)，功能是打印输出胜者。

```
def showWinner(self): # 输出胜者
    def showFont(screen, text, location_x, locaiton_y, height):
        font = pygame.font.SysFont(None, height)
        font_image = font.render(text, True, (255, 215, 0), (255, 255, 255))
          # 金黄色
        font_image_rect = font_image.get_rect()
        font_image_rect.x = location_x
        font_image_rect.y = locaiton_y
        screen.blit(font_image, font_image_rect)

    if self.winner == MAP_ENUM.player1:
        str = 'White Wins!'
    else:
        str = 'Black Wins!'
    showFont(self.screen, str, map_w / 5, screen_h / 8, 100)
    # 居上中，字号100 pygame.mouse.set_visible(True)
```

到此为止，本实例全部介绍完毕。执行程序后，先选择使用黑棋还是白棋，单击选择按钮后，即可进入游戏界面。黑棋获胜的界面效果如图 1-11 所示。

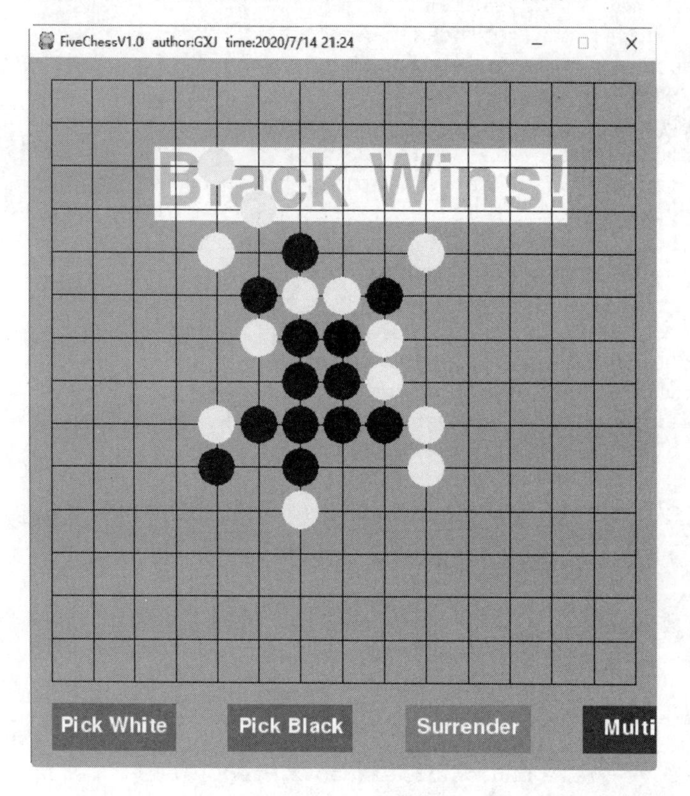


图 1-11　执行效果

# 第 2 章

## 在线商城系统（Django+Mezzanine+Cartridge 实现）

在信息时代的今天，网络已经成为人们工作和学习的一部分，不断充实和改变着人们的生活。其中网络购物已经成为人们生活中的重要组成部分，双十一、双十二、618等购物狂欢季已经逐渐深入人心，渐渐养成了全民网购的习惯。在本章的内容中，将详细讲解使用 Python 第三方库开发一个在线商城系统的知识，介绍 Python 使用 Django、Mezzanine 和 Cartridge 开发一个大型商城系统的过程。

## 2.1 项目介绍

本项目的客户是一家民营图书销售公司，为了扩大销售渠道，想开通网上商城，利用在线博客和电子商城来销售他们的图书。客户提出如下 3 点要求。

- 每个商品可以留言。
- 实现在线购物车处理和订单处理。
- 实现对产品、购物车和订单的管理功能。

本项目开发团队的具体职责如下。

- 项目经理：负责前期功能分析，选择第三方模块，策划构建系统模块，检查项目进度，质量检查。
- 软件工程师 PrA：配置系统文件，搭建数据库，实现数据访问层。
- 软件工程师 PrB：负责购物车处理模块、订单处理模块、商品评论模块、商品搜索模块的编码工作。
- 软件工程师 PrC：样式设计，系统测试，后期调试，并负责商品显示模块、商品分类模块、商品管理模块的编码工作。

整个项目的具体职责流程如图 2-1 所示。

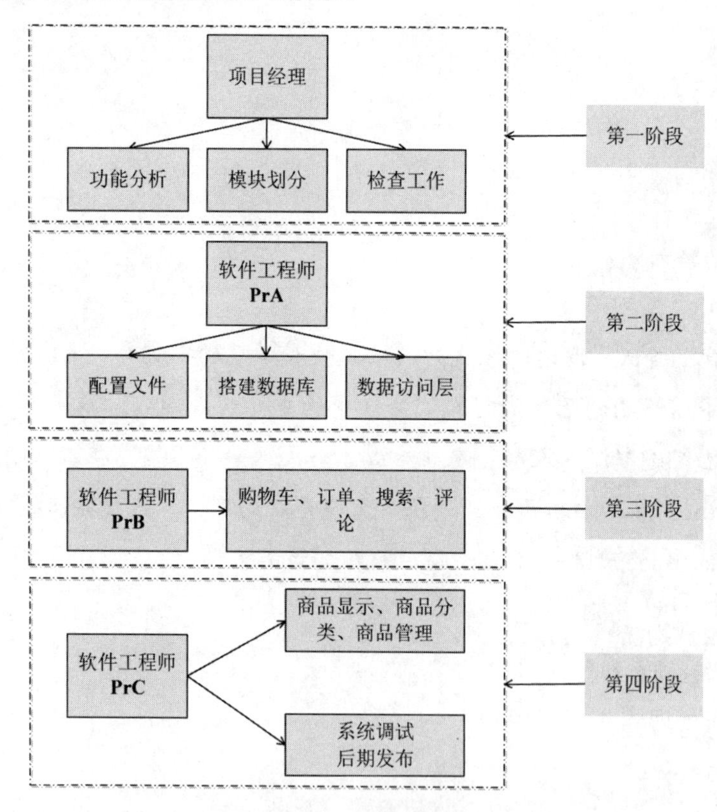


图 2-1　职责流程图

## 2.2 项目规划分析

在具体编码工作开始之前，需要进行项目规划分析方面的工作，为后期的编码工作打好基础。本节将详细讲解项目规划分析的知识。

### 2.2.1 电子商务的简要介绍

理论上，电子商务的范围很大，概括起来主要有两类，一类是 B2B，另一类是 B2C。B2B 的全称是 Business to Business，主要面向的是企业与企业，或是为大型的商业买卖提供的交易平台，企业通过这个平台进行采购、销售、结算等环节，可降低成本，提高效率。但这种平台对性能、安全和服务的要求比较高。B2C 的全称是 Business to Customer，它直接面向终端的大众消费者，其经营也有两种形式：一种类似大型超市，里面提供大量的货物商品，消费者可以浏览、挑选商品，直接在线结账付款，如当当网上书店、卓越网上商城等，都是采用 B2C 中的这种形式；另一种是类似城市里面的大商场，如华联等，在这个商场里面有许多柜台或专柜，都在卖自己的东西，消费者可以根据自己的需求直接到相应柜台购买商品，然后去商场服务台结账，柜台按类别或经营范围来划分，如新浪网的电子商城，就是采用 B2C 中的这种形式。不管是 B2B 还是 B2C，其基本模式是相同的，即浏览查看商品，然后下订单，双方确认后付款交货，完成交易。

电子商城类的网站由于经常涉及输入商品信息，所以有必要开发一套 CMS(Content Managment System)系统，即信息发布系统。CMS 系统由后台人工输入信息，然后系统自动将信息整理保存到数据库，而用户在前台浏览到的均为系统自动产生的网页，所有的过程都无须手工制作 HTML 网页而自动进行信息发布及管理。CMS 系统又可分为两大类：第一类是将内容生成静态网页，如一些新闻站点；第二类是从数据库实时读取数据。本实例的实现属于第一类。

### 2.2.2 在线博客+商城系统构成模块

(1) 博客系统模块。

为了提高用户体验，可以在系统中发布和产品相关的日志信息，例如商品评测、新品发布和商品试用体验等。

(2) 会员处理模块。

为了方便用户购买图书，提高系统人气，系统中设立了会员功能。成为系统会员后，可以对自己的资料进行管理，并且可以集中管理自己的订单。

(3) 购物车处理模块。

作为网上商城系统必不可少的环节，为满足用户的购物需求，本系统设立了购物车功能。用户可以把需要的商品放到购物车中保存，提交在线订单后即可完成在线商品的购买。

(4) 商品查询模块。

为了方便用户购买商品，系统设立了商品快速查询模块，供用户根据商品的信息快速找到自己需要的商品。

(5) 订单处理模块。

为方便商家处理用户的购买信息，系统设立了订单处理功能。通过该功能，可以及时处理用户的订单信息，使用户尽快买到自己的商品。

(6) 商品分类模块。

为了便于用户对商品进行浏览，系统将商品划分为不同的类别，以便用户迅速找到自己需要的商品。

(7) 商品管理模块。

为方便系统的升级和维护，建立专用的商品管理模块以实现商品的添加、删除和修改功能，满足系统更新的需求。

上述应用模块的具体运行流程如图 2-2 所示。

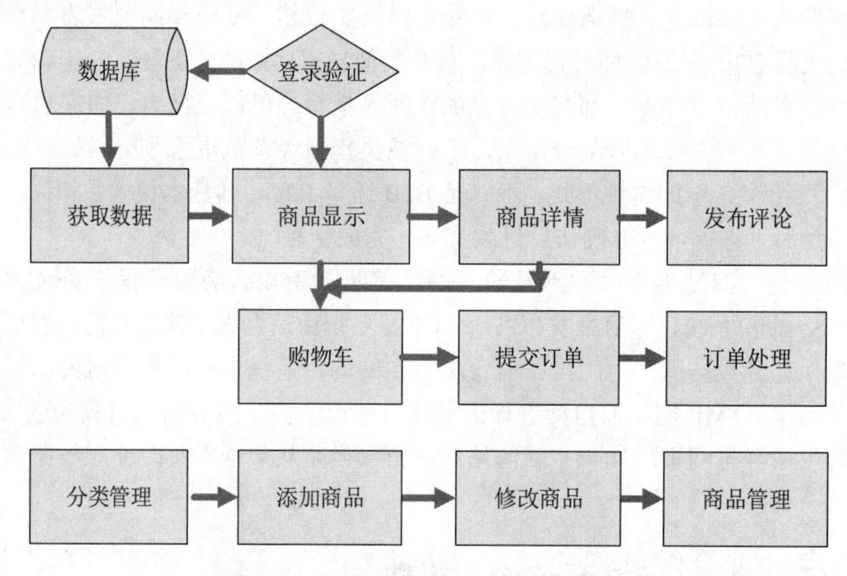


图 2-2　在线商城系统模块运行流程图

## 2.3　规划项目文件

扫码观看视频讲解

在开发一个大型的应用程序时，规划项目文件是一个非常重要的前期准备工作，是关系到整个项目的实现流程是否能顺利完成的关键。本节将根据严格的市场需求分析，规划出本项目的文件结构。

为整个项目规划具体实现文件后，各构成模块文件的具体说明如下。

- 系统配置文件：功能是对项目程序进行总体配置。
- 路径导航模块：功能是设置 URL 路径的导航链接。
- 商品显示模块：功能是将系统内的商品逐一显示出来。

- 购物车处理模块：功能是将满意的商品放在购物车内。
- 订单处理模块：功能是实现对系统内购物订单的处理。
- 商品评论模块：功能是供用户对系统内的某商品发布评论。
- 商品搜索模块：功能是使用户迅速地搜索出自己需要的商品。
- 商品分类模块：功能是将系统内的商品类别以指定形式显示出来。
- 系统管理模块：功能是对系统内数据进行管理维护。

**注 意**

在此声明规划阶段的重要性。开发者需要先分析网络中的一些在线购物系统，这样基本功能就了解得差不多了。任何购物系统都需要几个核心功能：商品展示、购物车处理、订单处理。只要设计好上述必需的核心功能，在此基础上进行扩充即可。

## 2.4　使用第三方库 Mezzanine 和 Cartridge

扫码观看视频讲解

为了提高开发效率，本项目将使用第三方库 Mezzanine 和 Cartridge。本节将简要介绍 Mezzanine 和 Cartridge 库的基本用法。

### 2.4.1　使用库 Mezzanine

Mezzanine 是一款著名的开源、基于 Django 的 CMS 系统 (Content Management System) 框架。其实可以将任何一个网站看作一个特定的内容管理系统，只不过每个网站发布和管理的具体内容不一样，例如携程发布的是航班、酒店和用户的订单信息，而淘宝发布的是商品和用户的订单信息。下面将详细讲述框架 Mezzanine 的使用知识。

在安装 Mezzanine 之前，需要先确保已经安装了 Django，然后使用如下命令安装 Mezzanine：

```
pip install mezzanine
```

接下来使用 Mezzanine 快速创建一个 CMS 内容管理系统，具体实现流程如下。

(1) 使用如下命令创建一个 Mezzanine 工程，项目名是 testing：

```
mezzanine-project testing
```

(2) 使用如下命令进入项目目录：

```
cd testing
```

(3) 使用如下命令创建一个数据库：

```
python manage.py createdb
```

在这个过程，需要填写如下基本信息。

- 域名和端口：默认为 http://127.0.0.1:8000/。

- ▶ 默认的超级管理员账号和密码。
- ▶ 默认主页。

(4) 使用如下命令启动这个项目：

```
python manage.py runserver
```

当显示如下所示的信息时，说明成功运行了新建的 Mezzanine 项目 testing：

```
              .....
          _d^^^^^^^^^b_
       .d''           ``b.
     .p'                `q.
    .d'                   `b.
   .d'                     `b.   * Mezzanine 4.2.3
   ::                       ::   * Django 1.10.8
  ::    M E Z Z A N I N E    ::    * Python 3.6.0
   ::                       ::   * SQLite 3.2.2
   `p.                     .q'   * Windows 10
    `p.                   .q'
     `b.                 .d'
       `q..           ..p'
          ^q........p^
              ''''

Performing system checks...

System check identified no issues (0 silenced).
April 17, 2018 - 14:07:33
Django version 1.10.8, using settings 'testing.settings'
Starting development server at http://127.0.0.1:8000/
```

在浏览器的地址栏中输入 http://127.0.0.1:8000/，来到系统主页，如图 2-3 所示。

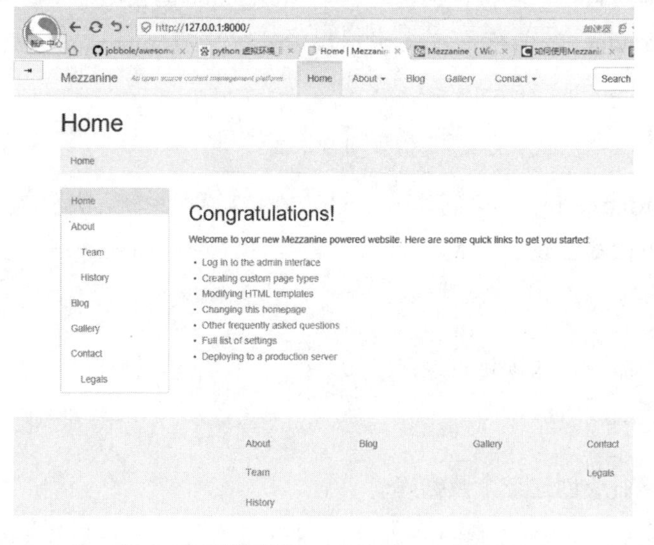


图 2-3　系统主页 http://127.0.0.1:8000/

(5) 后台管理首页是 http://127.0.0.1:8000/admin，如图 2-4 所示。在登录后台管理首页时，使用在创建数据库时设置的管理员账号。

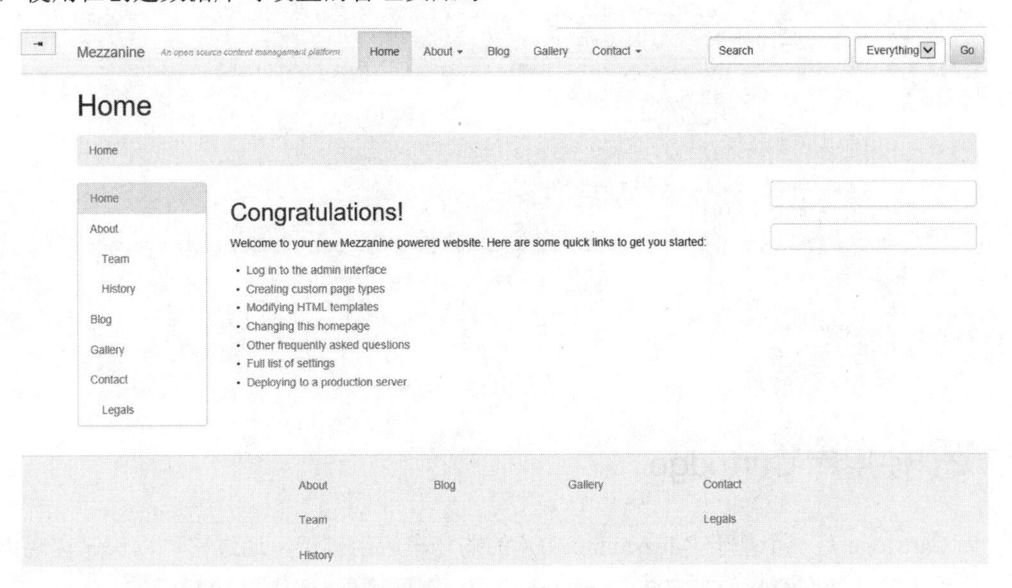


图 2-4　后台管理首页 http://127.0.0.1:8000/admin

后台管理系统的主要功能如下所示。

- 进入 Content > Pages：配置导航、页脚信息。
- 进入 Content > Blog posts：添加分类、发布文章。
- 进入 Site > Settings：配置网站 Site Title、Tagline。

(6) 系统主页默认为显示 Home 页面，如果想让博客的列表作为主页，只须将文件 url.py 中的代码行 un-comment 修改为如下内容：

```
url("^$", "mezzanine.blog.views.blog_post_list", name="home")
```

也就是将文件 url.py 中的如下代码注释掉：

```
#url("^$", direct_to_template, {"template": "index.html"}, name="home")
```

然后将文件 url.py 中的如下代码取消注释：

```
url("^$", "mezzanine.blog.views.blog_post_list", name="home")
```

(7) 如果想去掉导航栏中的 Search 输入框可选项，需要添加如下所示的配置项：

```
SEARCH_MODEL_CHOICES = []
```

如果想去掉左侧菜单连接和页脚，则需要添加如下所示的配置项：

```
PAGE_MENU_TEMPLATES = ( (1, "Top navigation bar",
"pages/menus/dropdown.html"), )
```

(8) Mezzanine 默认支持 4 种数据库，分别是 postgresql_psycopg2、MySQL、SQLite3 和 Oracle，在默认情况下使用 SQLite3。我们可以在文件 local_settings.py 中的如下代码段中修改设置：

```
DATABASES = {
    "default": {
        # Ends with "postgresql_psycopg2", "mysql", "sqlite3" or "oracle".
        "ENGINE": "django.db.backends.sqlite3",
        # DB name or path to database file if using sqlite3.
        "NAME": "dev.db",
        # Not used with sqlite3.
        "USER": "",
        # Not used with sqlite3.
        "PASSWORD": "",
        # Set to empty string for localhost. Not used with sqlite3.
        "HOST": "",
        # Set to empty string for default. Not used with sqlite3.
        "PORT": "",
    }
}
```

## 2.4.2 使用库 Cartridge

库 Cartridge 是一个基于 Mezzanine 构建的购物车应用框架，通过它可以快速实现电子商务应用中的购物车程序。在安装 Cartridge 之前，需要确保已经安装 Mezzanine，然后使用如下命令安装 Cartridge：

```
pip install Cartridge
```

接下来便可以使用 Cartridge 快速创建一个购物车应用程序，具体实现流程如下所示。

(1) 使用如下命令创建一个 Cartridge 项目，项目名称是 car：

```
mezzanine-project -a cartridge car
```

(2) 使用如下命令进入项目目录：

```
cd car
```

(3) 使用如下命令创建一个数据库，默认数据库类型是 SQLite3：

```
python manage.py createdb --noinput
```

在这个过程中，需要填写系统默认的管理员账号信息，其中用户名默认为 admin，密码默认为 default。

(4) 使用如下命令启动这个项目：

```
python manage.py runserver
```

当显示如下所示的信息时，说明成功运行新建的 Cartridge 项目 car：

```
             .....
         _d^^^^^^^^^b_
      .d''           ``b.
    .p'                `q.
   .d'                   `b.
  .d'                     `b.   * Mezzanine 4.2.3
  ::                       ::   * Django 1.10.8
```

```
  ::   M E Z Z A N I N E   ::  * Python 3.6.0
   ::                     ::   * SQLite 3.2.2
    `p.                 .q'   * Windows 10
     `p.               .q'
      `b.             .d'
        `q..        ..p'
          ^q........p^
              ''''

Performing system checks...

System check identified no issues (0 silenced).
April 17, 2018 - 21:05:02
Django version 1.10.8, using settings 'car.settings'
Starting development server at http://127.0.0.1:8000/
Quit the server with CTRL-BREAK.
```

在浏览器的地址栏中输入 http://127.0.0.1:8000/，来到系统主页，如图 2-5 所示。

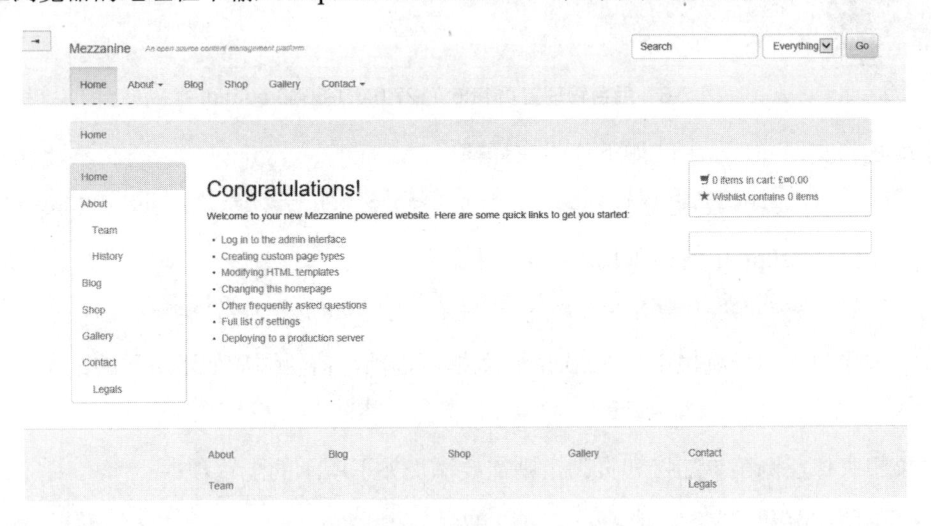


图 2-5　系统主页 http://127.0.0.1:8000/

(5) 后台管理首页是 http://127.0.0.1:8000/admin，如图 2-6 所示。在登录后台管理首页时，使用在创建数据库时提供的默认账号和密码。

跟传统的 Django 和 Mezzanine 项目相比，Cartridge 提供了和电子商务功能密切相关的模块，具体说明如下所示。

- Products：实现商品管理功能。
- Product options：设置商品规格信息，包括颜色、尺寸和其他规格信息。
- Discount codes：设置商品折扣信息。
- Sales：设置销售信息。
- Orders：实现订单管理功能。

(6) 系统主页默认为 Home 页面，如果想让博客的列表主页作为主页，只须将文件 url.py 中的代码行 un-comment 修改为如下内容：

```
url("^$", "mezzanine.blog.views.blog_post_list", name="home")
```

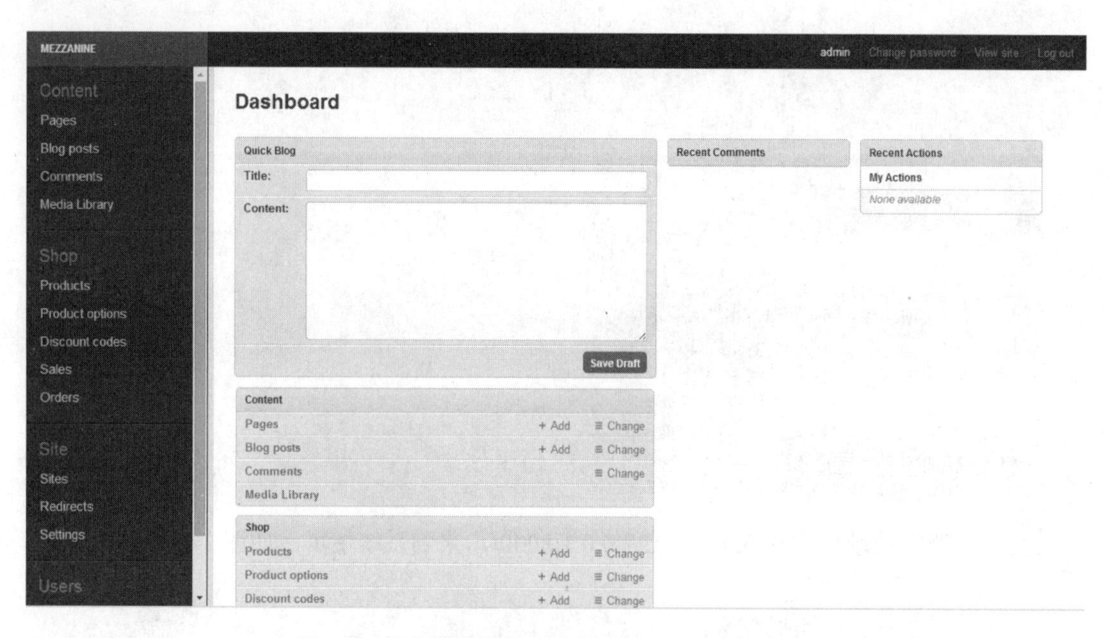


图 2-6 后台管理首页 http://127.0.0.1:8000/admin

也就是将文件 url.py 中的如下代码注释掉：

```
#url("^$", direct_to_template, {"template": "index.html"}, name="home")
```

然后将文件 url.py 中的如下代码取消注释：

```
url("^$", "mezzanine.blog.views.blog_post_list", name="home")
```

(7) 如果想去掉导航栏中的 Search 输入框可选项，需要添加如下所示的配置项：

```
SEARCH_MODEL_CHOICES = []
```

如果想去掉左侧菜单连接和页脚，则需要添加如下所示的配置项：

```
PAGE_MENU_TEMPLATES = ( (1, "Top navigation bar",
"pages/menus/dropdown.html"), )
```

(8) Mezzanine 默认支持 4 种数据库，分别是 postgresql_psycopg2、MySQL、SQLite3 和 Oracle，在默认情况下使用 SQLite3。我们可以在文件 local_settings.py 中的如下代码段中修改设置：

```
DATABASES = {
    "default": {
        # Ends with "postgresql_psycopg2", "mysql", "sqlite3" or "oracle".
        "ENGINE": "django.db.backends.sqlite3",
        # DB name or path to database file if using sqlite3.
        "NAME": "dev.db",
        # Not used with sqlite3.
        "USER": "",
        # Not used with sqlite3.
        "PASSWORD": "",
        # Set to empty string for localhost. Not used with sqlite3.
        "HOST": "",
        # Set to empty string for default. Not used with sqlite3.
```

```
        "PORT": "",
    }
}
```

## 2.5　实现基本功能

本节将详细介绍使用第三方库 Mezzanine 和 Cartridge 实现本系统基本功能的过程，主要包括实现项目配置、后台模块、博客模块和商品展示模块等功能。

### 2.5.1　项目配置

(1) 使用如下命令创建一个 Mezzanine 项目，项目名称是 bookshop：

```
mezzanine-project bookshop
```

(2) 在配置文件 settings.py 的 INSTALLED_APPS 中安装库 Mezzanine 模块和库 Cartridge 模块：

```
INSTALLED_APPS = (
   "django.contrib.admin",
   "django.contrib.auth",
   "django.contrib.contenttypes",
   "django.contrib.redirects",
   "django.contrib.sessions",
   "django.contrib.sites",
   "django.contrib.sitemaps",
   "django.contrib.staticfiles",
   "mezzanine.boot",
   "mezzanine.conf",
   "mezzanine.core",
   "mezzanine.generic",
   "mezzanine.pages",
   "cartridge.shop",
   "mezzanine.blog",
   "mezzanine.forms",
   "mezzanine.galleries",
   "mezzanine.twitter",
   # "mezzanine.accounts",
   # "mezzanine.mobile",
)
```

(3) 文件 urls.py 实现 URL 链接页面的路径导航功能，主要实现代码如下所示：

```
urlpatterns += [
   url("^shop/", include("cartridge.shop.urls")),
   url("^account/orders/$", order_history, name="shop_order_history"),
   url("^$", direct_to_template, {"template": "index.html"}, name="home"),
url("^", include("mezzanine.urls")),
]
```

## 2.5.2 后台模块

(1) 通过如下所示的命令更新系统数据库：

```
python manage.py migrate
```

(2) 通过如下所示的命令新建一个管理员账户：

```
python manage.py createsuperuser
```

在浏览器的地址栏中输入 http://127.0.0.1:8000/admin/，来到后台登录页面，输入上面创建的用户名和密码登录后台，后台管理主页效果如图 2-7 所示。

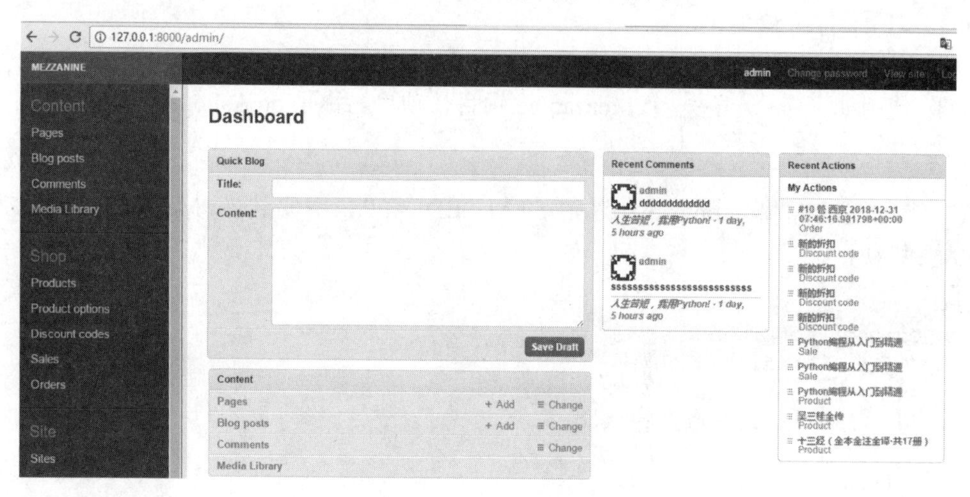


图 2-7 后台管理主页

因为我们使用了第三方库 Cartridge，所以在后台会显示商城模块的功能，例如添加商品页面的效果如图 2-8 所示。

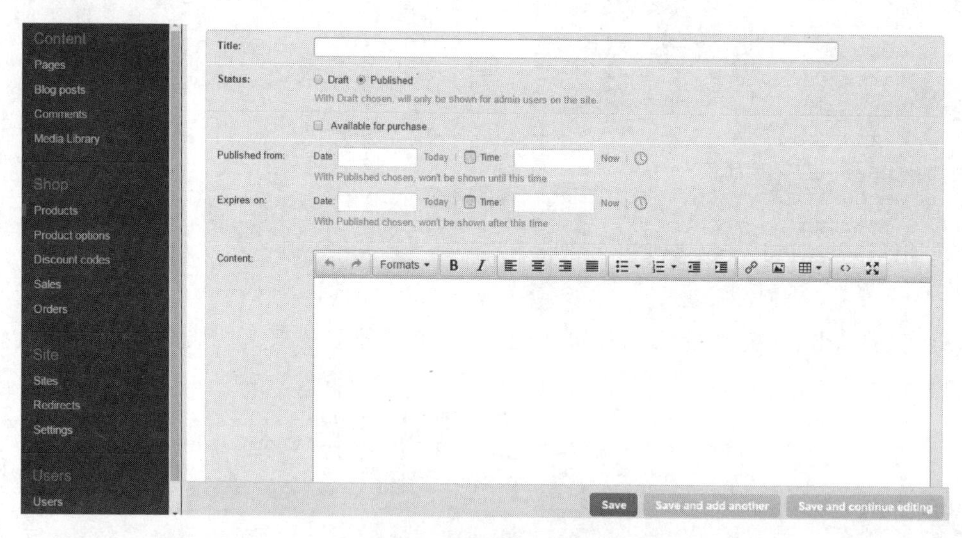


图 2-8 添加商品页面

## 2.5.3 博客模块

(1) 模板文件 blog_post_list.html 的功能是列表显示系统内的博客信息，主要实现代码如下所示：

```
{% block title %}
{% if page %}
{% editable page.title %}{{ page.title }}{% endeditable %}
{% else %}
{% trans "Blog 日志" %}
{% endif %}
{% endblock %}

{% block breadcrumb_menu %}
{{ block.super }}
{% if tag or category or year or month or author %}
<li>{% spaceless %}
{% if tag %}
    {% trans "Tag:" %} {{ tag }}
{% else %}{% if category %}
    {% trans "Category:" %} {{ category }}
{% else %}{% if year or month %}
    {% if month %}{{ month }}, {% endif %}{{ year }}
{% else %}{% if author %}
    {% trans "Author:" %} {{ author.get_full_name|default:author.username }}
{% endif %}{% endif %}{% endif %}{% endif %}
{% endspaceless %}
</li>
{% endif %}
{% endblock %}

{% block main %}

{% if tag or category or year or month or author %}
    {% block blog_post_list_filterinfo %}
    <p>
    {% if tag %}
        {% trans "Viewing posts tagged" %} {{ tag }}
    {% else %}{% if category %}
        {% trans "Viewing posts for the category" %} {{ category }}
    {% else %}{% if year or month %}
        {% trans "Viewing posts from" %} {% if month %}{{ month }}, {% endif %}
        {{ year }}
    {% else %}{% if author %}
        {% trans "Viewing posts by" %}
        {{ author.get_full_name|default:author.username }}
    {% endif %}{% endif %}{% endif %}{% endif %}
    {% endblock %}
    </p>
{% else %}
    {% if page %}
    {% block blog_post_list_pagecontent %}
    {% if page.get_content_model.content %}
```

```
        {% editable page.get_content_model.content %}
        {{ page.get_content_model.content|richtext_filters|safe }}
        {% endeditable %}
    {% endif %}
    {% endblock %}
    {% endif %}
{% endif %}

{% for blog_post in blog_posts.object_list %}
{% block blog_post_list_post_title %}
{% editable blog_post.title %}
<h2>
    <a href="{{ blog_post.get_absolute_url }}">{{ blog_post.title }}</a>
</h2>
{% endeditable %}
{% endblock %}
{% block blog_post_list_post_metainfo %}
{% editable blog_post.publish_date %}
<h6 class="post-meta">
    {% trans "发布者" %}:
    {% with blog_post.user as author %}
    <a href="{% url "blog_post_list_author"
author %}">{{ author.get_full_name|default:author.username }}</a>
    {% endwith %}
    {% with blog_post.categories.all as categories %}
    {% if categories %}
    {% trans "所属类别: " %}
    {% for category in categories %}
    <a href="{% url "blog_post_list_category"
category.slug %}">{{ category }}</a>{% if not forloop.last %}, {% endif %}
    {% endfor %}
    {% endif %}
    {% endwith %}
    {% blocktrans with
sometime=blog_post.publish_date|timesince %}{{ sometime }} ago{%
endblocktrans %}
</h6>
{% endeditable %}
{% endblock %}

{% if settings.BLOG_USE_FEATURED_IMAGE and blog_post.featured_image %}
{% block blog_post_list_post_featured_image %}
<a href="{{ blog_post.get_absolute_url }}">
    <img class="img-thumbnail pull-left" src="{{ MEDIA_URL }}{% thumbnail
blog_post.featured_image 90 90 %}">
</a>
{% endblock %}
{% endif %}

{% block blog_post_list_post_content %}
{% editable blog_post.content %}
{{ blog_post.description_from_content|safe }}
{% endeditable %}
{% endblock %}
```

```
{% block blog_post_list_post_links %}
<div class="blog-list-detail">
    {% keywords_for blog_post as tags %}
    {% if tags %}
    <ul class="list-inline tags">
    {% trans "Tags" %}:
    {% spaceless %}
    {% for tag in tags %}
    <li><a href="{% url "blog_post_list_tag" tag.slug %}"
class="tag">{{ tag }}</a>{% if not forloop.last %}, {% endif %}</li>
    {% endfor %}
    {% endspaceless %}
    </ul>
    {% endif %}
    <p>
    <a href="{{ blog_post.get_absolute_url }}">{% trans "详情" %}</a>
    {% if blog_post.allow_comments %}
    /
    {% if settings.COMMENTS_DISQUS_SHORTNAME %}
```

在浏览器的地址栏中输入 http://127.0.0.1:8000/blog/后，会显示系统内的博客列表，如图 2-9 所示。

图 2-9　博客列表

(2) 模板文件 blog_post_detail.html 的功能是显示某一条博客的详细信息，主要实现代码如下所示：

```
{% block breadcrumb_menu %}
{{ block.super }}
<li class="active">{{ blog_post.title }}</li>
{% endblock %}

{% block main %}

{% block blog_post_detail_postedby %}
```

```
{% editable blog_post.publish_date %}
<h6 class="post-meta">
   {% trans "发布者" %}:
   {% with blog_post.user as author %}
   <a href="{% url "blog_post_list_author"
author %}">{{ author.get_full_name|default:author.username }}</a>
   {% endwith %}
   {% blocktrans with
sometime=blog_post.publish_date|timesince %}{{ sometime }}之前发布{%
endblocktrans %}
</h6>
{% endeditable %}
{% endblock %}
{% block blog_post_detail_commentlink %}
<p>
   {% if blog_post.allow_comments %}
      {% if settings.COMMENTS_DISQUS_SHORTNAME %}
         (<a href="{{ blog_post.get_absolute_url }}#disqus_thread"
            data-disqus-identifier="{% disqus_id_for blog_post %}">{%
spaceless %}
            {% trans "评论" %}
         {% endspaceless %}</a>)
      {% else %}(<a href="#comments">{% spaceless %}
         {% blocktrans count
comments_count=blog_post.comments_count %}{{ comments_count }} comment{%
plural %}{{ comments_count }} comments{% endblocktrans %}
         {% endspaceless %}</a>)
      {% endif %}
   {% endif %}
</p>
{% endblock %}

{% block blog_post_detail_featured_image %}
{% if settings.BLOG_USE_FEATURED_IMAGE and blog_post.featured_image %}
<p><img class="img-responsive" src="{{ MEDIA_URL }}{% thumbnail
blog_post.featured_image 600 0 %}"></p>
{% endif %}
{% endblock %}

{% if settings.COMMENTS_DISQUS_SHORTNAME %}
{% include "generic/includes/disqus_counts.html" %}
{% endif %}

{% block blog_post_detail_content %}
{% editable blog_post.content %}
{{ blog_post.content|richtext_filters|safe }}
{% endeditable %}
{% endblock %}

{% block blog_post_detail_keywords %}
{% keywords_for blog_post as tags %}
{% if tags %}
{% spaceless %}
<ul class="list-inline tags">
   <li>{% trans "Tags" %}:</li>
```

```
    {% for tag in tags %}
    <li><a href="{% url "blog_post_list_tag" tag.slug %}">{{ tag }}</a>{% if not
forloop.last %}, {% endif %}</li>
    {% endfor %}
</ul>
{% endspaceless %}
{% endif %}
{% endblock %}

{% block blog_post_detail_rating %}
<div class="panel panel-default rating">
    <div class="panel-body">
    {% rating_for blog_post %}
    </div>
</div>
{% endblock %}

{% block blog_post_detail_sharebuttons %}
{% set_short_url_for blog_post %}
<a class="btn btn-sm share-twitter" target="_blank"
href="https://twitter.com/intent/tweet?url={{ blog_post.short_url|urlencode
 }}&amp;te xt={{ blog_post.title|urlencode }}">{% trans "Share on
Twitter" %}</a>
<a class="btn btn-sm share-facebook" target="_blank"
href="https://www.facebook.com/sharer/sharer.php?u={{ request.build_absolut
e_uri }}">{% trans "Share on Facebook" %}</a>
{% endblock %}

{% block blog_post_previous_next %}
<ul class="pager">
{% with blog_post.get_previous_by_publish_date as previous %}
{% if previous %}
<li class="previous">
    <a href="{{ previous.get_absolute_url }}">&larr; {{ previous }}</a>
</li>
{% endif %}
{% endwith %}
{% with blog_post.get_next_by_publish_date as next %}
{% if next %}
<li class="next">
    <a href="{{ next.get_absolute_url }}">{{ next }} &rarr;</a>
</li>
{% endif %}
{% endwith %}
</ul>
{% endblock %}

{% block blog_post_detail_related_posts %}
{% if related_posts %}
<div id="related-posts">
<h3>{% trans 'Related posts' %}</h3>
<ul class="list-unstyled">
{% for post in related_posts %}
    <li><a href="{{ post.get_absolute_url }}">{{ post.title }}</a></li>
{% endfor %}
```

```
</ul>
</div>
{% endif %}
```

为了节省服务器的开支，本系统使用静态技术生成每一个博客详情页面，例如某篇博客的标题是“本站郑重承诺，所有商品，假一赔十”，则在浏览器的地址栏中输入“http://127.0.0.1:8000/blog/本站郑重承诺所有商品假一赔十/”后，会显示这篇博客的详细信息，如图 2-10 所示。

图 2-10　博客详情页面

## 2.5.4　商品展示模块

在后台添加一个商品后，在前台可以显示这个商品的详细信息。在模板目录 shop 中保存了商品展示模块的实现文件。

模板文件 product.html 的功能是展示某个商品的详细信息，包括名称、图片、售价、评分和购买数量，主要实现代码如下所示：

```
{% block breadcrumb_menu %}
{{ block.super }}
<li>{{ product.title }}</li>
{% endblock %}

{% block title %}
{% editable product.title %}{{ product.title }}{% endeditable %}
{% endblock %}

{% block main %}

{% if images %}
{% spaceless %}
```

```
<ul id="product-images-large" class="list-unstyled list-inline">
   {% for image in images %}
   <li id="image-{{ image.id }}-large"{% if not
forloop.first %}style="display:none;"{% endif %}>
      <a class="product-image-large" href="{{ MEDIA_URL }}{{ image.file }}">
         <img alt="{{ image.description }}" src="{{ MEDIA_URL }}{% thumbnail
image.file 0 300 %}" class="img-thumbnail img-responsive col-xs-12">
      </a>
   </li>
   {% endfor %}
</ul>

{% if images|length != 1 %}
<ul id="product-images-thumb" class="list-unstyled list-inline">
   {% for image in images %}
   <li>
      <a class="thumbnail" id="image-{{ image.id }}"
href="{{ MEDIA_URL }}{{ image.file }}">
         <img alt="{{ image.description }}" src="{{ MEDIA_URL }}{% thumbnail
image.file 75 75 %}">
      </a>
   </li>
   {% endfor %}
</ul>
{% endif %}

{% endspaceless %}
{% endif %}

{% editable product.content %}
{{ product.content|richtext_filters|safe }}
{% endeditable %}

{% if product.available and has_available_variations %}
<ul id="variations" class="list-unstyled">
   {% for variation in variations %}
   <li id="variation-{{ variation.sku }}"
      {% if not variation.default %}style="display:none;"{% endif %}>
      {% if variation.has_price %}
         {% if variation.on_sale %}
            <span class="old-price">{{ variation.unit_price|currency }}</span>
            {% trans "售价:" %}
         {% endif %}
         <span class="price">{{ variation.price|currency }}</span>
      {% else %}
         {% if has_available_variations %}
         <span class="error-msg">
         {% trans "所选的选项当前不可用" %}
         </span>
         {% endif %}
      {% endif %}
   </li>
   {% endfor %}
</ul>
```

```
{% errors_for add_product_form %}

<form method="post" id="add-cart" class="shop-form">
    {% fields_for add_product_form %}
    <div class="form-actions">
        <input type="submit" class="btn btn-primary btn-lg pull-right"
name="add_cart" value="{% trans "直接购买" %}">
        {% if settings.SHOP_USE_WISHLIST %}
        <input type="submit" class="btn btn-default btn-lg pull-left"
name="add_wishlist" value="{% trans "先保存再购买" %}">
        {% endif %}
    </div>
</form>
{% else %}
<p class="error-msg">{% trans "当前产品不可购买." %}</p>
{% endif %}

{% if settings.SHOP_USE_RATINGS %}
<div class="panel panel-default rating">
    <div class="panel-body">{% rating_for product %}</div>
</div>
{% endif %}

{% if settings.SHOP_USE_RELATED_PRODUCTS and related_products %}
<h2>{% trans "相关产品" %}</h2>
<div class="row related-products">
    {% for product in related_products %}
    <div class="col-xs-6 col-sm-4 col-md-3 product-thumb">
        <a class="thumbnail" href="{{ product.get_absolute_url }}">
            {% if product.image %}
            <img src="{{ MEDIA_URL }}{% thumbnail product.image 90 90 %}">
            {% endif %}
            <div class="caption">
            <h6>{{ product }}</h6>
            <div class="price-info">
            {% if product.has_price %}
                {% if product.on_sale %}
                <span class="old-price">{{ product.unit_price|currency }}</span>
                {% trans "售价:" %}
                {% endif %}
                <span class="price">{{ product.price|currency }}</span>
            {% else %}
                <span class="coming-soon">{% trans "马上" %}</span>
            {% endif %}
            </div>
            </div>
        </a>
    </div>
    {% endfor %}
```

商品展示页面的执行效果如图 2-11 所示。

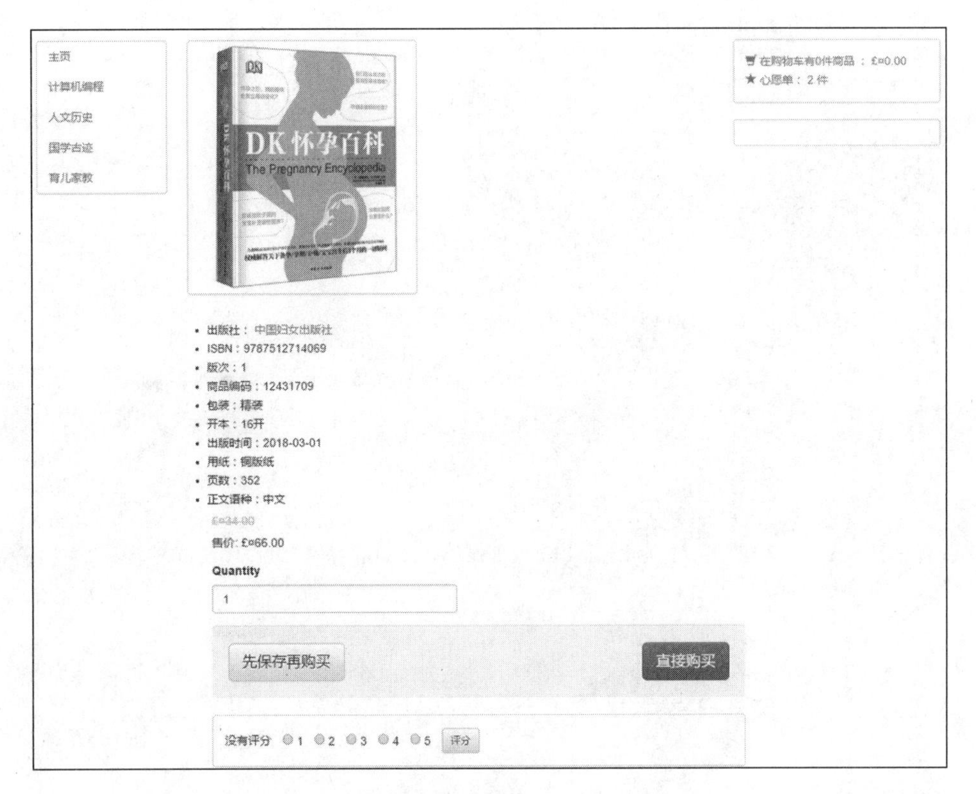


图 2-11　商品展示页面

## 2.6　在线购物

扫码观看视频讲解

在线商城系统的最大特色便是实现在线购物。基于在线购物模块的重要性，所以本系统的购物功能将在本节单独讲解。因为第三方库 Cartridge 为我们提供了完整的购物车功能，所以整个开发过程非常简单。下面将按照本系统的购物流程来讲解在线购物模块的实现过程。

### 2.6.1　购物车页面

单击某商品下面的“直接购买”按钮后，来到购物车页面。模板文件 cart.html 实现了购物车页面，在此页面中显示有购物车中的商品名称、购买数量、单价、总价、“删除”按钮和折扣码信息，主要实现代码如下所示：

```
{{ cart_formset.management_form }}
<table class="table table-striped">
    <thead>
    <tr>
        <th colspan="2" class="left">{% trans "商品" %}</th>
        <th>{% trans "单价" %}</th>
        <th class="center">{% trans "数量" %}</th>
```

```
        <th>{% trans "总价" %}</th>
        <th class="center">{% trans "删除?" %}</th>
    </tr>
    </thead>
    <tbody>
    {% for form in cart_formset.forms %}
    {% with form.instance as item %}
    <tr>
        <td width="30">
            {{ form.id }}
            {% if item.image %}
            <a href="{{ item.get_absolute_url }}">
                <img alt="{{ item.description }}" src="{{ MEDIA_URL }}{% thumbnail
item.image 30 30 %}">
            </a>
            {% endif %}
        </td>
        <td class="left">
            <a href="{{ item.get_absolute_url }}">{{ item.description }}</a>
        </td>
        <td>{{ item.unit_price|currency }}</td>
        <td class="quantity">{{ form.quantity }}</td>
        <td>{{ item.total_price|currency }}</td>
        <td class="center">{{ form.DELETE }}</td>
    </tr>
    {% endwith %}
    {% endfor %}
    <tr>
        <td colspan="5">{% order_totals %}</td>
        <td> </td>
    </tr>
    </tbody>
</table>

<div class="form-actions">
    <a href="{% url "shop_checkout" %}" class="btn btn-primary btn-lg
pull-right">
        {% if request.session.order.step %}{% trans "去结账" %}{% else %}{% trans
"去结账" %}{% endif %}
    </a>
    <input type="submit" name="update_cart" class="btn btn-default btn-lg
pull-left" value="{% trans "更新购物车" %}">
</div>
</form>

{% if discount_form %}
<form method="post" class="discount-form col-md-12 text-right">
    {% fields_for discount_form %}
    <input type="submit" class="btn btn-default" value="{% trans "支付" %}">
</form>
{% endif %}

{% if settings.SHOP_USE_UPSELL_PRODUCTS %}
{% with request.cart.upsell_products as upsell_products %}
{% if upsell_products %}
```

```
<h2>{% trans "You may also like:" %}</h2>
<div class="row">
    {% for product in upsell_products %}
    <div class="col-xs-6 col-sm-4 col-md-3 product-thumb">
        <a class="thumbnail" href="{{ product.get_absolute_url }}">
            {% if product.image %}
            <img src="{{ MEDIA_URL }}{% thumbnail product.image 90 90 %}">
            {% endif %}
            <div class="caption">
            <h6>{{ product }}</h6>
            <div class="price-info">
            {% if product.has_price %}
                {% if product.on_sale %}
                <span class="old-price">{{ product.unit_price|currency }}</span>
                {% trans "On sale:" %}
                {% endif %}
                <span class="price">{{ product.price|currency }}</span>
            {% else %}
                <span class="coming-soon">{% trans "Coming soon" %}</span>
            {% endif %}
            </div>
            </div>
        </a>
    </div>
    {% endfor %}
</div>
{% endif %}
{% endwith %}
{% endif %}
```

购物车页面的执行效果如图 2-12 所示。

图 2-12　购物车页面

## 2.6.2　订单详情页面

单击“去结账”按钮后，来到订单详情页面 checkout.html。在此页面中设置收货人的详细联系信息。文件 checkout.html 的主要实现代码如下所示：

```
{% block meta_title %}{% trans "结账" %} - {{ step_title }}{% endblock %}
```

```
{% block title %}{% trans "订单支付" %} - {% trans "步" %} {{ step }} {% trans
"/" %} {{ steps|length }}{% endblock %}
{% block body_id %}checkout{% endblock %}

{% block extra_head %}
<script>
var _gaq = [['_trackPageview', '{{ request.path }}{{ step_url }}/']];
$(function() {$('.middle :input:visible:enabled:first').focus();});
</script>
{% endblock %}

{% block breadcrumb_menu %}
{% for step in steps %}
<li>
   {% if step.title == step_title %}
   <strong>{{ step.title }}</strong>
   {% else %}
   {{ step.title }}
   {% endif %}
</li>
{% endfor %}
<li>{% trans "完成" %}</li>
{% endblock %}

{% block main %}

{% block before-form %}{% endblock %}
<div class="row">
<form method="post" class="col-md-8 checkout-form">
   {% csrf_token %}

   {% block fields %}{% endblock %}

   {% block nav-buttons %}
      {% if request.cart.has_items %}
         <div class="form-actions">
            <input type="submit" class="btn btn-lg btn-primary pull-right"
value="{% trans "下一步" %}">
            {% if not CHECKOUT_STEP_FIRST %}
            <input type="submit" class="btn btn-lg btn-default pull-left"
name="back" value="{% trans "后退" %}">
            {% endif %}
         </div>
      {% else %}
         <p>{% trans "你的购物车为空." %}</p>
         <p>{% trans "会话超时." %}</p>
         <p>{% trans "给您带来的不便，我们深表歉意。" %}</p>
         <br>
         <p><a class="btn btn-lg btn-primary" href="{% url "page" "shop" %}">{%
trans "继续购物" %}</a></p>
      {% endif %}
   {% endblock %}

</form>
```

```
{% if request.cart.has_items %}
<div class="col-md-4">
   <div class="panel panel-default checkout-panel">
   <div class="panel-body">
   <ul class="media-list">
   {% for item in request.cart %}
   <li class="media">
      {% if item.image %}
      <img class="pull-left" alt="{{ item.description }}"
src="{{ MEDIA_URL }}{% thumbnail item.image 30 30 %}">
      {% endif %}
      <div class="media-body">
         {{ item.quantity }} x {{ item.description }}
         <span class="price">{{ item.total_price|currency }}</span>
      </div>
   </li>
   {% endfor %}
   </ul>
   {% order_totals %}
   <br style="clear:both;">
   <a class="btn btn-default" href="{% url "shop_cart" %}">{% trans "修改购物
车" %}</a>
   </div>
   </div>
</div>
{% endif %}
```

订单详情页面的执行效果如图 2-13 所示。

图 2-13　订单详情页面

### 2.6.3 在线支付页面

单击“下一步”按钮，来到在线支付页面 payment_fields.html，在此页面设置使用的银行卡信息。文件 payment_fields.html 的具体实现代码如下所示：

```
{% load i18n mezzanine_tags %}
<fieldset>
    <legend>{% trans "支付信息" %}</legend>
    {% fields_for form.card_name_field %}
    {% fields_for form.card_type_field %}
    {% with form.card_expiry_fields as card_expiry_fields %}
    <div class="form-group card-expiry-fields{% if
card_expiry_fields.errors.card_expiry_year %} error{% endif %}">
        <label class="control-label">{% trans "您的银行卡已经过期" %}</label>
        {% fields_for card_expiry_fields %}
    </div>
    <div class="clearfix"></div>
    {% endwith %}
    {% fields_for form.card_fields %}
</fieldset>
```

在线支付页面的执行效果如图 2-14 所示。

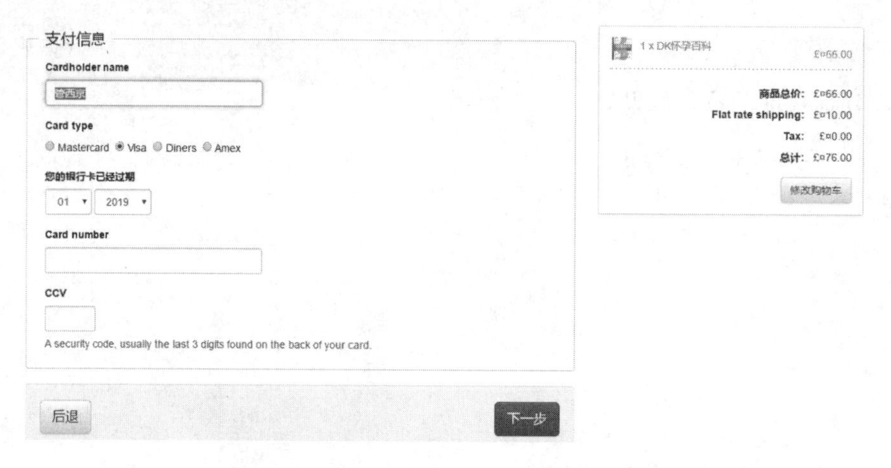


图 2-14 在线支付页面

### 2.6.4 订单确认页面

单击“下一步”按钮，来到订单确认页面 confirmation.html，在此页面显示订单信息、快递信息和购物车信息，是购物者完成购物前的最后确认工作。文件 confirmation.html 的主要实现代码如下所示：

```
<div class="confirmation col-md-6">
    <div class="panel panel-default">
    <div class="panel-body">
    <h3>{% trans "订单信息" %}</h3>
    <ul class="list-unstyled">
```

```
        {% for field, value in form.billing_detail_fields.values %}
        <li><label>{{ field }}:</label> {{ value }}</li>
        {% endfor %}

    </ul>
    </div>
    </div>
</div>

<div class="confirmation col-md-6">
    <div class="panel panel-default">
    <div class="panel-body">
    <h3>{% trans "快递信息" %}</h3>
    <ul class="list-unstyled">

        {% for field, value in form.shipping_detail_fields.values %}
        <li><label>{{ field }}:</label> {{ value }}</li>
        {% endfor %}

        {% for field, value in form.additional_instructions_field.values %}
        <li><label>{{ field }}:</label> {{ value }}</li>
        {% endfor %}

    </ul>
    </div>
    </div>
</div>
{% if settings.SHOP_PAYMENT_STEP_ENABLED %}
{% comment %}
<br style="clear:both;">
<div class="confirmation col-md-6">
    <div class="panel panel-default">
    <div class="panel-body">
    <h3>{% trans "支付信息" %}</h3>
    <ul class="list-unstyled">

        {% for field, value in form.card_name_field.values %}
        <li><label>{{ field }}:</label> {{ value }}</li>
        {% endfor %}

        {% for field, value in form.card_type_field.values %}
        <li><label>{{ field }}:</label> {{ value }}</li>
        {% endfor %}

        <li>
            {% with form.card_expiry_fields.values as expiry_fields %}
            {% with expiry_fields.next as month_field %}
            <label>{{ month_field.0 }}:</label>
            {{ month_field.1 }}/{{ expiry_fields.next.1 }}
            {% endwith %}
            {% endwith %}
        </li>

        {% for field, value in form.card_fields.values %}
```

```
        <li><label>{{ field }}:</label> {{ value }}</li>
        {% endfor %}

    </ul>
    </div>
    </div>
</div>
```

订单确认页面的执行效果如图 2-15 所示。

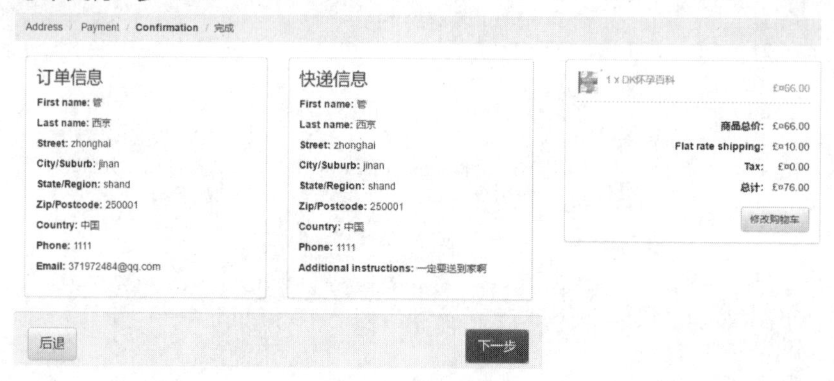


图 2-15　订单确认页面

## 2.6.5　订单完成发送提醒邮件

单击“下一步”按钮，来到订单完成页面 complete.html，在此页面不但可以查看订单信息，而且可以发送一封提醒邮件到会员邮箱。文件 complete.html 的主要实现代码如下所示：

```
{% block title %}{% trans "完成订单" %}{% endblock %}

{% block breadcrumb_menu %}
{% for step in steps %}
<li>{{ step.title }}</li>
{% endfor %}
<li><strong>{% trans "完成" %}</strong></li>
{% endblock %}


{% block main %}
<p>{% trans "感谢您的购物，您的订单已完成." %}</p>
<p>{% trans "我们已通过电子邮件向您发送了订单信息." %}</p>
<p>{% trans "您也可以使用以下链接之一查看订单信息." %}</p>
<br>
<form class="order-complete-form" method="post" action="{% url
"shop_invoice_resend" order.id %}?next={{ request.path }}">
    {% csrf_token %}
    {% if has_pdf %}
    <a class="btn btn-primary" href="{% url "shop_invoice"
order.id %}?format=pdf">{% trans "下载 PDF 格式的订单" %}</a>
```

```
    {% endif %}
    <input type="submit" class="btn btn-default" value="{% trans "重新发送提醒邮件" %}">
</form>
{% endblock %}
```

在文件 settings.py 中设置邮件服务器的信息，主要代码如下所示：

```
EMAIL_HOST='smtp.qq.com'
EMAIL_HOST_PASSWORD=''
EMAIL_HOST_USER='729××××××@qq.com'
EMAIL_PORT=25

EMAIL_SUBJECT_PREFIX='[Django] '
EMAIL_USE_TLS=True
```

订单完成页面的执行效果如图 2-16 所示。

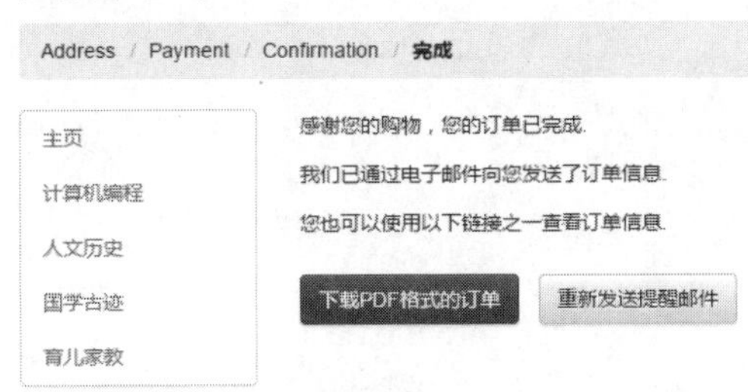


图 2-16　订单完成页面

在发送的提醒邮件中，会显示订单的详细信息，如图 2-17 所示。

**您的订单信息**

| 订单详情 | 快递详情 |
| --- | --- |
| First name: 管<br>Last name: 西京<br>Street: zhonghai<br>City/Suburb: jinan<br>State/Region: shand<br>Zip/Postcode: 250001<br>Country: 中国<br>Phone: 1111<br>Email: 371972484@qq.com | First name: 管<br>Last name: 西京<br>Street: zhonghai<br>City/Suburb: jinan<br>State/Region: shand<br>Zip/Postcode: 250001<br>Country: 中国<br>Phone: 1111 |

**订单商品**

| 商品 | 单价 | 数量 | 价格 |
| --- | --- | --- | --- |
| DK怀孕百科 | £¤66.00 | 1 | £¤66.00 |
| | | | 商品总价: £¤66.00<br>Flat rate shipping: £¤10.00<br>Tax: £¤0.00<br>总计: £¤76.00 |

图 2-17　提醒邮件中的订单信息

# 第 3 章

# 房产价格数据可视化分析系统

## （网络爬虫+ MySQL+pylab 实现）

房产的价格现在已经成为人们最关注的对象之一，些许的风吹草动都会引起大家的注意。本章将详细讲解使用 Python 语言采集主流网站中国内主流城市房价信息的过程，包括新房价格、二手房价格和房租价格，并进一步分析这些采集的数据，将房产的价格用更加直观的图表形式展示出来。

## 3.1 背景介绍

随着房价的不断升高，人们对房价的关注度也越来越高，房产投资者希望通过房价数据预判房价走势，从而进行有效的投资，获取收益；因结婚、小孩上学等需要买房的民众，希望通过房价数据寻找买房的最佳时机，以最合适的价格购买能满足需要的房产。

在当前市场环境下，因为房价水平牵动了大多数人的心，所以各大房产网站都上线了“查房价”相关的功能模块，以满足购房者或计划购房者关注房价行情的需求，从而实现增加产品活跃度、促进购房转化的目的。

各个房产网市场的用户群大都相同，但主要房源资源和营销方式有所差异。然而，房产网巨头公司的房源，由于有品牌与质量的优势正快速扩张，市场上的推广费用也越来越贵。而购房者迫切希望通过分析找到最精确的房价查询系统，在这个时候，推出一款能够完整展示房产信息的软件变得愈发重要。

## 3.2 需求分析

本项目将提供国内主流城市、每个区域、每个小区的房价成交情况、关注情况、发展走势，乃至每个小区的解读/评判，以解决用户购房没有价格依据、无从选择购房时机的问题；满足用户及时了解房价行情、以最合适价格购买最合适位置房产的需求。

通过使用本系统，可以产生如下所示的价值。

- 增加活跃：由于对房价的关注是中长期性质的，不断更新的行情数据可以增加用户的活跃度。
- 促进转化：使用房价数据在用户购房时帮助推荐合适的位置与价格，可以提高用户的咨询率与成交率。
- 减少流失：若没有此功能，会导致一些购房观望者无从得知房价变化，而选择最终离开。

## 3.3 模块架构

本房产价格数据可视化系统的基本模块架构如图 3-1 所示。

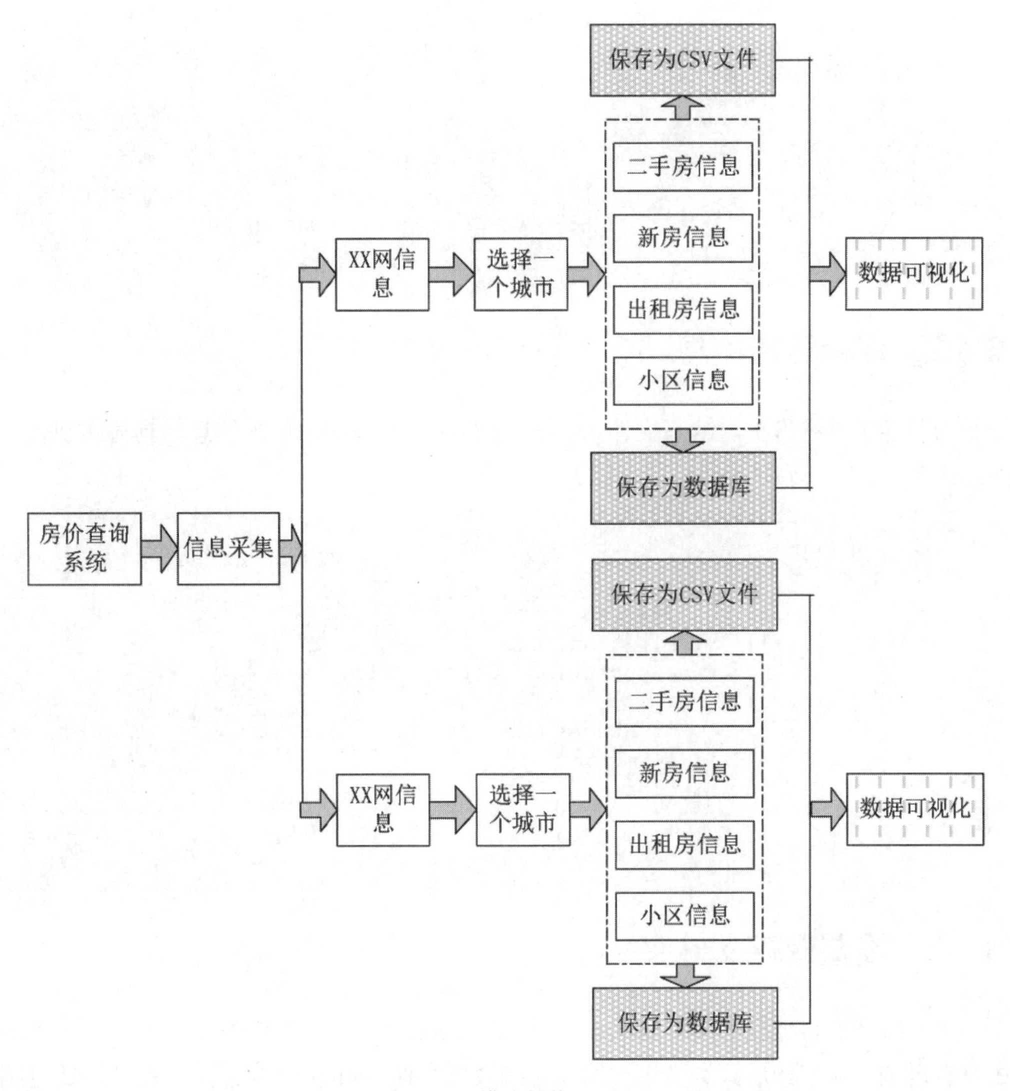


图 3-1　模块架构

## 3.4　系统设置

在开发一个大型应用程序时，需要模块化开发经常用到的系统设置模块。本节将详细讲解实现本项目系统模块的过程。

### 3.4.1　选择版本

因为在当前市面中同时存在 Python 2 和 Python 3 版本，所以本系统分别推出了对应的两个实现版本。编写文件 version.py，供用户选择使用不同的 Python 版本，具体实现代码如下所示：

```
import sys
```

```
if sys.version_info < (3, 0):   # 如果小于 Python 3
    PYTHON_3 = False
else:
    PYTHON_3 = True

if not PYTHON_3:   # 如果不是 Python 3
    reload(sys)
    sys.setdefaultencoding("utf-8")
```

## 3.4.2 保存日志信息

为了便于系统维护，编写文件 log.py 保存使用本系统的日志信息，具体实现代码如下所示：

```
import logging
from lib.utility.path import LOG_PATH

logger = logging.getLogger(__name__)
logger.setLevel(level=logging.INFO)
handler = logging.FileHandler(LOG_PATH + "/log.txt")
handler.setLevel(logging.INFO)
formatter = logging.Formatter('%(asctime)s - %(levelname)s - %(message)s')
handler.setFormatter(formatter)
logger.addHandler(handler)

if __name__ == '__main__':
    pass
```

## 3.4.3 设置保存文件夹

本系统能够将抓取的房价信息保存到本地 CSV 文件中，保存 CSV 文件的文件夹的命名机制有日期、城市和房源类型等。编写系统设置文件 path.py，功能是根据不同的机制创建对应的文件夹来保存 CSV 文件。文件 path.py 的具体实现代码如下所示：

```
def get_root_path():
    file_path = os.path.abspath(inspect.getfile(sys.modules[__name__]))
    parent_path = os.path.dirname(file_path)
    lib_path = os.path.dirname(parent_path)
    root_path = os.path.dirname(lib_path)
    return root_path


def create_data_path():
    root_path = get_root_path()
    data_path = root_path + "/data"
    if not os.path.exists(data_path):
        os.makedirs(data_path)
    return data_path


def create_site_path(site):
```

```
    data_path = create_data_path()
    site_path = data_path + "/" + site
    if not os.path.exists(site_path):
        os.makedirs(site_path)
    return site_path


def create_city_path(site, city):
    site_path = create_site_path(site)
    city_path = site_path + "/" + city
    if not os.path.exists(city_path):
        os.makedirs(city_path)
    return city_path


def create_date_path(site, city, date):
    city_path = create_city_path(site, city)
    date_path = city_path + "/" + date
    if not os.path.exists(date_path):
        os.makedirs(date_path)
    return date_path


# const for path
ROOT_PATH = get_root_path()
DATA_PATH = ROOT_PATH + "/data"
SAMPLE_PATH = ROOT_PATH + "/sample"
LOG_PATH = ROOT_PATH + "/log"

if __name__ == "__main__":
    create_date_path("lianjia", "sh", "20160912")
    create_date_path("beike", "bj", "20160912")
```

### 3.4.4　设置爬取城市

本系统能够爬取国内主流一线、二线城市的房价。编写文件 city.py，设置要爬取的城市，实现城市缩写和城市名的映射。如果想爬取其他已有城市数据的话，需要把相关城市的信息放入到文件 city.py 的字典中。文件 city.py 的具体实现代码如下所示：

```
cities = {
    'bj': '北京',
    'cd': '成都',
    'cq': '重庆',
    'cs': '长沙',
    'dg': '东莞',
    'dl': '大连',
    'fs': '佛山',
    'gz': '广州',
    'hz': '杭州',
    'hf': '合肥',
    'jn': '济南',
    'nj': '南京',
    'qd': '青岛',
```

```
    'sh': '上海',
    'sz': '深圳',
    'su': '苏州',
    'sy': '沈阳',
    'tj': '天津',
    'wh': '武汉',
    'xm': '厦门',
    'yt': '烟台',
}


lianjia_cities = cities
beike_cities = cities


def create_prompt_text():
    """
    根据已有城市中英文对照表拼接选择提示信息
    :return: 拼接好的字串
    """
    city_info = list()
    count = 0
    for en_name, ch_name in cities.items():
        count += 1
        city_info.append(en_name)
        city_info.append(": ")
        city_info.append(ch_name)
        if count % 4 == 0:
            city_info.append("\n")
        else:
            city_info.append(", ")
    return 'Which city do you want to crawl?\n' + ''.join(city_info)


def get_chinese_city(en):
    """
    将拼音名转中文城市名
    :param en: 拼音
    :return: 中文
    """
    return cities.get(en, None)


def get_city():
    city = None
    # 允许用户通过命令直接指定
    if len(sys.argv) < 2:
        print("Wait for your choice.")
        # 让用户选择爬取哪个城市的价格数据
        prompt = create_prompt_text()
        # 判断 Python 版本
        if not PYTHON_3:  # 如果不是 Python 3
            city = raw_input(prompt)
        else:
```

```
        city = input(prompt)
    elif len(sys.argv) == 2:
        city = str(sys.argv[1])
        print("City is: {0}".format(city))
    else:
        print("At most accept one parameter.")
        exit(1)

    chinese_city = get_chinese_city(city)
    if chinese_city is not None:
        message = 'OK, start to crawl ' + get_chinese_city(city)
        print(message)
        logger.info(message)
    else:
        print("No such city, please check your input.")
        exit(1)
    return city


if __name__ == '__main__':
    print(get_chinese_city("sh"))
```

## 3.4.5 处理区县信息

(1) 因为每个城市都有不同的行政区，所以编写文件 area.py 处理区县信息，具体实现代码如下所示：

```
def get_district_url(city, district):
    """
    拼接指定城市的区县 url
    :param city: 城市
    :param district: 区县
    :return:
    """
    return "http://{0}.{1}.com/xiaoqu/{2}".format(city, SPIDER_NAME, district)


def get_areas(city, district):
    """
    通过城市和区县名获得下级板块名
    :param city: 城市
    :param district: 区县
    :return: 区县列表
    """
    page = get_district_url(city, district)
    areas = list()
    try:
        headers = create_headers()
        response = requests.get(page, timeout=10, headers=headers)
        html = response.content
        root = etree.HTML(html)
        links = root.xpath(DISTRICT_AREA_XPATH)
```

```
        # 针对超级链接标签 a 中的子标签 list 进行处理
        for link in links:
            relative_link = link.attrib['href']
            # 去掉最后的"/"
            relative_link = relative_link[:-1]
            # 获取最后一节
            area = relative_link.split("/")[-1]
            # 去掉区县名,防止重复
            if area != district:
                chinese_area = link.text
                chinese_area_dict[area] = chinese_area
                # print(chinese_area)
                areas.append(area)
        return areas
    except Exception as e:
        print(e)
```

(2) 编写文件 district.py，获取各个区县的详细信息，具体实现代码如下所示：

```
chinese_city_district_dict = dict()      # 城市代码和中文名映射
chinese_area_dict = dict()               # 板块代码和中文名映射
area_dict = dict()


def get_chinese_district(en):
    """
    拼音区县名转中文区县名
    :param en: 英文
    :return: 中文
    """
    return chinese_city_district_dict.get(en, None)


def get_districts(city):
    """
    获取各城市的区县中英文对照信息
    :param city: 城市
    :return: 英文区县名列表
    """
    url = 'https://{0}.{1}.com/xiaoqu/'.format(city, SPIDER_NAME)
    headers = create_headers()
    response = requests.get(url, timeout=10, headers=headers)
    html = response.content
    root = etree.HTML(html)
    elements = root.xpath(CITY_DISTRICT_XPATH)
    en_names = list()
    ch_names = list()
    for element in elements:
        link = element.attrib['href']
        en_names.append(link.split('/')[-2])
        ch_names.append(element.text)

        # 打印区县英文和中文名列表
    for index, name in enumerate(en_names):
        chinese_city_district_dict[name] = ch_names[index]
```

```
        # print(name + ' -> ' + ch_names[index])
    return en_names
```

## 3.5 破解反爬机制

扫码观看视频讲解

市面中的很多站点都设立了反爬机制，防止站点内的信息被爬取。本节将详细讲解破解反爬机制的过程。

### 3.5.1 定义爬虫基类

编写文件 base_spider.py，定义爬虫基类，具体过程是首先设置随机延迟，防止爬虫被禁止；然后设置要爬取的目标站点(下面代码默认抓取的是×壳)；最后获取城市列表来选择将要爬取的目标城市。文件 base_spider.py 的具体实现代码如下所示。

```
thread_pool_size = 50

# 防止爬虫被禁，设定随机延迟
# 如果不想延迟，就设定 False
# 具体时间可以修改 random_delay，由于为多线程，建议数值大于 10
RANDOM_DELAY = False
LIANJIA_SPIDER = "lianjia"
BEIKE_SPIDER = "ke"
# SPIDER_NAME = LIANJIA_SPIDER
SPIDER_NAME = BEIKE_SPIDER


class BaseSpider(object):
    @staticmethod
    def random_delay():
        if RANDOM_DELAY:
            time.sleep(random.randint(0, 16))

    def __init__(self, name):
        self.name = name
        if self.name == LIANJIA_SPIDER:
            self.cities = lianjia_cities
        elif self.name == BEIKE_SPIDER:
            self.cities = beike_cities
        else:
            self.cities = None
        # 准备日期信息，爬到的数据存放到日期相关文件夹下
        self.date_string = get_date_string()
        print('Today date is: %s' % self.date_string)

        self.total_num = 0  # 总的小区个数，用于统计
        print("Target site is {0}.com".format(SPIDER_NAME))
        self.mutex = threading.Lock()  # 创建锁

    def create_prompt_text(self):
        """
```

```
        根据已有城市中英文对照表拼接选择提示信息
        :return: 拼接好的字串
        """
        city_info = list()
        count = 0
        for en_name, ch_name in self.cities.items():
            count += 1
            city_info.append(en_name)
            city_info.append(": ")
            city_info.append(ch_name)
            if count % 4 == 0:
                city_info.append("\n")
            else:
                city_info.append(", ")
        return 'Which city do you want to crawl?\n' + ''.join(city_info)

    def get_chinese_city(self, en):
        """
        拼音名转中文城市名
        :param en: 拼音
        :return: 中文
        """
        return self.cities.get(en, None)
```

## 3.5.2 浏览器用户代理

编写文件 headers.py，实现浏览器用户代理功能，具体实现代码如下所示：

```
USER_AGENTS = [
    "Mozilla/4.0 (compatible; MSIE 6.0; Windows NT 5.1; SV1; AcooBrowser; .NET
CLR 1.1.4322; .NET CLR 2.0.50727)",
    "Mozilla/4.0 (compatible; MSIE 7.0; Windows NT 6.0; Acoo Browser; SLCC1; .NET
CLR 2.0.50727; Media Center PC 5.0; .NET CLR 3.0.04506)",
    "Mozilla/4.0 (compatible; MSIE 7.0; AOL 9.5; AOLBuild 4337.35; Windows NT
5.1; .NET CLR 1.1.4322; .NET CLR 2.0.50727)",
    "Mozilla/5.0 (Windows; U; MSIE 9.0; Windows NT 9.0; en-US)",
    "Mozilla/5.0 (compatible; MSIE 9.0; Windows NT 6.1; Win64; x64;
Trident/5.0; .NET CLR 3.5.30729; .NET CLR 3.0.30729; .NET CLR 2.0.50727; Media
Center PC 6.0)",
    "Mozilla/5.0 (compatible; MSIE 8.0; Windows NT 6.0; Trident/4.0; WOW64;
Trident/4.0; SLCC2; .NET CLR 2.0.50727; .NET CLR 3.5.30729; .NET CLR
3.0.30729; .NET CLR 1.0.3705; .NET CLR 1.1.4322)",
    "Mozilla/4.0 (compatible; MSIE 7.0b; Windows NT 5.2; .NET CLR 1.1.4322; .NET
CLR 2.0.50727; InfoPath.2; .NET CLR 3.0.04506.30)",
    "Mozilla/5.0 (Windows; U; Windows NT 5.1; zh-CN) AppleWebKit/53.15 (KHTML,
like Gecko, Safari/419.3) Arora/0.3 (Change: 287 c9dfb30)",
    "Mozilla/5.0 (X11; U; Linux; en-US) AppleWebKit/527+ (KHTML, like Gecko,
Safari/419.3) Arora/0.6",
    "Mozilla/5.0 (Windows; U; Windows NT 5.1; en-US; rv:1.8.1.2pre)
Gecko/20070215 K-Ninja/2.1.1",
    "Mozilla/5.0 (Windows; U; Windows NT 5.1; zh-CN; rv:1.9) Gecko/20080705
Firefox/3.0 Kapiko/3.0",
    "Mozilla/5.0 (X11; Linux i686; U;) Gecko/20070322 Kazehakase/0.4.5",
```

```
    "Mozilla/5.0 (X11; U; Linux i686; en-US; rv:1.9.0.8) Gecko
Fedora/1.9.0.8-1.fc10 Kazehakase/0.5.6",
    "Mozilla/5.0 (Windows NT 6.1; WOW64) AppleWebKit/535.11 (KHTML, like Gecko)
Chrome/17.0.963.56 Safari/535.11",
    "Mozilla/5.0 (Macintosh; Intel Mac OS X 10_7_3) AppleWebKit/535.20 (KHTML,
like Gecko) Chrome/19.0.1036.7 Safari/535.20",
    "Opera/9.80 (Macintosh; Intel Mac OS X 10.6.8; U; fr) Presto/2.9.168
Version/11.52",
]

def create_headers():
    headers = dict()
    headers["User-Agent"] = random.choice(USER_AGENTS)
    headers["Referer"] = "http://www.{0}.com".format(SPIDER_NAME)
    return headers
```

### 3.5.3　在线 IP 代理

编写文件 proxy.py，功能是模拟专业在线代理中的 IP 地址，具体实现代码如下所示：

```
def spider_proxyip(num=10):
    try:
        url = 'http://www.网站域名.com/nt/1'
        req = requests.get(url, headers=create_headers())
        source_code = req.content
        print(source_code)
        soup = BeautifulSoup(source_code, 'lxml')
        ips = soup.findAll('tr')

        for x in range(1, len(ips)):
            ip = ips[x]
            tds = ip.findAll("td")
            proxy_host = "{0}://".format(tds[5].contents[0]) +\
            tds[1].contents[0] + ":" + tds[2].contents[0]
            proxy_temp = {tds[5].contents[0]: proxy_host}
            proxys_src.append(proxy_temp)
            if x >= num:
                break
    except Exception as e:
        print("spider_proxyip exception:")
        print(e)
```

## 3.6　爬虫抓取信息

扫码观看视频讲解

本系统的核心是爬虫抓取房价信息，本节将详细讲解爬取不同类型房价信息的过程。

### 3.6.1　设置解析元素

编写文件 xpath.py，功能是根据要爬取的目标网站设置要抓取的 HTML 元素，具体实现代码如下所示。

```
from lib.spider.base_spider import SPIDER_NAME, LIANJIA_SPIDER, BEIKE_SPIDER

if SPIDER_NAME == LIANJIA_SPIDER:
    ERSHOUFANG_QU_XPATH = '//*[@id="filter-options"]/dl[1]/dd/div/a'
    ERSHOUFANG_BANKUAI_XPATH = '//*[@id="filter-options"]/dl[1]/dd/div[2]/a'
    XIAOQU_QU_XPATH = '//*[@id="filter-options"]/dl[1]/dd/div/a'
    XIAOQU_BANKUAI_XPATH = '//*[@id="filter-options"]/dl[1]/dd/div[2]/a'
    DISTRICT_AREA_XPATH = '//div[3]/div[1]/dl[2]/dd/div/div[2]/a'
    CITY_DISTRICT_XPATH = '///div[3]/div[1]/dl[2]/dd/div/div/a'
elif SPIDER_NAME == BEIKE_SPIDER:
    ERSHOUFANG_QU_XPATH = '//*[@id="filter-options"]/dl[1]/dd/div/a'
    ERSHOUFANG_BANKUAI_XPATH = '//*[@id="filter-options"]/dl[1]/dd/div[2]/a'
    XIAOQU_QU_XPATH = '//*[@id="filter-options"]/dl[1]/dd/div/a'
    XIAOQU_BANKUAI_XPATH = '//*[@id="filter-options"]/dl[1]/dd/div[2]/a'
    DISTRICT_AREA_XPATH = '//div[3]/div[1]/dl[2]/dd/div/div[2]/a'
    CITY_DISTRICT_XPATH = '///div[3]/div[1]/dl[2]/dd/div/div/a'
```

## 3.6.2 爬取二手房信息

(1) 编写文件 ershou_spider.py，定义爬取二手房数据的爬虫派生类，具体实现流程如下所示。

- 编写函数 collect_area_ershou_data()，功能是获取每个板块下所有的二手房信息，并且将这些信息写入到 CSV 文件中保存。对应代码如下所示：

```
def collect_area_ershou_data(self, city_name, area_name, fmt="csv"):
    """
      :param city_name: 城市
    :param area_name: 板块
    :param fmt: 保存文件格式
    :return: None
    """
    district_name = area_dict.get(area_name, "")
    csv_file = self.today_path + "/{0}_{1}.csv".format(district_name, area_name)
    with open(csv_file, "w") as f:
        # 开始获得需要的板块数据
        ershous = self.get_area_ershou_info(city_name, area_name)
        # 锁定，多线程读写
        if self.mutex.acquire(1):
            self.total_num += len(ershous)
            # 释放
            self.mutex.release()
        if fmt == "csv":
            for ershou in ershous:
                # print(date_string + "," + xiaoqu.text())
                f.write(self.date_string + "," + ershou.text() + "\n")
    print("Finish crawl area: " + area_name + ", save data to : " + csv_file)
```

- 编写函数 get_area_ershou_info()，功能是通过爬取页面获得城市指定板块的二手房信息，对应代码如下所示：

```
@staticmethod
def get_area_ershou_info(city_name, area_name):
```

```
"""
:param city_name: 城市
:param area_name: 板块
:return: 二手房数据列表
"""
total_page = 1
district_name = area_dict.get(area_name, "")
# 中文区县
chinese_district = get_chinese_district(district_name)
# 中文板块
chinese_area = chinese_area_dict.get(area_name, "")

ershou_list = list()
page = 'http://{0}.{1}.com/ershoufang/{2}/'.format(city_name,
   SPIDER_NAME, area_name)
print(page)  # 打印板块页面地址
headers = create_headers()
response = requests.get(page, timeout=10, headers=headers)
html = response.content
soup = BeautifulSoup(html, "lxml")

# 获得总的页数，查找总页码的元素信息
try:
   page_box = soup.find_all('div', class_='page-box')[0]
   matches = re.search('.*"totalPage":(\d+),.*', str(page_box))
   total_page = int(matches.group(1))
except Exception as e:
   print("\tWarning: only find one page for {0}".format(area_name))
   print(e)

# 从第一页开始,一直遍历到最后一页
for num in range(1, total_page + 1):
   page = 'http://{0}.{1}.com/ershoufang/{2}/pg{3}'.format(city_name,
      SPIDER_NAME, area_name, num)
   print(page)  # 打印每一页的地址
   headers = create_headers()
   BaseSpider.random_delay()
   response = requests.get(page, timeout=10, headers=headers)
   html = response.content
   soup = BeautifulSoup(html, "lxml")

   # 获得有小区信息的panel
   house_elements = soup.find_all('li', class_="clear")
   for house_elem in house_elements:
      price = house_elem.find('div', class_="totalPrice")
      name = house_elem.find('div', class_='title')
      desc = house_elem.find('div', class_="houseInfo")
      pic = house_elem.find('a', class_="img").find('img', class_="lj-lazy")

      # 继续清理数据
      price = price.text.strip()
      name = name.text.replace("\n", "")
      desc = desc.text.replace("\n", "").strip()
      pic = pic.get('data-original').strip()
      # print(pic)
```

```
            # 作为对象保存
            ershou = ErShou(chinese_district, chinese_area, name, price, desc, pic)
            ershou_list.append(ershou)
    return ershou_list
```

▶ 编写函数 start(self)，功能是根据获取的城市参数来爬取这个城市的二手房信息，对应的实现代码如下所示：

```
def start(self):
    city = get_city()
    self.today_path = create_date_path("{0}/ershou".format(SPIDER_NAME),
                        city, self.date_string)

    t1 = time.time()  # 开始计时

    # 获得城市有多少区列表, district: 区县
    districts = get_districts(city)
    print('City: {0}'.format(city))
    print('Districts: {0}'.format(districts))

    # 获得每个区的板块, area: 板块
    areas = list()
    for district in districts:
        areas_of_district = get_areas(city, district)
        print('{0}: Area list:  {1}'.format(district, areas_of_district))
        # 用list的extend方法L1.extend(L2)，该方法将参数L2的全部元素添加到L1的尾部
        areas.extend(areas_of_district)
        # 使用一个字典来存储区县和板块的对应关系, 例如{'beicai': 'pudongxinqu', }
        for area in areas_of_district:
            area_dict[area] = district
    print("Area:", areas)
    print("District and areas:", area_dict)

    # 准备线程池用到的参数
    nones = [None for i in range(len(areas))]
    city_list = [city for i in range(len(areas))]
    args = zip(zip(city_list, areas), nones)
    # areas = areas[0: 1]   # For debugging

    # 针对每个板块写一个文件,启动一个线程来操作
    pool_size = thread_pool_size
    pool = threadpool.ThreadPool(pool_size)
    my_requests = threadpool.makeRequests(self.collect_area_ershou_data, args)
    [pool.putRequest(req) for req in my_requests]
    pool.wait()
    pool.dismissWorkers(pool_size, do_join=True)  # 完成后退出

    # 计时结束，统计结果
    t2 = time.time()
    print("Total crawl {0} areas.".format(len(areas)))
    print("Total cost {0} second to crawl {1} data items.".format(t2 - t1,
          self.total_num))
```

(2) 编写文件 ershou.py，功能是爬取指定城市的二手房信息，具体实现代码如下所示：

```
from lib.spider.ershou_spider import *

if __name__ == "__main__":
    spider = ErShouSpider(SPIDER_NAME)
    spider.start()
```

执行文件 ershou.py 后，会先提示用户选择一个要爬取的城市：

```
Today date is: 20190212
Target site is ke.com
Wait for your choice.
Which city do you want to crawl?
bj: 北京, cd: 成都, cq: 重庆, cs: 长沙
dg: 东莞, dl: 大连, fs: 佛山, gz: 广州
hz: 杭州, hf: 合肥, jn: 济南, nj: 南京
qd: 青岛, sh: 上海, sz: 深圳, su: 苏州
sy: 沈阳, tj: 天津, wh: 武汉, xm: 厦门
yt: 烟台,
```

输入一个城市的两个字母标识，例如输入 bj 并按 Enter 键后，会爬取当天北京市的二手房信息，并将爬取到的信息保存到 CSV 文件中，如图 3-2 所示。

data\ke\ershou\bj\20190211

| 名称 | 修改日期 | 类型 | 大小 |
|---|---|---|---|
| changping_anningzhuang1.csv | 2019/2/11 20:46 | Microsoft Excel ... | 0 KB |
| changping_aolinpikegongyuan11.csv | 2019/2/11 20:46 | Microsoft Excel ... | 0 KB |
| changping_lishuiqiao1.csv | 2019/2/11 20:44 | Microsoft Excel ... | 0 KB |
| chaoyang_andingmen.csv | 2019/2/11 20:44 | Microsoft Excel ... | 48 KB |
| chaoyang_anzhen1.csv | 2019/2/11 20:46 | Microsoft Excel ... | 86 KB |
| chaoyang_baiziwan.csv | 2019/2/11 20:41 | Microsoft Excel ... | 0 KB |
| chaoyang_beigongda.csv | 2019/2/11 20:44 | Microsoft Excel ... | 27 KB |
| chaoyang_beiyuan2.csv | 2019/2/11 20:41 | Microsoft Excel ... | 0 KB |
| chaoyang_cbd.csv | 2019/2/11 20:47 | Microsoft Excel ... | 2 KB |
| chaoyang_changying.csv | 2019/2/11 20:41 | Microsoft Excel ... | 0 KB |
| chaoyang_chaoqing.csv | 2019/2/11 20:41 | Microsoft Excel ... | 0 KB |
| chaoyang_chaoyanggongyuan.csv | 2019/2/11 20:46 | Microsoft Excel ... | 66 KB |
| chaoyang_chaoyangmenwai1.csv | 2019/2/11 20:45 | Microsoft Excel ... | 66 KB |
| chaoyang_dashanzi.csv | 2019/2/11 20:43 | Microsoft Excel ... | 17 KB |
| chaoyang_dawanglu.csv | 2019/2/11 20:42 | Microsoft Excel ... | 0 KB |
| chaoyang_dingfuzhuang.csv | 2019/2/11 20:47 | Microsoft Excel ... | 65 KB |
| chaoyang_dongba.csv | 2019/2/11 20:47 | Microsoft Excel ... | 83 KB |
| chaoyang_dongdaqiao.csv | 2019/2/11 20:45 | Microsoft Excel ... | 45 KB |
| chaoyang_dongzhimen.csv | 2019/2/11 20:47 | Microsoft Excel ... | 92 KB |
| chaoyang_dougezhuang.csv | 2019/2/11 20:46 | Microsoft Excel ... | 63 KB |
| chaoyang_fatou.csv | 2019/2/11 20:46 | Microsoft Excel ... | 50 KB |
| chaoyang_ganluyuan.csv | 2019/2/11 20:48 | Microsoft Excel ... | 5 KB |
| chaoyang_gaobeidian.csv | 2019/2/11 20:45 | Microsoft Excel ... | 39 KB |
| chaoyang_gongti.csv | 2019/2/11 20:47 | Microsoft Excel ... | 62 KB |
| chaoyang_guangqumen.csv | 2019/2/11 20:48 | Microsoft Excel ... | 77 KB |
| chaoyang_guanzhuang.csv | 2019/2/11 20:46 | Microsoft Excel ... | 42 KB |
| chaoyang_guozhan1.csv | 2019/2/11 20:47 | Microsoft Excel ... | 77 KB |
| chaoyang_hepingli.csv | 2019/2/11 20:48 | Microsoft Excel ... | 75 KB |
| chaoyang_hongmiao.csv | 2019/2/11 20:47 | Microsoft Excel ... | 52 KB |

图 3-2　爬取到的二手房信息被保存到 CSV 文件中

### 3.6.3 爬取楼盘信息

(1) 编写文件 loupan_spider.py，定义爬取楼盘数据的爬虫派生类，具体实现流程如下所示。

- 编写函数 collect_city_loupan_data()，功能是将指定城市的新房楼盘数据爬取下来，并将爬取的信息保存到 CSV 文件中。对应的实现代码如下所示：

```
def collect_city_loupan_data(self, city_name, fmt="csv"):
    """
    :param city_name: 城市
    :param fmt: 保存文件格式
    :return: None
    """
    csv_file = self.today_path + "/{0}.csv".format(city_name)
    with open(csv_file, "w") as f:
        # 开始获得需要的板块数据
        loupans = self.get_loupan_info(city_name)
        self.total_num = len(loupans)
        if fmt == "csv":
            for loupan in loupans:
                f.write(self.date_string + "," + loupan.text() + "\n")
    print("Finish crawl: " + city_name + ", save data to : " + csv_file)
```

- 编写函数 get_loupan_info()，功能是爬取指定目标城市的新房楼盘信息，对应代码如下所示：

```
@staticmethod
def get_loupan_info(city_name):
    """
    :param city_name: 城市
    :return: 新房楼盘信息列表
    """
    total_page = 1
    loupan_list = list()
    page = 'http://{0}.fang.{1}.com/loupan/'.format(city_name, SPIDER_NAME)
    print(page)
    headers = create_headers()
    response = requests.get(page, timeout=10, headers=headers)
    html = response.content
    soup = BeautifulSoup(html, "lxml")

    # 获得总的页数
    try:
        page_box = soup.find_all('div', class_='page-box')[0]
        matches = re.search('.*data-total-count="(\d+)".*', str(page_box))
        total_page = int(math.ceil(int(matches.group(1)) / 10))
    except Exception as e:
        print("\tWarning: only find one page for {0}".format(city_name))
        print(e)

    print(total_page)
    # 从第一页开始,一直遍历到最后一页
```

```
    headers = create_headers()
    for i in range(1, total_page + 1):
        page = 'http://{0}.fang.{1}.com/loupan/pg{2}'.format(city_name,
               SPIDER_NAME, i)
        print(page)
        BaseSpider.random_delay()
        response = requests.get(page, timeout=10, headers=headers)
        html = response.content
        soup = BeautifulSoup(html, "lxml")

        # 获得有小区信息的panel
        house_elements = soup.find_all('li', class_="resblock-list")
        for house_elem in house_elements:
            price = house_elem.find('span', class_="number")
            total = house_elem.find('div', class_="second")
            loupan = house_elem.find('a', class_='name')

            # 继续清理数据
            try:
                price = price.text.strip()
            except Exception as e:
                price = '0'

            loupan = loupan.text.replace("\n", "")

            try:
                total = total.text.strip().replace(u'总价', '')
                total = total.replace(u'/套起', '')
            except Exception as e:
                total = '0'

            print("{0} {1} {2} ".format(
                loupan, price, total))

            # 作为对象保存
            loupan = LouPan(loupan, price, total)
            loupan_list.append(loupan)
    return loupan_list
```

- 编写函数 start(self)，功能是根据获取的城市参数来爬取这个城市的新房楼盘信息，对应的实现代码如下所示：

```
def start(self):
    city = get_city()
    print('Today date is: %s' % self.date_string)
    self.today_path = create_date_path("{0}/loupan".format(SPIDER_NAME),
                      city, self.date_string)

    t1 = time.time()  # 开始计时
    self.collect_city_loupan_data(city)
    t2 = time.time()  # 计时结束，统计结果

    print("Total crawl {0} loupan.".format(self.total_num))
    print("Total cost {0} second ".format(t2 - t1))
```

(2) 编写文件 loupan.py，功能是爬取指定城市的新房楼盘信息，具体实现代码如下所示：

```
from lib.spider.loupan_spider import *

if __name__ == "__main__":
    spider = LouPanBaseSpider(SPIDER_NAME)
    spider.start()
```

执行文件 loupan.py 后，会先提示用户选择一个要爬取信息的城市：

```
Today date is: 20190212
Target site is ke.com
Wait for your choice.
Which city do you want to crawl?
bj: 北京, cd: 成都, cq: 重庆, cs: 长沙
dg: 东莞, dl: 大连, fs: 佛山, gz: 广州
hz: 杭州, hf: 合肥, jn: 济南, nj: 南京
qd: 青岛, sh: 上海, sz: 深圳, su: 苏州
sy: 沈阳, tj: 天津, wh: 武汉, xm: 厦门
yt: 烟台,
```

输入一个城市的两个字母标识，例如输入 jn 并按 Enter 键后，会爬取当天济南市的新房楼盘信息，并将爬取到的信息保存到 CSV 文件中，如图 3-3 所示。

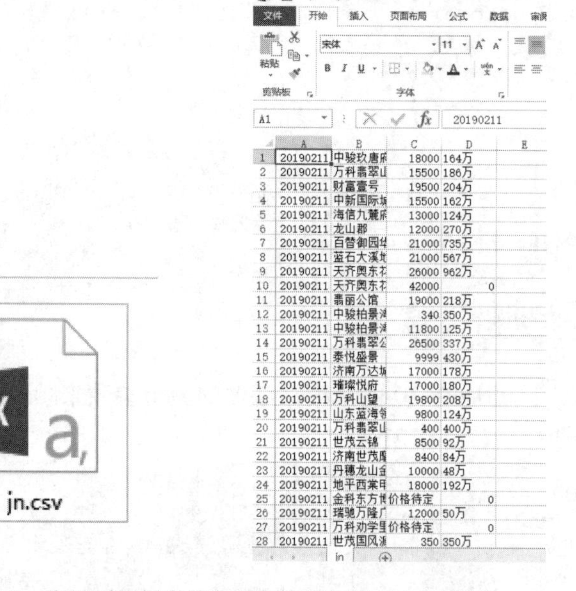


| | A | B | C | D | E |
|---|---|---|---|---|---|
| 1 | 20190211 | 中骏玖唐 | 18000 | 164万 | |
| 2 | 20190211 | 万科翡翠 | 15500 | 186万 | |
| 3 | 20190211 | 财富壹号 | 19500 | 204万 | |
| 4 | 20190211 | 中新国际 | 15500 | 162万 | |
| 5 | 20190211 | 海信九麓 | 13000 | 124万 | |
| 6 | 20190211 | 龙山郡 | 12000 | 270万 | |
| 7 | 20190211 | 百替御园 | 21000 | 735万 | |
| 8 | 20190211 | 蓝石大溪 | 21000 | 567万 | |
| 9 | 20190211 | 天齐奥东 | 26000 | 962万 | |
| 10 | 20190211 | 天齐奥东 | 42000 | 0 | |
| 11 | 20190211 | 翡丽公馆 | 19000 | 218万 | |
| 12 | 20190211 | 中骏柏景 | 340 | 350万 | |
| 13 | 20190211 | 中骏柏景 | 11800 | 125万 | |
| 14 | 20190211 | 万科翡翠 | 26500 | 337万 | |
| 15 | 20190211 | 泰悦盛景 | 9999 | 430万 | |
| 16 | 20190211 | 济南万达 | 17000 | 178万 | |
| 17 | 20190211 | 璀璨悦府 | 17000 | 180万 | |
| 18 | 20190211 | 万科山望 | 19800 | 208万 | |
| 19 | 20190211 | 山东蓝海 | 9800 | 124万 | |
| 20 | 20190211 | 万科翡翠 | 400 | 400万 | |
| 21 | 20190211 | 世茂云锦 | 8500 | 92万 | |
| 22 | 20190211 | 济南世茂 | 8400 | 84万 | |
| 23 | 20190211 | 丹穗龙山 | 10000 | 48万 | |
| 24 | 20190211 | 地平西棠 | 18000 | 192万 | |
| 25 | 20190211 | 金科东方 | 价格待定 | 0 | |
| 26 | 20190211 | 瑞驰万隆 | 12000 | 50万 | |
| 27 | 20190211 | 万科劝学 | 价格待定 | 0 | |
| 28 | 20190211 | 世茂国风 | 350 | 350万 | |

图 3-3　爬取到的新房楼盘信息被保存到 CSV 文件中

## 3.6.4　爬取小区信息

(1) 编写文件 xiaoqu_spider.py，定义爬取小区数据的爬虫派生类，具体实现流程如下所示：

- 编写函数 collect_area_xiaoqu_data()，功能是获取每个板块下的所有小区的信息，

并且将这些信息写入到 CSV 文件中进行保存。对应代码如下所示:

```
def collect_area_xiaoqu_data(self, city_name, area_name, fmt="csv"):
    """
    :param city_name: 城市
    :param area_name: 板块
    :param fmt: 保存文件格式
    :return: None
    """
    district_name = area_dict.get(area_name, "")
    csv_file = self.today_path + "/{0}_{1}.csv".format(district_name,
              area_name)
    with open(csv_file, "w") as f:
        # 开始获得需要的板块数据
        xqs = self.get_xiaoqu_info(city_name, area_name)
        # 锁定
        if self.mutex.acquire(1):
            self.total_num += len(xqs)
            # 释放
            self.mutex.release()
        if fmt == "csv":
            for xiaoqu in xqs:
                f.write(self.date_string + "," + xiaoqu.text() + "\n")
    print("Finish crawl area: " + area_name + ", save data to : " + csv_file)
    logger.info("Finish crawl area: " + area_name + ", save data to : " +
              csv_file)
```

▶ 编写函数 get_xiaoqu_info(),功能是获取指定小区的详细信息,对应代码如下所示:

```
@staticmethod
def get_xiaoqu_info(city, area):
    total_page = 1
    district = area_dict.get(area, "")
    chinese_district = get_chinese_district(district)
    chinese_area = chinese_area_dict.get(area, "")
    xiaoqu_list = list()
    page = 'http://{0}.{1}.com/xiaoqu/{2}/'.format(city, SPIDER_NAME, area)
    print(page)
    logger.info(page)

    headers = create_headers()
    response = requests.get(page, timeout=10, headers=headers)
    html = response.content
    soup = BeautifulSoup(html, "lxml")

    # 获得总的页数
    try:
        page_box = soup.find_all('div', class_='page-box')[0]
        matches = re.search('.*"totalPage":(\d+),.*', str(page_box))
        total_page = int(matches.group(1))
    except Exception as e:
        print("\tWarning: only find one page for {0}".format(area))
        print(e)
```

```
    # 从第一页开始,一直遍历到最后一页
    for i in range(1, total_page + 1):
        headers = create_headers()
        page = 'http://{0}.{1}.com/xiaoqu/{2}/pg{3}'.format(city,
               SPIDER_NAME, area, i)
        print(page)  # 打印板块页面地址
        BaseSpider.random_delay()
        response = requests.get(page, timeout=10, headers=headers)
        html = response.content
        soup = BeautifulSoup(html, "lxml")

        # 获得有小区信息的panel
        house_elems = soup.find_all('li', class_="xiaoquListItem")
        for house_elem in house_elems:
            price = house_elem.find('div', class_="totalPrice")
            name = house_elem.find('div', class_='title')
            on_sale = house_elem.find('div', class_="xiaoquListItemSellCount")

            # 继续清理数据
            price = price.text.strip()
            name = name.text.replace("\n", "")
            on_sale = on_sale.text.replace("\n", "").strip()

            # 作为对象保存
            xiaoqu = XiaoQu(chinese_district, chinese_area, name, price, on_sale)
            xiaoqu_list.append(xiaoqu)
    return xiaoqu_list
```

- 编写函数 start(self)，功能是根据获取的城市参数来爬取这个城市的小区信息，对应的实现代码如下所示：

```
def start(self):
    city = get_city()
    self.today_path = create_date_path("{0}/xiaoqu".format(SPIDER_NAME),
                      city, self.date_string)
    t1 = time.time()  # 开始计时

    # 获得城市有多少区列表, district: 区县
    districts = get_districts(city)
    print('City: {0}'.format(city))
    print('Districts: {0}'.format(districts))

    # 获得每个区的板块, area: 板块
    areas = list()
    for district in districts:
        areas_of_district = get_areas(city, district)
        print('{0}: Area list:  {1}'.format(district, areas_of_district))
        # 用list的extend方法L1.extend(L2)，该方法将参数L2的全部元素添加到L1的尾部
        areas.extend(areas_of_district)
        # 使用一个字典来存储区县和板块的对应关系, 例如{'beicai': 'pudongxinqu', }
```

```
        for area in areas_of_district:
            area_dict[area] = district
    print("Area:", areas)
    print("District and areas:", area_dict)

    # 准备线程池用到的参数
    nones = [None for i in range(len(areas))]
    city_list = [city for i in range(len(areas))]
    args = zip(zip(city_list, areas), nones)
    # areas = areas[0: 1]

    # 针对每个板块写一个文件,启动一个线程来操作
    pool_size = thread_pool_size
    pool = threadpool.ThreadPool(pool_size)
    my_requests = threadpool.makeRequests(self.collect_area_xiaoqu_data, args)
    [pool.putRequest(req) for req in my_requests]
    pool.wait()
    pool.dismissWorkers(pool_size, do_join=True)  # 完成后退出

    # 计时结束，统计结果
    t2 = time.time()
    print("Total crawl {0} areas.".format(len(areas)))
    print("Total cost {0} second to crawl {1} data items.".format(t2 - t1,
self.total_num))
```

(2) 编写文件 xiaoqu.py，功能是爬取指定城市的小区信息，具体实现代码如下所示：

```
from lib.spider.xiaoqu_spider import *

if __name__ == "__main__":
   spider = XiaoQuBaseSpider(SPIDER_NAME)
   spider.start()
```

执行文件 xiaoqu.py 后，会先提示用户选择一个要爬取信息的城市：

```
Today date is: 20190212
Target site is ke.com
Wait for your choice.
Which city do you want to crawl?
bj: 北京, cd: 成都, cq: 重庆, cs: 长沙
dg: 东莞, dl: 大连, fs: 佛山, gz: 广州
hz: 杭州, hf: 合肥, jn: 济南, nj: 南京
qd: 青岛, sh: 上海, sz: 深圳, su: 苏州
sy: 沈阳, tj: 天津, wh: 武汉, xm: 厦门
yt: 烟台,
```

输入一个城市的两个字母标识，例如输入 jn 并按 Enter 键后，会爬取当天济南市的小区信息，并将爬取到的信息保存到 CSV 文件中，如图 3-4 所示。

data\ke\xiaoqu\jn\20190211

| 名称 | 修改日期 | 类型 | 大小 |
|---|---|---|---|
| changqing_changqing1111.csv | 2019/2/11 21:28 | Microsoft Excel ... | 4 KB |
| changqing_changqingdaxuecheng.csv | 2019/2/11 21:27 | Microsoft Excel ... | 2 KB |
| gaoxin_aotizhonglu.csv | 2019/2/11 21:27 | Microsoft Excel ... | 1 KB |
| gaoxin_aotizhongxin.csv | 2019/2/11 21:27 | Microsoft Excel ... | 2 KB |
| gaoxin_guojihuizhanzhongxin.csv | 2019/2/11 21:27 | Microsoft Excel ... | 1 KB |
| gaoxin_hanyu.csv | 2019/2/11 21:27 | Microsoft Excel ... | 3 KB |
| gaoxin_jingshidonglu.csv | 2019/2/11 21:27 | Microsoft Excel ... | 1 KB |
| gaoxin_kanghonglu.csv | 2019/2/11 21:27 | Microsoft Excel ... | 1 KB |
| gaoxin_qiluruanjianyuan.csv | 2019/2/11 21:27 | Microsoft Excel ... | 2 KB |
| gaoxin_shijidadao.csv | 2019/2/11 21:27 | Microsoft Excel ... | 1 KB |
| huaiyin_baimashan.csv | 2019/2/11 21:27 | Microsoft Excel ... | 3 KB |
| huaiyin_chalujie.csv | 2019/2/11 21:27 | Microsoft Excel ... | 3 KB |
| huaiyin_daguanyuan.csv | 2019/2/11 21:27 | Microsoft Excel ... | 5 KB |
| huaiyin_dikoulu.csv | 2019/2/11 21:27 | Microsoft Excel ... | 6 KB |
| huaiyin_duandian.csv | 2019/2/11 21:27 | Microsoft Excel ... | 3 KB |
| huaiyin_ertongyiyuan.csv | 2019/2/11 21:27 | Microsoft Excel ... | 1 KB |
| huaiyin_huochezhan3.csv | 2019/2/11 21:27 | Microsoft Excel ... | 3 KB |
| huaiyin_jianshelu2.csv | 2019/2/11 21:27 | Microsoft Excel ... | 4 KB |
| huaiyin_jingshixilu.csv | 2019/2/11 21:27 | Microsoft Excel ... | 1 KB |
| huaiyin_kuangshanqu1.csv | 2019/2/11 21:27 | Microsoft Excel ... | 3 KB |
| huaiyin_lashan.csv | 2019/2/11 21:27 | Microsoft Excel ... | 4 KB |
| huaiyin_liancheng.csv | 2019/2/11 21:27 | Microsoft Excel ... | 2 KB |
| huaiyin_liuzhangshanlu.csv | 2019/2/11 21:27 | Microsoft Excel ... | 1 KB |
| huaiyin_meilihu.csv | 2019/2/11 21:27 | Microsoft Excel ... | 2 KB |
| huaiyin_nanxinzhuang.csv | 2019/2/11 21:27 | Microsoft Excel ... | 3 KB |

图 3-4　爬取到的小区信息被保存到 CSV 文件中

### 3.6.5　爬取租房信息

(1) 编写文件 zufang_spider.py，定义爬取租房数据的爬虫派生类，具体实现流程如下所示。

- 编写函数 collect_area_zufang_data()，功能是获取每个板块下的所有租房信息，并且将这些信息写入 CSV 文件中进行保存。对应的实现代码如下所示：

```
def collect_area_zufang_data(self, city_name, area_name, fmt="csv"):
    """
    :param city_name: 城市
    :param area_name: 板块
    :param fmt: 保存文件格式
    :return: None
    """
    district_name = area_dict.get(area_name, "")
    csv_file = self.today_path + "/{0}_{1}.csv".format(district_name, area_name)
    with open(csv_file, "w") as f:
        # 开始获得需要的板块数据
        zufangs = self.get_area_zufang_info(city_name, area_name)
        # 锁定
        if self.mutex.acquire(1):
            self.total_num += len(zufangs)
            # 释放
            self.mutex.release()
```

```
        if fmt == "csv":
            for zufang in zufangs:
                f.write(self.date_string + "," + zufang.text() + "\n")
    print("Finish crawl area: " + area_name + ", save data to : " + csv_file)
```

▶ 编写函数 get_area_zufang_info()，功能是获取指定城市指定板块的租房信息。对应的实现代码如下所示：

```
@staticmethod
def get_area_zufang_info(city_name, area_name):
    matches = None
    """
    :param city_name: 城市
    :param area_name: 板块
    :return: 出租房信息列表
    """
    total_page = 1
    district_name = area_dict.get(area_name, "")
    chinese_district = get_chinese_district(district_name)
    chinese_area = chinese_area_dict.get(area_name, "")
    zufang_list = list()
    page = 'http://{0}.{1}.com/zufang/{2}/'.format(city_name, SPIDER_NAME,
           area_name)
    print(page)

    headers = create_headers()
    response = requests.get(page, timeout=10, headers=headers)
    html = response.content
    soup = BeautifulSoup(html, "lxml")

    # 获得总的页数
    try:
        if SPIDER_NAME == "lianjia":
            page_box = soup.find_all('div', class_='page-box')[0]
            matches = re.search('.*"totalPage":(\d+),.*', str(page_box))
        elif SPIDER_NAME == "ke":
            page_box = soup.find_all('div', class_='content__pg')[0]
            # print(page_box)
            matches = re.search('.*data-totalpage="(\d+)".*', str(page_box))
        total_page = int(matches.group(1))
        # print(total_page)
    except Exception as e:
        print("\tWarning: only find one page for {0}".format(area_name))
        print(e)

    # 从第一页开始,一直遍历到最后一页
    headers = create_headers()
    for num in range(1, total_page + 1):
        page = 'http://{0}.{1}.com/zufang/{2}/pg{3}'.format(city_name,
               SPIDER_NAME, area_name, num)
        print(page)
        BaseSpider.random_delay()
        response = requests.get(page, timeout=10, headers=headers)
        html = response.content
        soup = BeautifulSoup(html, "lxml")
```

```
# 获得有小区信息的panel
if SPIDER_NAME == "lianjia":
   ul_element = soup.find('ul', class_="house-lst")
   house_elements = ul_element.find_all('li')
else:
   ul_element = soup.find('div', class_="content__list")
   house_elements = ul_element.find_all('div',
                    class_="content__list--item")

if len(house_elements) == 0:
   continue
# else:
#    print(len(house_elements))

for house_elem in house_elements:
   if SPIDER_NAME == "lianjia":
      price = house_elem.find('span', class_="num")
      xiaoqu = house_elem.find('span', class_='region')
      layout = house_elem.find('span', class_="zone")
      size = house_elem.find('span', class_="meters")
   else:
      price = house_elem.find('span',
              class_="content__list--item-price")
      desc1 = house_elem.find('p',
              class_="content__list--item--title")
      desc2 = house_elem.find('p',
              class_="content__list--item--des")

   try:
      if SPIDER_NAME == "lianjia":
         price = price.text.strip()
         xiaoqu = xiaoqu.text.strip().replace("\n", "")
         layout = layout.text.strip()
         size = size.text.strip()
      else:
         # 继续清理数据
         price=price.text.strip().replace(" ", "").replace("元/月", "")
         # print(price)
         desc1 = desc1.text.strip().replace("\n", "")
         desc2 = desc2.text.strip().replace("\n", "").replace(" ", "")
         # print(desc1)

         infos = desc1.split(' ')
         xiaoqu = infos[0]
         layout = infos[1]
         descs = desc2.split('/')
         # print(descs[1])
         size = descs[1].replace("m²", "平米")

      # print("{0} {1} {2} {3} {4} {5} {6}".format(
      #  chinese_district, chinese_area, xiaoqu, layout, size, price))

      # 作为对象保存
```

```
                zufang = ZuFang(chinese_district, chinese_area, xiaoqu, layout,
                         size, price)
                zufang_list.append(zufang)
            except Exception as e:
                print("=" * 20 + " page no data")
                print(e)
                print(page)
                print("=" * 20)
        return zufang_list
```

- 编写函数 start(self)，功能是根据获取的城市参数来爬取这个城市的租房信息，对应的实现代码如下所示：

```
    def start(self):
        city = get_city()
        self.today_path = create_date_path("{0}/zufang".format(SPIDER_NAME),
                           city, self.date_string)
        # collect_area_zufang('sh', 'beicai')  # For debugging, keep it here
        t1 = time.time()  # 开始计时

        # 获得城市有多少区列表, district: 区县
        districts = get_districts(city)
        print('City: {0}'.format(city))
        print('Districts: {0}'.format(districts))

        # 获得每个区的板块, area: 板块
        areas = list()
        for district in districts:
            areas_of_district = get_areas(city, district)
            print('{0}: Area list:  {1}'.format(district, areas_of_district))
            # 用list的extend方法L1.extend(L2)，该方法将参数L2的全部元素添加到L1的尾部
            areas.extend(areas_of_district)
            # 使用一个字典来存储区县和板块的对应关系, 例如{'beicai': 'pudongxinqu', }
            for area in areas_of_district:
                area_dict[area] = district
        print("Area:", areas)
        print("District and areas:", area_dict)

        # 准备线程池用到的参数
        nones = [None for i in range(len(areas))]
        city_list = [city for i in range(len(areas))]
        args = zip(zip(city_list, areas), nones)
        # areas = areas[0: 1]

        # 针对每个板块写一个文件,启动一个线程来操作
        pool_size = thread_pool_size
        pool = threadpool.ThreadPool(pool_size)
        my_requests = threadpool.makeRequests(self.collect_area_zufang_data, args)
        [pool.putRequest(req) for req in my_requests]
        pool.wait()
        pool.dismissWorkers(pool_size, do_join=True)  # 完成后退出

        # 计时结束，统计结果
        t2 = time.time()
        print("Total crawl {0} areas.".format(len(areas)))
```

```
    print("Total cost {0} second to crawl {1} data items.".format(t2 - t1,
        self.total_num))
```

(2) 编写文件 zufang.py，功能是爬取指定城市的租房信息，具体实现代码如下所示：

```
from lib.spider.zufang_spider import *

if __name__ == "__main__":
    spider = ZuFangBaseSpider(SPIDER_NAME)
    spider.start()
```

执行文件 zufang.py 后，会先提示用户选择一个要爬取信息的城市：

```
Today date is: 20190212
Target site is ke.com
Wait for your choice.
Which city do you want to crawl?
bj: 北京, cd: 成都, cq: 重庆, cs: 长沙
dg: 东莞, dl: 大连, fs: 佛山, gz: 广州
hz: 杭州, hf: 合肥, jn: 济南, nj: 南京
qd: 青岛, sh: 上海, sz: 深圳, su: 苏州
sy: 沈阳, tj: 天津, wh: 武汉, xm: 厦门
yt: 烟台,
```

输入一个城市的两个字母标识，例如输入 jn 并按 Enter 键后，会爬取当天济南市的租房信息，并将爬取到的信息保存到 CSV 文件中，如图 3-5 所示。

| | | | |
|---|---|---|---|
| gaoxin_aotizhongxin.csv | 2019/2/11 20:53 | Microsoft Excel ... | 7 KB |
| gaoxin_hanyu.csv | 2019/2/11 20:54 | Microsoft Excel ... | 0 KB |
| gaoxin_jingshidonglu.csv | 2019/2/11 20:57 | Microsoft Excel ... | 18 KB |
| gaoxin_kanghonglu.csv | 2019/2/11 20:57 | Microsoft Excel ... | 5 KB |
| gaoxin_qiluruanjianyuan.csv | 2019/2/11 20:55 | Microsoft Excel ... | 0 KB |
| huaiyin_baimashan.csv | 2019/2/11 20:56 | Microsoft Excel ... | 0 KB |
| huaiyin_chalujie.csv | 2019/2/11 20:57 | Microsoft Excel ... | 4 KB |
| huaiyin_daguanyuan.csv | 2019/2/11 20:56 | Microsoft Excel ... | 0 KB |
| huaiyin_dikoulu.csv | 2019/2/11 20:56 | Microsoft Excel ... | 0 KB |
| huaiyin_duandian.csv | 2019/2/11 20:56 | Microsoft Excel ... | 0 KB |
| huaiyin_ertongyiyuan.csv | 2019/2/11 20:56 | Microsoft Excel ... | 0 KB |
| huaiyin_huochezhan3.csv | 2019/2/11 20:56 | Microsoft Excel ... | 0 KB |
| huaiyin_jianshelu2.csv | 2019/2/11 20:56 | Microsoft Excel ... | 0 KB |
| huaiyin_jingshixilu.csv | 2019/2/11 20:57 | Microsoft Excel ... | 1 KB |
| huaiyin_kuangshanqu1.csv | 2019/2/11 20:56 | Microsoft Excel ... | 0 KB |
| huaiyin_lashan.csv | 2019/2/11 20:56 | Microsoft Excel ... | 0 KB |
| huaiyin_liancheng.csv | 2019/2/11 20:56 | Microsoft Excel ... | 0 KB |
| huaiyin_liuzhangshanlu.csv | 2019/2/11 20:57 | Microsoft Excel ... | 1 KB |
| huaiyin_meilihu.csv | 2019/2/11 20:57 | Microsoft Excel ... | 0 KB |
| huaiyin_nanxinzhuang.csv | 2019/2/11 20:57 | Microsoft Excel ... | 0 KB |
| huaiyin_quanjingwolongqu.csv | 2019/2/11 20:57 | Microsoft Excel ... | 0 KB |
| huaiyin_shengliyiyuan.csv | 2019/2/11 20:57 | Microsoft Excel ... | 0 KB |
| huaiyin_shigou.csv | 2019/2/11 20:57 | Microsoft Excel ... | 0 KB |
| huaiyin_tianqiaogongyeyuan.csv | 2019/2/11 20:57 | Microsoft Excel ... | 1 KB |
| huaiyin_wangguanzhuang.csv | 2019/2/11 20:57 | Microsoft Excel ... | 0 KB |
| huaiyin_xikezhan.csv | 2019/2/11 20:57 | Microsoft Excel ... | 0 KB |

图 3-5　爬取到的租房信息被保存到 CSV 文件中

## 3.7　数据可视化

扫码观看视频讲解

在爬取到房价数据后，我们可以对 CSV 文件实现可视化分析。但是为了更加方便的操作，可以将抓取的数据保存到数据库中，然后提取数据库中的数据进行数据分析。本节将详细讲解将数据保存到数据库并进行数据分析的过程。

### 3.7.1　爬取数据并保存到数据库

编写文件 xiaoqu_to_db.py，功能是爬取指定城市的小区房价数据并保存到数据库中，我们可以选择存储的数据库类型有 MySQL、MongoDB、JSON、CSV 和 Excel，默认的存储方式是 MySQL。文件 xiaoqu_to_db.py 的具体实现流程如下所示。

(1) 创建提示语句，询问用户将要爬取的目标城市，对应实现代码如下所示：

```
pymysql.install_as_MySQLdb()
def create_prompt_text():
    city_info = list()
    num = 0
    for en_name, ch_name in cities.items():
        num += 1
        city_info.append(en_name)
        city_info.append(": ")
        city_info.append(ch_name)
        if num % 4 == 0:
            city_info.append("\n")
        else:
            city_info.append(", ")
    return 'Which city data do you want to save ?\n' + ''.join(city_info)
```

(2) 设置数据库类型，根据不同的存储类型执行相应的写入操作，对应实现代码如下所示：

```
if __name__ == '__main__':
    # 设置目标数据库
    ##################################
    # mysql/mongodb/excel/json/csv
    database = "mysql"
    # database = "mongodb"
    # database = "excel"
    # database = "json"
    # database = "csv"
    ##################################
    db = None
    collection = None
    workbook = None
    csv_file = None
    datas = list()
```

```
    if database == "mysql":
        import records
        db = records.Database('mysql://root:66688888@localhost/
            lianjia?charset=utf8', encoding='utf-8')
    elif database == "mongodb":
        from pymongo import MongoClient
        conn = MongoClient('localhost', 27017)
        db = conn.lianjia  # 连接lianjia数据库，没有则自动创建
        collection = db.xiaoqu  # 使用xiaoqu集合，没有则自动创建
    elif database == "excel":
        import xlsxwriter
        workbook = xlsxwriter.Workbook('xiaoqu.xlsx')
        worksheet = workbook.add_worksheet()
    elif database == "json":
        import json
    elif database == "csv":
        csv_file = open("xiaoqu.csv", "w")
        line = "{0};{1};{2};{3};{4};{5};{6}\n".format('city_ch', 'date',
'district', 'area', 'xiaoqu', 'price', 'sale')
        csv_file.write(line)
```

(3) 准备日期信息，将爬到的数据保存到对应日期的相关文件夹下，实现代码如下所示：

```
    city = get_city()
    date = get_date_string()
    # 获得 csv 文件路径
    # date = "20180331"   # 指定采集数据的日期
    # city = "sh"         # 指定采集数据的城市
    city_ch = get_chinese_city(city)
    csv_dir = "{0}/{1}/xiaoqu/{2}/{3}".format(DATA_PATH, SPIDER_NAME, city, date)

    files = list()
    if not os.path.exists(csv_dir):
        print("{0} does not exist.".format(csv_dir))
        print("Please run 'python xiaoqu.py' firstly.")
        print("Bye.")
        exit(0)
    else:
        print('OK, start to process ' + get_chinese_city(city))
    for csv in os.listdir(csv_dir):
        data_csv = csv_dir + "/" + csv
        # print(data_csv)
        files.append(data_csv)
```

(4) 清理数据，删除没有房源信息的小区，对应实现代码如下所示：

```
    # 清理数据
    count = 0
```

```
row = 0
col = 0
for csv in files:
    with open(csv, 'r') as f:
        for line in f:
            count += 1
            text = line.strip()
            try:
                # 如果小区名里面没有逗号，那么总共是 6 项
                if text.count(',') == 5:
                    date, district, area, xiaoqu, price, sale = text.split(',')
                elif text.count(',') < 5:
                    continue
                else:
                    fields = text.split(',')
                    date = fields[0]
                    district = fields[1]
                    area = fields[2]
                    xiaoqu = ','.join(fields[3:-2])
                    price = fields[-2]
                    sale = fields[-1]
            except Exception as e:
                print(text)
                print(e)
                continue
            sale = sale.replace(r'套在售二手房', '')
            price = price.replace(r'暂无', '0')
            price = price.replace(r'元/m2', '')
            price = int(price)
            sale = int(sale)
            print("{0} {1} {2} {3} {4} {5}".format(date, district, area, xiaoqu,
                    price, sale))
```

(5) 将爬取到的房价数据添加到数据库或 JSON、Excel、CSV 文件中，对应实现代码如下所示：

```
            # 写入mysql数据库
            if database == "mysql":
                db.query('INSERT INTO xiaoqu (city, date, district, area,
                            xiaoqu, price, sale) ''VALUES(:city, :date,
                            :district, :area, :xiaoqu, :price, :sale)', city=city_ch,
                             date=date, district=district, area=area, xiaoqu=
                            xiaoqu, price=price, sale=sale)
            # 写入mongodb数据库
            elif database == "mongodb":
                data = dict(city=city_ch, date=date, district=district, area=area,
                       xiaoqu=xiaoqu, price=price, sale=sale)
                collection.insert(data)
            elif database == "excel":
                if not PYTHON_3:
                    worksheet.write_string(row, col, city_ch)
                    worksheet.write_string(row, col + 1, date)
```

```
                worksheet.write_string(row, col + 2, district)
                worksheet.write_string(row, col + 3, area)
                worksheet.write_string(row, col + 4, xiaoqu)
                worksheet.write_number(row, col + 5, price)
                worksheet.write_number(row, col + 6, sale)
            else:
                worksheet.write_string(row, col, city_ch)
                worksheet.write_string(row, col + 1, date)
                worksheet.write_string(row, col + 2, district)
                worksheet.write_string(row, col + 3, area)
                worksheet.write_string(row, col + 4, xiaoqu)
                worksheet.write_number(row, col + 5, price)
                worksheet.write_number(row, col + 6, sale)
            row += 1
        elif database == "json":
            data = dict(city=city_ch, date=date, district=district,
                    area=area, xiaoqu=xiaoqu, price=price, sale=sale)
            datas.append(data)
        elif database == "csv":
            line = "{0};{1};{2};{3};{4};{5};{6}\n".format(city_ch, date,
                    district, area, xiaoqu, price, sale)
            csv_file.write(line)

    # 写入，并且关闭句柄
    if database == "excel":
        workbook.close()
    elif database == "json":
        json.dump(datas, open('xiaoqu.json', 'w'), ensure_ascii=False, indent=2)
    elif database == "csv":
        csv_file.close()

    print("Total write {0} items to database.".format(count))
```

执行代码后，会提示用户选择一个目标城市：

```
Wait for your choice.
Which city do you want to crawl?
bj: 北京, cd: 成都, cq: 重庆, cs: 长沙
dg: 东莞, dl: 大连, fs: 佛山, gz: 广州
hz: 杭州, hf: 合肥, jn: 济南, nj: 南京
qd: 青岛, sh: 上海, sz: 深圳, su: 苏州
sy: 沈阳, tj: 天津, wh: 武汉, xm: 厦门
yt: 烟台,
```

假设输入 jn 并按 Enter 键，则会将济南市的小区信息保存到数据库中。因为在上述代码中设置的默认存储方式是 MySQL，所以会将爬取到的数据保存到 MySQL 数据库中，如图 3-6 所示。

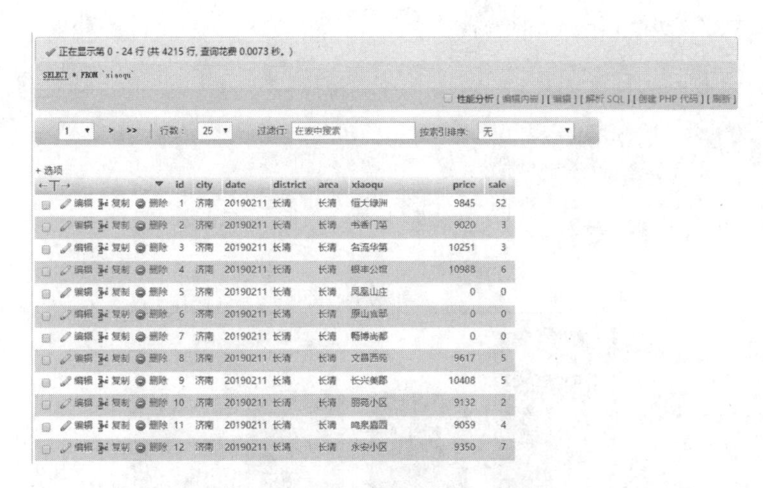
正在显示第 0 - 24 行 (共 4215 行, 查询花费 0.0073 秒。)

SELECT * FROM `xiaoqu`

| id | city | date | district | area | xiaoqu | price | sale |
|---|---|---|---|---|---|---|---|
| 1 | 济南 | 20190211 | 长清 | 长清 | 恒大绿洲 | 9845 | 52 |
| 2 | 济南 | 20190211 | 长清 | 长清 | 书香门第 | 9020 | 3 |
| 3 | 济南 | 20190211 | 长清 | 长清 | 名流华第 | 10251 | 3 |
| 4 | 济南 | 20190211 | 长清 | 长清 | 银丰公馆 | 10988 | 6 |
| 5 | 济南 | 20190211 | 长清 | 长清 | 凤凰山庄 | 0 | 0 |
| 6 | 济南 | 20190211 | 长清 | 长清 | 原山首府 | 0 | 0 |
| 7 | 济南 | 20190211 | 长清 | 长清 | 畅博尚都 | 0 | 0 |
| 8 | 济南 | 20190211 | 长清 | 长清 | 文昌西苑 | 9617 | 5 |
| 9 | 济南 | 20190211 | 长清 | 长清 | 长兴美郡 | 10408 | 5 |
| 10 | 济南 | 20190211 | 长清 | 长清 | 丽苑小区 | 9132 | 2 |
| 11 | 济南 | 20190211 | 长清 | 长清 | 鸣泉嘉园 | 9059 | 4 |
| 12 | 济南 | 20190211 | 长清 | 长清 | 永安小区 | 9350 | 7 |

图 3-6　保存到 MySQL 中的数据

**注　意**

MySQL 数据库的数据库结构，通过导入源码目录中的 SQL 文件 lianjia_xiaoqu.sql 创建。

## 3.7.2　可视化济南市房价最贵的 4 个小区

编写文件 pricetubiao.py，功能是提取 MySQL 数据库中的数据，可视化展示当日济南市房价最贵的 4 个小区。文件 pricetubiao.py 的具体实现代码如下所示：

```
import pymysql
from pylab import *
mpl.rcParams["font.sans-serif"] = ["SimHei"]
mpl.rcParams["axes.unicode_minus"] = False

##获取一个数据库连接，注意如果是 UTF-8 类型的，需要指定数据库
db = pymysql.connect(host="localhost", user='root', passwd="66688888",
port=3306, db="lianjia", charset='utf8')
cursor = db.cursor()  # 获取一个游标
sql = "select xiaoqu,price from xiaoqu where price!=0 order by price desc LIMIT 4"
cursor.execute(sql)
result = cursor.fetchall()  # result 为元组

# 将元组数据存进列表中
xiaoqu = []
price = []
for x in result:
    xiaoqu.append(x[0])
    price.append(x[1])

# 直方图
plt.bar(range(len(price)), price, color='steelblue', tick_label=xiaoqu)
plt.xlabel("小区名")
plt.ylabel("价格")
```

```
plt.title("济南房价 Top 4 小区")
for x, y in enumerate(price):
   plt.text(x - 0.1, y + 1, '%s' % y)
plt.show()
cursor.close()  # 关闭游标
db.close()  # 关闭数据库
```

执行效果如图 3-7 所示。

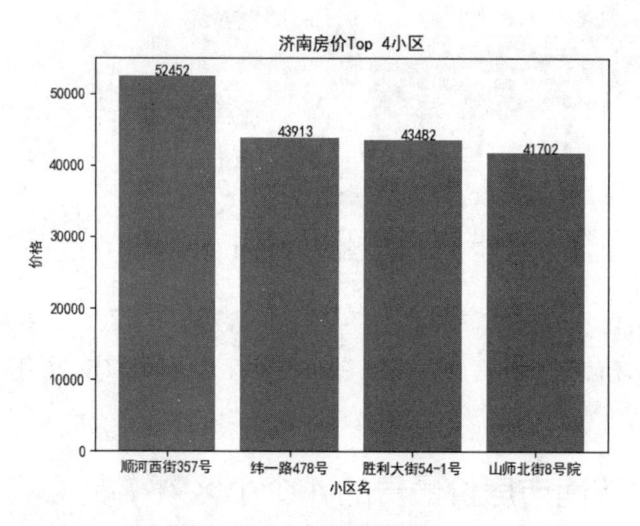


图 3-7　执行效果

## 3.7.3　可视化济南市主要行政区的房价均价

编写文件 gequ.py，功能是提取 MySQL 数据库中的数据，可视化展示当日济南市主要行政区的房价均价。文件 gequ.py 的具体实现代码如下所示：

```
import pymysql
from pylab import *
mpl.rcParams["font.sans-serif"] = ["SimHei"]
mpl.rcParams["axes.unicode_minus"] = False
plt.figure(figsize=(10, 6))
##获取一个数据库连接，注意如果是 UTF-8 类型的，需要指定数据库
db = pymysql.connect(host="localhost", user='root', passwd="66688888",
port=3306, db="lianjia", charset='utf8')
cursor = db.cursor()  # 获取一个游标

sql = "select district,avg(price) as avgsprice from xiaoqu where price!=0 group
by district"


cursor.execute(sql)
result = cursor.fetchall()  # result 为元组

# 将元组数据存进列表中
district = []
avgsprice = []
```

```
for x in result:
    district.append(x[0])
    avgsprice.append(x[1])

# 直方图
plt.bar(range(len(avgsprice)), avgsprice, color='steelblue',
tick_label=district)
plt.xlabel("行政区")
plt.ylabel("平均价格")
plt.title("济南市主要行政区房价均价")
for x, y in enumerate(avgsprice):
    plt.text(x - 0.5, y + 2, '%s' % y)
plt.show()
cursor.close()  # 关闭游标
db.close()  # 关闭数据库
```

执行效果如图 3-8 所示。

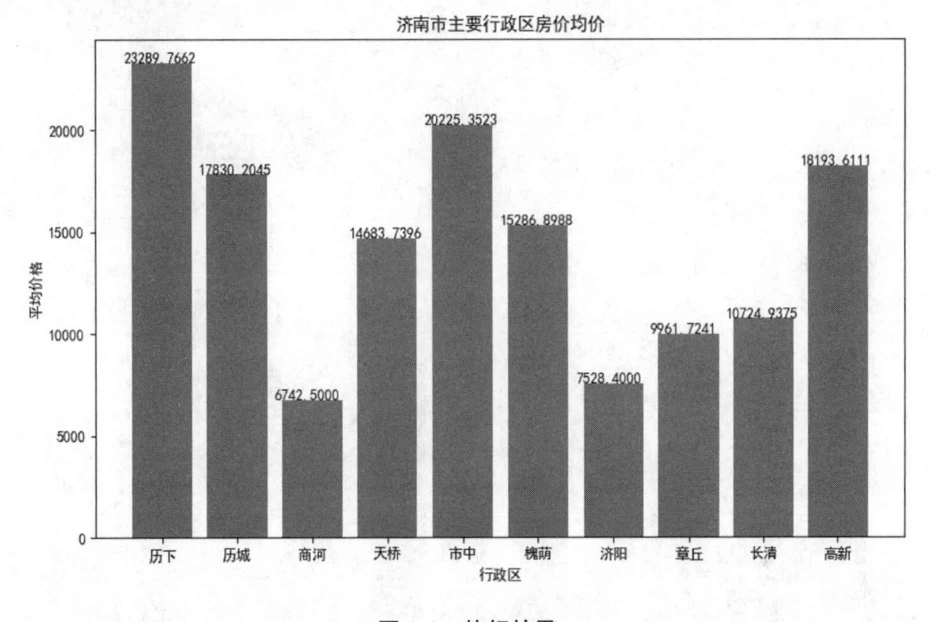


图 3-8　执行效果

## 3.7.4　可视化济南市主要行政区的房源数量

编写文件 fangyuanshuliang.py，功能是提取分析 MySQL 数据库中的数据，可视化展示当日济南市主要行政区的房源数量。文件 fangyuanshuliang.py 的具体实现代码如下所示：

```
##获取一个数据库连接，注意如果是 UTF-8 类型的，需要指定数据库
db = pymysql.connect(host="localhost", user='root', passwd="66688888",
port=3306, db="lianjia", charset='utf8')
cursor = db.cursor()  # 获取一个游标
sql = "SELECT district,sum(sale) as bili FROM xiaoqu where price!=0 and sale>=1
group by district"
cursor.execute(sql)
result = cursor.fetchall()  # result 为元组
```

```
# 将元组数据存进列表中
district = []
bili = []


for x in result:
    district.append(x[0])
    bili.append(x[1])


print(district)
print(bili)

# 直方图
plt.bar(range(len(bili)), bili, color='steelblue', tick_label=district)
plt.xlabel("行政区")
plt.ylabel("房源数量")
plt.title("济南市主要行政区房源数量(套)")
for x, y in enumerate(bili):
    plt.text(x - 0.2, y + 100, '%s' % y)
plt.show()

cursor.close()  # 关闭游标
db.close()  # 关闭数据库
```

执行效果如图 3-9 所示。

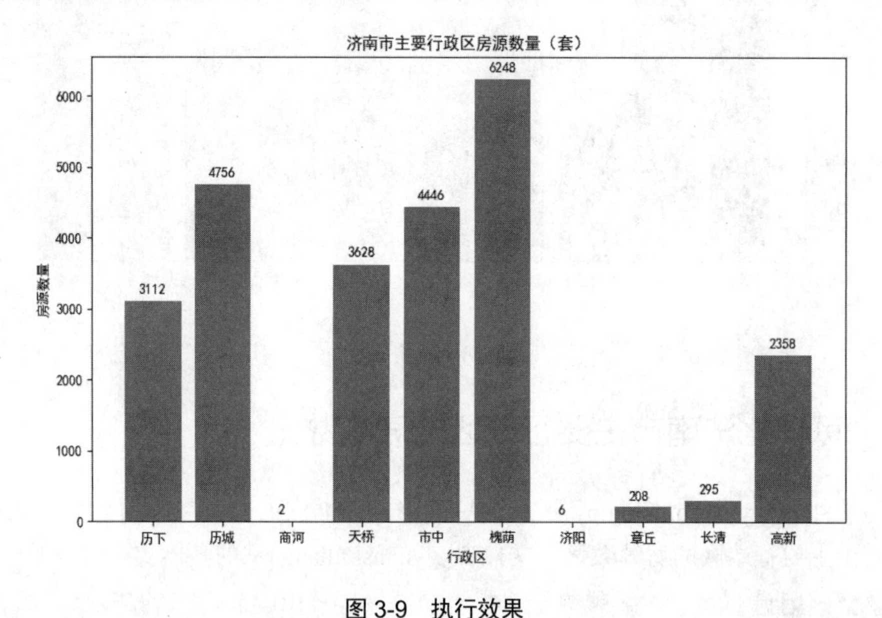


图 3-9 执行效果

## 3.7.5 可视化济南市各区的房源数量所占百分比

编写文件 bing.py，功能是提取分析 MySQL 数据库中的数据，可视化展示当日济南市各区房源数量所占百分比。为了使饼形图的界面美观，只是切片选取了济南市 7 个区的数据。文件 bing.py 的具体实现代码如下所示：

```
import pymysql
from pylab import *
mpl.rcParams["font.sans-serif"] = ["SimHei"]
mpl.rcParams["axes.unicode_minus"] = False
plt.figure(figsize=(9, 7))
##获取一个数据库连接，注意如果是 UTF-8 类型的，需要指定数据库
db = pymysql.connect(host="localhost", user='root', passwd="66688888",
port=3306, db="lianjia", charset='utf8')
cursor = db.cursor()  # 获取一个游标

sql = "select district,sum(sale) as quzongji from xiaoqu where price!=0 group
by district"
cursor.execute(sql)
result = cursor.fetchall()  # result 为元组
quzongji = []
district = []
for x in result:
    district.append(x[0])
    quzongji.append(x[1])

print(district)
print(quzongji)

sql1 = "select district,sum(sale) as quanbu from xiaoqu where price!=0"
cursor.execute(sql1)
result = cursor.fetchall()  # result 为元组

# 将元组数据存进列表中
quanbu = []
for x in result:
    quanbu.append(x[1])
print(quanbu)

import numpy as np

a = np.array(quzongji)
c = (a / quanbu)*100
print(c)


matplotlib.rcParams['font.sans-serif'] = ['SimHei']
matplotlib.rcParams['axes.unicode_minus'] = False

label_list = district[:7]    # 各部分标签
size = c[:7]    # 各部分大小
color = ["red", "green", "blue", "cyan", "magenta", "yellow", "black"]
# 各部分颜色
explode = [0, 0, 0.2, 0, 0, 0, 0.2]   # 各部分突出值
patches, l_text, p_text = plt.pie(size, explode=explode, colors=color,
labels=label_list, labeldistance=1.1, autopct="%1.2f%%", shadow=False,
                    startangle=90, pctdistance=0.7)
plt.axis("equal")    # 设置横轴和纵轴大小相等，这样饼才是圆的
```

```
plt.legend()
plt.show()

cursor.close()  # 关闭游标
db.close()  # 关闭数据库
```

执行效果如图 3-10 所示。

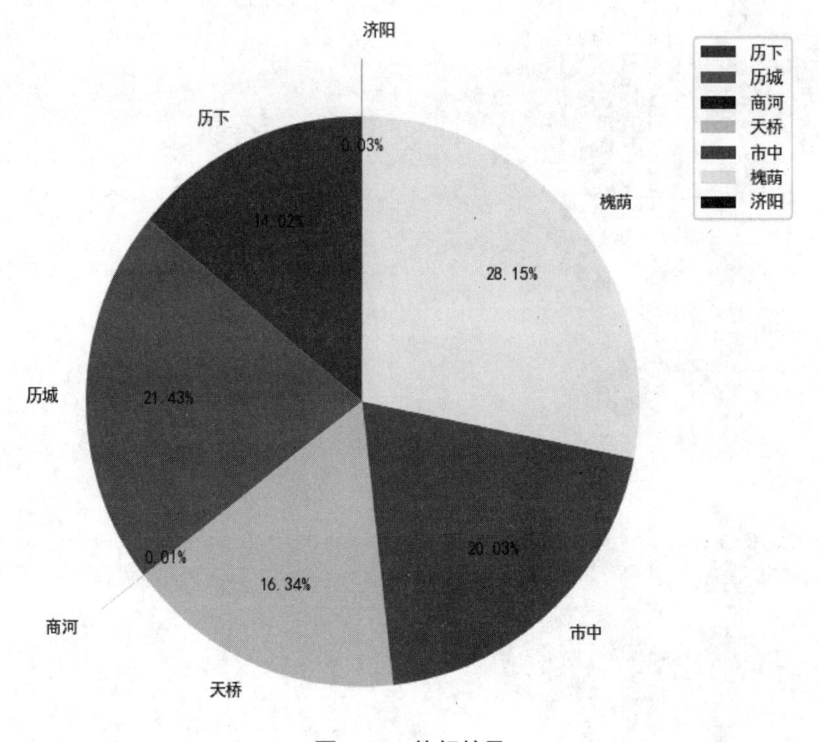


图 3-10　执行效果

# 第 4 章

# 招聘信息实时数据分析系统

## （网络爬虫+Flask+Highcharts+MySQL 实现）

21 世纪什么最重要？人才最重要。在现实世界中，各用人单位为了取得更好的发展，纷纷通过各种途径招纳人才。在众多招纳人才的途径中，招聘网成为了最重要的渠道之一。无论对于用人单位还是应聘者，分析招聘网的招聘信息是十分重要的。本章将通过一个综合实例的实现过程，详细讲解爬虫抓取某知名招聘网中招聘信息的方法，并讲解可视化分析招聘数据的过程。

扫码观看视频讲解

## 4.1 系统背景介绍

在当今社会环境下，招聘已然成为人力资源管理的热点，猎头公司、招聘网站、人才测评等配套服务机构应运而生，其核心在于为企业提供人才信息渠道。这些专业机构为企业提供专业服务，猎头公司、招聘网站解决的是“符合企业要求的人才在哪里”的问题，人才测评公司解决的是“这个人到底有何素质、适合做什么”的问题。随着企业用人需求的弹性化和不断动态变化，需要企业内部专业经理与外部专业机构解决“企业到底需要什么样的人”这一问题。

随着时代的发展，很多公司在招聘时都会收到成千上万的简历。如何挑选合适的应聘者成为公司比较棘手的事情，这给招聘单位的人事部门带来相当大的工作负担。与其他传统的人才中介相比，网上招聘具有低成本、容量大、速度快和强调个性化服务等优势。随着新增岗位的源源不断涌现，即使广泛存在的岗位对人的要求也变得模糊起来。对于用人单位的 HR 来说，及时了解招聘行情是自己最基本的业务范畴。而对于应聘者来说，根据自己的情况选择合适待遇的用人单位是自己的首要应聘目的。在这个时候，将招聘网中的招聘信息进行可视化处理就变得十分重要了。下面以某公司招聘 Python 开发工程师为例进行说明：

- 开发公司的 HR 可以可视化分析招聘网中和 Python 相关的招聘信息，了解不同学历和不同工作经验对应的薪资水平。
- Python 应聘者通过可视化分析招聘网中和 Python 相关的招聘信息，了解不同学历和不同工作经验对应的薪资水平。

目前可视化招聘信息已经成为人力资源经理关注的焦点。在特定的发展阶段、特定的文化背景下，面对变动的市场环境和弹性的岗位要求，企业到底需要什么样的人，为不同层次的人才提供什么样的待遇是他们格外关注的问题。基于目前招聘信息可视化需求分析的重要性，很多专业招聘网和猎头机构热衷于用可视化图表展示人才供求状况，可视化分析招聘信息大有可为。

## 4.2 系统架构分析

扫码观看视频讲解

本项目将首先用爬虫抓取某知名招聘网中招聘信息，然后可视化分析抓取到的招聘信息，为用人单位 HR 和应聘者提供强有力的数据支撑。

本项目的功能模块如图 4-1 所示，各个模块的具体说明如下所示。

(1) 系统设置。

设置使用 MySQL 数据库保存用爬虫抓取到的数据，然后用 Flask Web 框架提取数据库中的数据，用网页的形式可视化展示招聘信息。

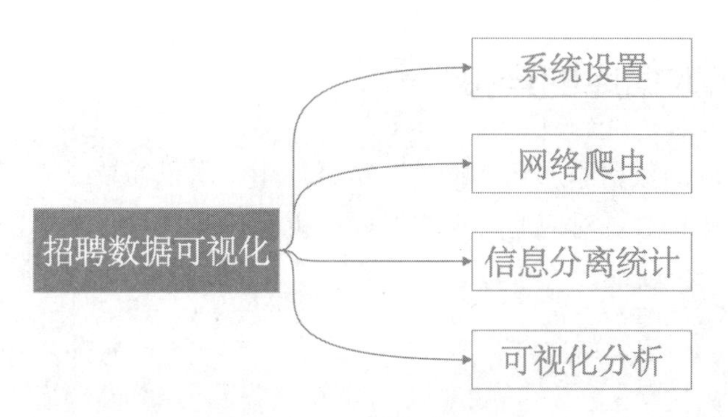


图 4-1　系统模块架构

(2) 网络爬虫。

根据用户输入的关键字，使用网络爬虫技术抓取招聘网中的招聘信息，并将抓取到的招聘信息添加到 MySQL 数据库中。

(3) 信息分离统计。

提取 MySQL 数据库中保存的爬虫数据，分别根据“工作地区”“工作经验”“薪资水平”和“学历水平”提取并分离招聘信息。

(4) 可视化分析。

提取 MySQL 数据库中保存的爬虫数据，然后使用开源框架 Highcharts 绘制柱状图和饼状图，可视化展示招聘数据信息。

## 4.3　系统设置

扫码观看视频讲解

本项目使用 MySQL 数据库保存用爬虫抓取到的数据，编写程序文件 config.py，设置连接 MySQL 数据库的参数，具体实现代码如下所示：

```
HOST = '127.0.0.1'
PORT = '3306'
USERNAME = 'root'
PASSWORD = '66688888'
DATABASE = 'u1'
DB_URI = 'mysql+pymysql://{username}:{password}@{host}:
{port}/{db}?charset= utf8mb4'.format(username=USERNAME, password=PASSWORD,
host=HOST, port=PORT, db=DATABASE)
```

## 4.4　网络爬虫

扫码观看视频讲解

本项目网络爬虫模块的实现文件是 data.py，功能是根据输入的关键字抓取招聘网中的招聘信息，并将抓取到的招聘信息添加到 MySQL 数据库中。

### 4.4.1 建立和数据库的连接

因为需要将爬取的数据添加到 MySQL 数据库中，所以需要导入在配置文件 config.py 中设置的数据库连接参数。对应的实现代码如下所示：

```
host = config.HOST
post = config.PORT
username = config.USERNAME
password = config.PASSWORD
database = config.DATABASE

etree = html.etree
tlock=threading.Lock()


# 拿到游标
cursor = db.cursor()
```

### 4.4.2 设置 HTTP 请求头 User-Agent

User-Agent 会告诉网站服务器，访问者是通过什么工具来发送请求的：如果是爬虫，一般会拒绝；如果是用户浏览器，就会应答。通过设置 HTTP 请求头 User-Agent，可以确保能够爬取到数据。对应的实现代码如下所示：

```
user_agent = [

    # Firefox
    "Mozilla/5.0 (Windows NT 6.1; WOW64; rv:34.0) Gecko/20100101 Firefox/34.0",
    "Mozilla/5.0 (X11; U; Linux x86_64; zh-CN; rv:1.4.2.10) Gecko/20100922
Ubuntu/4.10 (maverick) Firefox/3.6.10",
    # Safari
    "Mozilla/5.0 (Windows NT 6.1; WOW64) AppleWebKit/534.57.2 (KHTML, like Gecko)
Version/5.1.7 Safari/534.57.2",
    # chrome
    "Mozilla/5.0 (Windows NT 6.1; WOW64) AppleWebKit/537.36 (KHTML, like Gecko)
Chrome/34.0.2171.71 Safari/537.36",
    "Mozilla/5.0 (X11; Linux x86_64) AppleWebKit/537.11 (KHTML, like Gecko)
Chrome/23.0.1271.64 Safari/537.11",
    "Mozilla/5.0 (Windows; U; Windows NT 6.1; en-US) AppleWebKit/534.16 (KHTML,
like Gecko) Chrome/4.0.648.133 Safari/534.16",
    # 360
    "Mozilla/5.0 (Windows NT 6.1; WOW64) AppleWebKit/537.36 (KHTML, like Gecko)
Chrome/30.0.1594.101 Safari/537.36",
    "Mozilla/5.0 (Windows NT 6.1; WOW64; Trident/7.0; rv:11.0) like Gecko",
    # 猎豹浏览器
    "Mozilla/5.0 (Windows NT 6.1; WOW64) AppleWebKit/537.1 (KHTML, like Gecko)
Chrome/21.0.1180.71 Safari/537.1 LBBROWSER",
    "Mozilla/5.0 (compatible; MSIE 4.0; Windows NT 6.1; WOW64; Trident/5.0;
SLCC2; .NET CLR 2.0.50727; .NET CLR 3.5.30729; .NET CLR 3.0.30729; Media Center
PC 6.0; .NET4.0C; .NET4.0E; LBBROWSER)",
```

```
    "Mozilla/4.0 (compatible; MSIE 6.0; Windows NT 5.1; SV1; QQDownload
732; .NET4.0C; .NET4.0E; LBBROWSER)",
    # QQ 浏览器
    "Mozilla/5.0 (compatible; MSIE 4.0; Windows NT 6.1; WOW64; Trident/5.0;
SLCC2; .NET CLR 2.0.50727; .NET CLR 3.5.30729; .NET CLR 3.0.30729; Media Center
PC 6.0; .NET4.0C; .NET4.0E; QQBrowser/7.0.3698.400)",
    "Mozilla/4.0 (compatible; MSIE 6.0; Windows NT 5.1; SV1; QQDownload
732; .NET4.0C; .NET4.0E)"
]

def get_user_agent():
    """随机获取一个请求头"""
    return {'User-Agent': random.choice(user_agent)}


def requst(url):
    """requests 到 url 的 HTML"""
    html = requests.get(url,headers=get_user_agent())
    html.encoding = 'gbk'
    return etree.HTML(html.text)
```

### 4.4.3　抓取信息

分别编写函数 def get_url(url)和 def get_data(urls)，根据设置的要抓取的 URL 地址，分别抓取目标 URL 中的每个招聘信息中的职位名称、公司、城市地区、经验、发布日期、学历、薪资、招聘详情等信息。对应的实现代码如下所示：

```
def get_url(url):
    data = requst(url)
    href = data.xpath('//*[@id="resultList"]/div/p/span/a/@href')
    print(len(href))
    return href

def get_data(urls):
    """获取 51ob 职位信息,并存入数据库"""
    list_all = []
    for url in urls:
        regjob = re.compile(r'https://(.*?)51job.com', re.S)
        it = re.findall(regjob, url)
        if it != ['jobs.']:
            print('不匹配')
            continue
        try:
            data = requst(url)
            # 职位名称
            titles = data.xpath('/html/body/div[3]/div[2]/div[2]/div/div[1]
/h1/@title')[0]
            # 公司
            company = data.xpath('/html/body/div[3]/div[2]/div[2]/div/div[1]
/p[1]/a[1]/@title')[0]

            ltype = data.xpath('/html/body/div[3]/div[2]/div[2]/div/div[1]/
p[2]/@title')[0]
```

```
            ltype_str = "".join(ltype.split())
            # print(ltype_str)
            ltype_list = ltype_str.split('|')
            #城市地区
            addres = ltype_list[0]
            #经验
            exper = ltype_list[1]
            # 发布日期
            if len(ltype_list)>=5:
                #学历
                edu = ltype_list[2]
                dateT = ltype_list[4]
            else:
                #学历
                edu = "没有要求"
                dateT = ltype_list[-1]
            # 薪资
            salary = data.xpath('/html/body/div[3]/div[2]/div[2]/div/div[1]
/strong/text()')
            if len(salary) == 0:
                salary_list = [0,0]
            else:
                salary_list = salary_alter(salary)[0]
            # 招聘详情
            contents = data.xpath('/html/body/div[3]/div[2]/div[3]/div[1]
/div')[0]
            content = contents.xpath('string(.)')
            # content = content.replace(' ','')
            content = "".join(content.split())

            list_all.append([titles,company,addres,salary_list[0],salary_list[1],
dateT,edu,exper,content])
            item = [titles, company, addres, salary_list[0], salary_list[1], dateT,
                    edu, exper, content]
            write_db(item)
        except:
            print('爬取失败')
```

## 4.4.4 将抓取的信息添加到数据库

编写函数 write_db(data)，功能是将抓取到的招聘信息添加到 MySQL 数据库中。对应的实现代码如下所示：

```
def write_db(data):
    """写入数据库"""
    print(data)
    try:
        tlock.acquire()
        # rows 变量得到数据库中被影响的数据行数。
        rescoun = cursor.execute(
            "insert into data (post,company,address,salary_min,salary_max,
dateT,edu,exper,content) values(%s,%s,%s,%s,%s,%s,%s,%s,%s)", data)
        # 向数据库提交
```

```
        db.commit()
        tlock.release()
        # 如果没有 commit()，库中字段已经向下移位但内容没有写进，可是自动生成的 ID 会自动增加
        print('成功')
        global db_item
        db_item = db_item + 1

    except:
        # 发生错误时回滚
        db.rollback()
        tlock.release()
        print('插入失败')
```

### 4.4.5　处理薪资数据

为了便于在可视化图表中展示薪资水平，向数据库中添加的是单位为元的数据，在图表中展示的薪资单位是千元。如果招聘信息没有工资数据，则向数据库添加 0。如果在招聘信息中提供的是年薪，则向数据库中添加除以 12 的月薪。对应的实现代码如下所示：

```
def salary_alter(salarys):   #[]
    salary_list = []
    for salary in salarys:
        # print(salary)
        if salary == '':
            a = [0,0]
        re_salary = re.findall('[\d+\.\d]*', salary)  # 提取数值--是文本值
        salary_min = float(re_salary[0])  # 将文本转化成数值型,带有小数，用 float()

        wan = lambda x,y : [x*10000,y*10000]
        qian = lambda x, y: [x * 1000, y * 1000]
        wanqian = lambda x,y,s :wan(x,y) if '万' in salary else qian(x,y)
        tian = lambda x, y, s: [x,y] if '元' in salary else qian(x, y)

        if '年' in salary:
            salary_max = float(re_salary[2])
            a = wanqian(salary_min,salary_max,salary)
            a[0] = round(a[0] / 12, 2)
            a[1] = round(a[1] / 12, 2)
        elif '月' in salary:
            salary_max = float(re_salary[2])
            a = wanqian(salary_min, salary_max, salary)
        elif '天' in salary:
            salary_max = float(re_salary[0])
            a = tian(salary_min, salary_max, salary)
            a[0] *= 31
            a[1] *= 31
        salary_list.append(a)
    return salary_list
```

### 4.4.6 清空数据库数据

为了保证可视化分析的数据具备时效性和准确性，在每一次新的爬虫之前要清空之前抓取的数据。对应的实现代码如下所示：

```
def data_clr():
    # 清空 data
    try:
        tlock.acquire()
        query = "truncate table `data`"
        cursor.execute(query)
        db.commit()
        tlock.release()
        print('data 原表已清空')
    except Exception as aa:
        print(aa)
        print('无 data 表！')
```

### 4.4.7 执行爬虫程序

设置要爬取的 URL 地址，开始执行爬虫程序。对应的实现代码如下所示：

```
def main(kw, city,startpage):
    print(kw)
    print(city)
    city_id = get_cityid(city)
    print(city_id)
    data_clr()
    page = startpage
    global db_item
    db_item = 0
    while(db_item <= 200):
        url = "https://search.51job.com/list/{},000000,0000,00,9,99,{},2, \
{}.html".format(city_id,kw,page)
        print(url)
        # url = 'https://search.51job.com/jobsearch/search_result.php'
        a =get_url(url)
        print(a)
        get_data(a)
        page = page + 1
        print(db_item)
        # time.sleep(0.2)
```

## 4.5 信息分离统计

扫码观看视频讲解

提取在 MySQL 数据库中保存的爬虫数据，分别根据“工作地区”“工作经验”“薪资水平”和“学历水平”提取招聘信息并进行统计，最终的统计结果将作为后面可视化图表的素材数据。

### 4.5.1 根据“工作经验”分析数据

编写程序文件 jinyan.py，根据用人单位的“工作经验”要求提取并统计招聘信息，具体实现代码如下所示：

```
def get_edu():
    try:
        cursor.execute("select exper from data ")
        salary = cursor.fetchall()
        # 向数据库提交
        db.commit()
        return salary
    except:
        # 发生错误时回滚
        db.rollback()
        print("查询失败")
        return 0

def jinyanfun():
    edu = get_edu()
    data = []
    for i in edu:
        # print(type(i))
        year = re.findall(r"\d+", i[0])
        if len(year)==1:
            data.append(year[0]+'年工作经验')
        elif len(year)==2:
            data.append(year[0]+'-'+year[1]+'年工作经验')
        elif len(year)==0:
            data.append('无工作经验')

    data = DataFrame(data)
    da = data[0].value_counts()
    # print(da)
    list_all = []
    for (i, j) in zip(da.index, da):
        print(j,i)
        list_all.append([i,j])
    # print(type(da))
    return list_all


if __name__ == '__main__':
    a = jinyanfun()
    print(a)
```

执行代码后，会基于当前抓取到的招聘信息提取“工作经验”列的数据，并进行数据统计：

```
76 3-4 年工作经验
56 2 年工作经验
37 1 年工作经验
```

```
33 无工作经验
14 5-7 年工作经验
1 3 年工作经验
[['3-4 年工作经验', 76], ['2 年工作经验', 56], ['1 年工作经验', 37], ['无工作经验', 33],
['5-7 年工作经验', 14], ['3 年工作经验', 1]]
```

## 4.5.2 根据“工作地区”分析数据

编写程序文件 map.py，根据用人单位的“工作地区”要求提取并统计招聘信息，具体实现代码如下所示：

```
def get_xinzi():
    try:
        cursor.execute("select address from data ")
        salary = cursor.fetchall()
        # 向数据库提交
        db.commit()
        return salary
    except:
        # 发生错误时回滚
        db.rollback()
        print("查询失败")
        return 0
a = get_xinzi()

list = []
for i in a:
    test = i[0].split('-')
    list.append(test[0])
    # print(test)

data = DataFrame(list)
da = data[0].value_counts()


for (i,j) in zip(da.index,da):
    pass
    print(j,i)
print(type(da))
```

执行代码后，会基于当前抓取到的招聘信息提取“工作地区”列的数据，并进行数据统计：

```
169 广州
14 佛山
12 东莞
8 珠海
4 中山
3 深圳
1 南昌
1 汕头
1 上海
1 惠州
```

```
1 澄迈
1 韶关
1 广东省
```

### 4.5.3 根据“薪资水平”分析数据

编写程序文件 xinzi.py，根据用人单位的“薪资水平”要求提取并统计招聘信息。在本项目中将工资分为如下 8 个档次：

- 小于 5000 元。
- 5000～8000 元。
- 8000～11000 元。
- 11000～14000 元。
- 14000～17000 元。
- 17000～20000 元。
- 20000～23000 元。
- 高于 23000 元。

文件 xinzi.py 的具体实现代码如下所示：

```
def get_xinzi():
    try:
        cursor.execute("select salary_min,salary_max from data ")
        salary = cursor.fetchall()
        # 向数据库提交
        db.commit()
        return salary
    except:
        # 发生错误时回滚
        db.rollback()
        print("查询失败")
        return 0

def xinzi():
    a = get_xinzi()
    data = []
    for i in a:
        data.append((int(i[0])+int(i[1]))/2)

    fenzu=pd.cut(data,[0,5000,8000,11000,14000,17000,20000,23000,9000000000],
right=False)
    pinshu=fenzu.value_counts()
    # print(pinshu)
    list = []
    for i in pinshu:
        # print(i)
        list.append(i)

    list_all = [
        ['小于5k', list[0]],
        ['5k~8k', list[1]],
```

```
        ['8k~11k',list[2]],
        ['11k~14k',list[3]],
        ['14k-17k', list[4]],
        ['17k-20k', list[5]],
        ['20k-23k', list[6]],
        ['23K~', list[7]]
    ]
    return list

if __name__ == '__main__':
    a = xinzi()
    print(a)
```

执行代码后，会基于当前抓取到的招聘信息提取“薪资水平”列的数据，并进行数据统计。

```
[11, 34, 47, 53, 31, 14, 15, 12]
```

### 4.5.4 根据“学历水平”分析数据

编写程序文件xueli.py，根据用人单位的“学历水平”要求提取并统计招聘信息，具体实现代码如下所示：

```
def get_edu():
    try:
        cursor.execute("select edu from data ")
        salary = cursor.fetchall()
        # 向数据库提交
        db.commit()
        return salary
    except:
        # 发生错误时回滚
        db.rollback()
        print("查询失败")
        return 0

def xuelifun():
    edu = get_edu()
    data = []
    for i in edu:
       data.append(i)
       # print(i)

    data = DataFrame(data)
    da = data[0].value_counts()
    # print(da)
    list_all = []
    for (i, j) in zip(da.index, da):
        print(j, i)
        if '招' in i:
            # print(i)
            continue
        list_all.append([i, j])
    # print(type(da))
```

```
    return list_all
if __name__ == '__main__':
    a = xuelifun()
    print(a)
    print(type(a))
```

执行代码后，会基于当前抓取到的招聘信息提取“学历水平”列的数据，并进行数据统计。

```
107 本科
73 大专
26 没有要求
5 硕士
2 中专
2 招若干人
1 招 4 人
1 招 2 人
[['本科', 107], ['大专', 73], ['没有要求', 26], ['硕士', 5], ['中专', 2]]
```

## 4.6　数据可视化

扫码观看视频讲解

提取在 MySQL 数据库中保存的爬虫数据，并基于前面分析统计步骤中得到的统计结果，使用开源框架 Highcharts 绘制柱状图和饼状图，使用 Flask 框架可视化展示招聘数据信息。

### 4.6.1　Flask Web 架构

编写程序文件 app.py，使用 Flask 框架创建一个 Web 项目，设置不同的 URL 参数对应的 HTML 模板文件。文件 app.py 的具体实现代码如下所示：

```
from flask import Flask  # 导入 Flask 模块
from flask import render_template #导入模板函数
import _thread,time

from flask import request
from servers.data import main
from servers import xinzi,xueli,jinyan
from flask_sqlalchemy import SQLAlchemy #导入 SQLAlchemy 模块，连接数据库
from sqlalchemy import or_
import config #导入配置文件
app = Flask(__name__) #Flask 初始化

app.jinja_env.auto_reload = True
app.config['TEMPLATES_AUTO_RELOAD'] = True
app.config.from_object(config) #初始化配置文件
db = SQLAlchemy(app) #获取配置参数，将和数据库相关的配置加载到 SQLAlchemy 对象中

from models.models import Data    #导入 Data 模块
```

```
#创建表和字段

@app.route('/')
def first():
    return render_template("input.html")

@app.route('/list') #定义路由
def list(): #定义list函数
    # data = Data.query.all()
    page = int(request.args.get('page', 1))
    per_page = int(request.args.get('per_page', 2))
    key = request.args.get('key', '')
    # Data.address.like("%{}%".format('java'))
    data = Data.query.filter(or_(Data.post.like("%{}%".format(key)),
                                 Data.company.like("%{}%".format(key)),
                                 Data.address.like("%{}%".format(key)),
                                 Data.salary_max.like("{}".format(key)),

Data.salary_min.like("{}".format(key)) )).paginate(page, 12, error_out=False)
    return render_template("data.html", datas=data,key=key)


@app.route('/search')
def search():
    kw =request.args.get("kw")
    city =request.args.get("city")

    # main(kw, city)
    # 创建两个线程
    try:
        _thread.start_new_thread(main, (kw, city, 1,))
        _thread.start_new_thread(main, (kw, city, 51,))
    except:
        print("Error: 无法启动线程")

    time.sleep(50)
    xz = xinzi.xinzi()
    xl = xueli.xuelifun()
    jy = jinyan.jinyanfun()
    return render_template('h.html', **locals())


#通过url传递信息
@app.route('/chart')
def charts():
    xz = xinzi.xinzi()
    xl = xueli.xuelifun()
    jy = jinyan.jinyanfun()
    return render_template('h.html',**locals())


#通过url传递信息
@app.route('/xinzi')
```

```
def xinzitest():
    a = xinzi.xinzi()
    print(a)
    return str(a)

@app.route('/xueli')
def xuelitest():
    data = xueli.xuelifun()
    print(data)

    return str(data)

if __name__ == '__main__':
    app.run()
```

## 4.6.2 Web 主页

在本 Flask Web 模块中，Web 主页对应的 HTML 模板文件是 input.html，功能是提供一个表单供用户分别输入岗位名称和搜索的省份，单击“搜索”按钮后会根据用户输入的岗位条件爬虫抓取目标网站中的招聘信息。文件 input.html 的具体实现代码如下所示：

```
{% extends "base.html" %}
{% block content %}

    <style>
    body{
        background: url('{{ url_for('static',filename='img/background.jpg') }}');
    }
    #sousuo{
        height: 500px;
    }
    </style>
    <div id="sousuo"style="text-align: center ; padding: 5px">
            <form method="get" action="/search">
                <div class="form">
                    <p>岗位名称</p>
                    <input class="form-name"placeholder="java 软件工程师"
name="kw" type="text" autofocus>
                </div><br>
                <div class="form" style="margin-top: 30px;">
                    <p>搜索的省份</p>
                    <input class="form-name"placeholder="例如:全国,江苏省"
name="city" type="text" autofocus>
                </div><br>
                <input type="submit" value="搜索" class="btn" />
            </form>
    </div>

{% endblock %}
```

执行代码后的 Web 主页效果如图 4-2 所示。

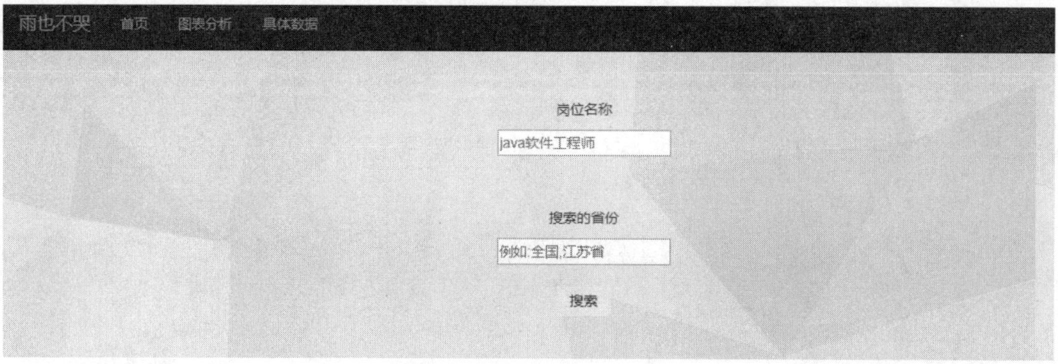

图 4-2 Web 主页

## 4.6.3 数据展示页面

在本 Flask Web 模块中，数据展示页面对应的 HTML 模板文件是 data.html，功能是获取在 MySQL 数据库中保存的招聘信息，并通过表格分页的形式展示这些招聘信息。文件 data.html 的具体实现代码如下所示：

```
{% extends "base.html" %}
{% block content %}
    <form method="get" action="/list">
        <div class="input-group col-md-3" style="margin-top:0px; margin-left:
75%; positon:relative">
            <input type="text" class="form-control" placeholder="请输入要搜索的内
容" name="key"/>
            <span class="input-group-btn">
               <button class="btn btn-info btn-search" style="margin-left:3px">
搜索</button>
            </span>
        </div>
    </form>


    <table class="table table-bordered">
    <tr>
        <th>职位</th>
        <th>公司</th>
        <th>城市</th>
        <th>最低薪资</th>
        <th>最高薪资</th>
        <th>发布日期</th>
    </tr>
        {% for a in datas.items %}
            <tr>
                <td>{{ a.post }}</td>
                <td>{{ a.company }}</td>
```

```
            <td>{{ a.address }}</td>
            <td>{{ a.salary_min }}</td>
            <td>{{ a.salary_max }}</td>
            <td>{{ a.dateT }}</td>
        </tr>
    {% endfor %}
    </table>


    <div style="text-align: center">
        <nav aria-label="Page navigation example">
        <ul class="pagination justify-content-center">

            {% if datas.has_prev %}
                <li class="page-item">
            <a class="page-link" href="/list?page={{ datas.prev_num }}"
aria-label="Previous">
                <span aria-hidden="true">&laquo;</span>
                <span class="sr-only">Previous</span>
            </a>
        </li>
            {% endif %}

            {% for i in datas.iter_pages() %}
                {% if i == None %}
                    <li class="page-item"><a class="page-link"
href="#">...</a></li>
                {% else %}
                    <li class="page-item"><a class="page-link"
href="/list?page={{ i }}&key={{ key }}">{{ i }}</a></li>
                {% endif %}
            {% endfor %}

            {% if datas.has_next %}
                <li class="page-item">
            <a class="page-link"
href="/list?page={{ datas.next_num }}&key={{ key }}" aria-label="Next">
                <span aria-hidden="true">&raquo;</span>
                <span class="sr-only">Next</span>
            </a>
        </li>
            {% endif %}
        </ul>
        </nav>
        当前页数：{{ datas.page }}
        总页数：{{ datas.pages }}
        一共有{{ datas.total }}条数据
        <br>
    </div>
{% endblock %}
```

执行代码后的数据展示页面效果如图 4-3 所示。

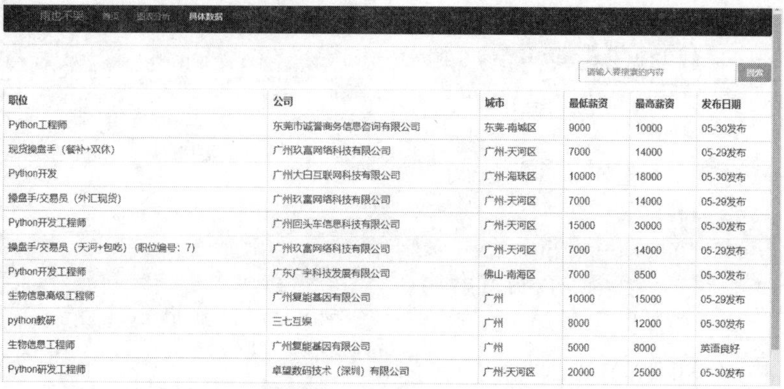

| 职位 | 公司 | 城市 | 最低薪资 | 最高薪资 | 发布日期 |
|---|---|---|---|---|---|
| Python工程师 | 东莞市诚誉商务信息咨询有限公司 | 东莞-南城区 | 9000 | 10000 | 05-30发布 |
| 现货操盘手（餐补+双休） | 广州玖富网络科技有限公司 | 广州-天河区 | 7000 | 14000 | 05-29发布 |
| Python开发 | 广州大白互联网科技有限公司 | 广州-海珠区 | 10000 | 18000 | 05-30发布 |
| 操盘手/交易员（外汇现货） | 广州玖富网络科技有限公司 | 广州-天河区 | 7000 | 14000 | 05-29发布 |
| Python开发工程师 | 广州回头车信息科技有限公司 | 广州-天河区 | 15000 | 30000 | 05-30发布 |
| 操盘手/交易员（天河+包吃）（职位编号：7） | 广州玖富网络科技有限公司 | 广州-天河区 | 7000 | 14000 | 05-29发布 |
| Python开发工程师 | 广东广宇科技发展有限公司 | 佛山-南海区 | 7000 | 8500 | 05-30发布 |
| 生物信息高级工程师 | 广州复能基因有限公司 | 广州 | 10000 | 15000 | 05-29发布 |
| python教研 | 三七互娱 | 广州 | 8000 | 12000 | 05-30发布 |
| 生物信息工程师 | 广州复能基因有限公司 | 广州 | 5000 | 8000 | 英语良好 |
| Python研发工程师 | 卓望数码技术（深圳）有限公司 | 广州-天河区 | 20000 | 25000 | 05-30发布 |

图 4-3　数据展示页面

## 4.6.4　数据可视化页面

在本 Flask Web 模块中，数据可视化页面对应的 HTML 模板文件是 h.html，功能是根据 MySQL 数据库中保存的招聘信息的统计结果，使用 Highcharts 绘制统计图表。文件 h.html 的具体实现代码如下所示：

```
{% extends "base.html" %}
{% block content %}
    <style>
    body{
        background:
url('{{ url_for('static',filename='img/background.jpg') }}');
    }
    </style>

    <div id="xinzi"></div>
    <table align="center">
    <tr>
        <td>
            <div id="xueli"></div>
        </td>
        <td>
            <div id="jingyan"></div>
        </td>
    </tr>
    </table>


    <script type="text/javascript">
    $(document).ready(function() {
        var chart = {
            type: 'column',
            backgroundColor: 'rgba(0,0,0,0)'
        };
        var title = {
            useHTML: true,
            style: {
```

```
            color: '#000',          //字体颜色
            fontSize: "29px",       //字体大小
            fontWeight: 'bold'
        },
        text: '工资分布图'
    };
    var subtitle = {
        text: '51job.com'
    };
    var xAxis = {
        categories:
['0~5k','5~8k','8k~11k','11k~14k','14k~17k','17k~20k','20k~23k','23K 以上'],
        crosshair: true
    };
    var yAxis = {
        min: 0,
        title: {
            text: '岗位数'
        }
    };
    var tooltip = {
        headerFormat: '<span
style="font-size:10px">{point.key}</span><table>',
        pointFormat: '<tr><td
style="color:{series.color};padding:0">{series.name}: </td>' +
            '<td style="padding:0"><b>{point.y:.1f} 个</b></td></tr>',
        footerFormat: '</table>',
        shared: true,
        useHTML: true
    };
    var plotOptions = {
        column: {
            pointPadding: 0.2,
            borderWidth: 0
        }
    };
    var credits = {
        enabled: false
    };

    var series= [{
        name: '岗位数',
        data: {{ xz|tojson }}
    }];

    var json = {};
    json.chart = chart;
    json.title = title;
    json.subtitle = subtitle;
    json.tooltip = tooltip;
    json.xAxis = xAxis;
    json.yAxis = yAxis;
    json.series = series;
    json.plotOptions = plotOptions;
    json.credits = credits;
    $('#xinzi').highcharts(json);
```

```
});
</script>
<script type="text/javascript">
  $(document).ready(function() {
   let chart = {
      plotBackgroundColor: null,
      plotBorderWidth: null,
      plotShadow: false,
      backgroundColor: 'rgba(0,0,0,0)'
   };
   let title = {
      useHTML: true,
       style: {
          color: '#000',       //字体颜色
          fontSize: "29px",   //字体大小
          fontWeight: 'bold'
        },
     text: '学历占比情况'
   };
   let tooltip = {
     pointFormat: '{series.name}: <b>{point.percentage:.1f}%</b>'
   };
   let plotOptions = {
     pie: {
       allowPointSelect: true,
       cursor: 'pointer',
       dataLabels: {
         enabled: true
       },
       showInLegend: true
     }
   };
   let series= [{
     type: 'pie',
     name: '学历',
     data: {{xl|tojson}}
   }];

   let json = {};
   json.chart = chart;
   json.title = title;
   json.tooltip = tooltip;
   json.series = series;
   json.plotOptions = plotOptions;
   $('#xueli').highcharts(json);
});
</script>
<script type="text/javascript">
$(document).ready(function() {
   let chart = {
      plotBackgroundColor: null,
      plotBorderWidth: null,
      plotShadow: false,
      backgroundColor: 'rgba(0,0,0,0)'
   };
   let title = {
      useHTML: true,
       style: {
```

```
          color: '#000',     //字体颜色
          "fontSize": "29px",   //字体大小
          fontWeight: 'bold'
        },
      text: '工作年限要求'
    };
    let tooltip = {
      pointFormat: '{series.name}: <b>{point.percentage:.1f}%</b>'
    };
    let plotOptions = {
      pie: {
        allowPointSelect: true,
        cursor: 'pointer',
        dataLabels: {
          enabled: true
        },
        showInLegend: true
      }
    };
    let series= [{
      type: 'pie',
      name: '工作年限要求',
      data: {{jy|tojson}}
    }];

    let json = {};
    json.chart = chart;
    json.title = title;
    json.tooltip = tooltip;
    json.series = series;
    json.plotOptions = plotOptions;
    $('#jinyan').highcharts(json);
  });
  </script>

{% endblock %}
```

执行代码后的数据可视化页面效果如图 4-4 所示。

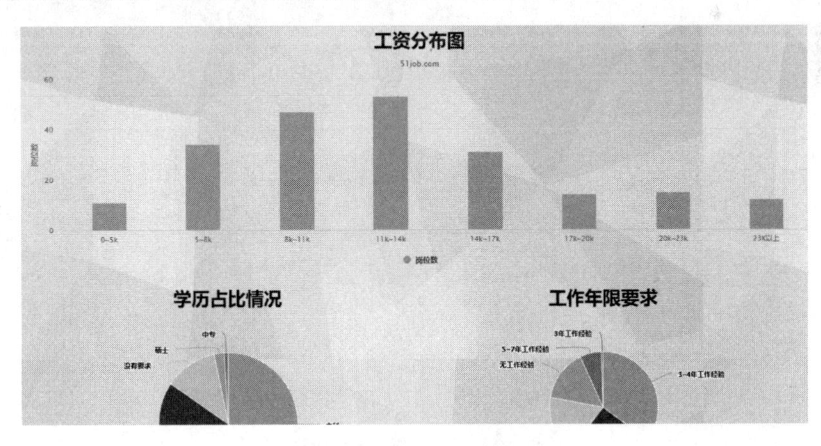


图 4-4　数据可视化页面

# 第 5 章

# 基于深度学习的 AI 人脸识别系统（Flask+OpenCV-Python+ Keras+Sklearn 实现）

近年来，随着人工智能技术的飞速发展，机器学习和深度学习技术已经摆在了人们的面前，逐渐成为程序员们的学习热点。本章将详细介绍使用深度学习技术开发人脸识别系统的知识，详细讲解使用 Flask+OpenCV-Python+ Keras+Sklearn 实现一个大型人工智能项目的过程。

## 5.1 人工智能基础

扫码观看视频讲解

人工智能就是我们平常所说的AI，全称是Artificial Intelligence。人工智能是研究、开发用于模拟、延伸和扩展人类智能的理论、方法、技术及应用系统的一门新的技术科学。本节将简要介绍人工智能技术的基本知识。

### 5.1.1 人工智能介绍

人类发展已有几千年，自从机器诞生以来，聪明的人类就开始试图让机器具有智能，也就是人工智能。人工智能是一门极富挑战性的科学，从事这项工作的人必须懂得计算机知识、心理学和哲学。人工智能是包括内容十分广泛的科学，它由不同领域的学科组成，如机器学习、计算机视觉，等等。总的说来，人工智能研究的一个主要目标是使机器能够胜任一些通常需要人类智能才能完成的复杂工作。

人工智能不是一个非常庞大的概念，单从字面上，应该理解为人类创造的智能。那么什么是智能呢？如果人类创造了一个机器人，这个机器人能有像人类一样甚至超过人类的推理、学习、感知、处理等这些能力，那么就可以将这个机器人称为是一个有智能的物体，也就是人工智能。

现在通常将人工智能分为弱人工智能和强人工智能。我们在电影里看到的一些人工智能大部分都是强人工智能，他们能像人类一样思考如何处理问题，甚至能在一定程度上做出比人类更好的决定，他们能自适应周围的环境，解决一些程序中没有遇到过的突发事件。但是在目前的现实世界中，大部分人工智能只是实现了弱人工智能，这能够让机器具备观察和感知的能力，在经过一定的训练后能计算一些人类不能计算的事情；但是它并没有自适应能力，也就是它不会处理突发的情况，只能处理程序中已经写好的、已经预测到的事情。

### 5.1.2 人工智能的发展历程

1950年，一位名叫马文·明斯基(人工智能之父)的大四学生与他的同学邓恩·埃德蒙一起，建造了世界上第一台神经网络计算机。同样是在1950年，被称为“计算机之父”的阿兰·图灵提出了一个举世瞩目的想法：图灵测试。按照图灵的设想，如果一台机器能够与人类开展对话而不能被辨别出机器身份，那么这台机器就具有智能。而就在这一年，图灵还大胆预言了真正具备智能机器的可行性。

时间跳转到了20世纪70年代，人工智能也步入了一段艰难险阻的岁月。对于人工智能的研究，由于科研人员对于难度估量过低和缺乏经费等原因，导致与美国国防高级研究计划署的合作计划失败，社会舆论的压力也开始慢慢压向人工智能这边，这也导致很多研究经费被转移到了其他项目上，让大家对人工智能的前景产生担忧。

人工智能产业面临衰落，但科技并不会因外界因素而停止发展，至 20 世纪 80 年代初期人工智能产业开始重新崛起。从 20 世纪 90 年代中期开始，随着 AI 技术尤其是神经网络技术的逐步发展，以及人们对 AI 逐渐抱有客观理性的认知，人工智能技术开始进入平稳发展时期。1997 年 5 月 11 日，IBM 的计算机系统“深蓝”战胜了国际象棋世界冠军卡斯帕罗夫，又一次在公众领域引发了现象级的 AI 话题讨论。这是人工智能发展的一个重要里程碑。

2006 年，Hinton 在神经网络的深度学习领域取得突破，人类又一次看到机器赶超人类的希望，这也是标志性的技术进步。紧接着谷歌、微软、百度等互联网巨头，还有众多的初创科技公司，纷纷加入人工智能产品的战场，掀起又一轮的智能化狂潮。

2016 年，Google 公司的 AlphaGo 战胜韩国棋手李世石，再度引发 AI 热潮。

## 5.1.3　和人工智能相关的几个重要概念

### 1. 监督学习

监督学习的任务是学习一个模型，这个模型可以处理任意的输入，并且针对每个输入都可以映射输出一个预测结果。这里的模型相当于数学中一个函数，输入就相当于数学中的 X，而预测的结果就相当于数学中的 Y。对于每一个 X，都可以通过一个映射函数映射出一个结果。

### 2. 非监督学习

非监督学习指直接对没有标记的训练数据进行建模学习。注意，这里的数据是没有标记的数据，它与监督学习的最基本区别是建模的数据一个有标记一个没有标记。例如聚类(将物理或抽象对象的集合分成由类似的对象组成的多个类的过程被称为聚类)就是一种典型的非监督学习，分类就是一种典型的监督学习。

### 3. 半监督学习

当我们拥有标记的数据很少，未被标记的数据很多，但是人工标注又比较昂贵的时候，可以根据一些条件(查询算法)查询(query)一些数据，让专家进行标记。这是半监督学习与其他算法的本质区别。

### 4. 主动学习

当使用一些传统的监督学习方法做分类处理的时候，通常是训练样本的规模越大，分类的效果就越好。但是在现实生活的很多场景中，标记样本的获取是很困难的，这需要领域内的专家进行人工标注，所花费的时间成本和经济成本都是很大的。而且，如果训练样本的规模过于庞大，训练的时间花费也会比较多。那么问题来了：有没有一种有效办法，能够使用较少的训练样本来获得性能较好的分类器呢？主动学习(Active Learning)为我们提供了这种可能。主动学习通过一定的算法查询最有用的未标记样本，并交由专家进行标记，然后用查询到的样本训练分类模型来提高模型的精确度。

在人类的学习过程中，通常利用已有的经验来学习新的知识，又依靠获得的知识来总结和积累经验，经验与知识不断交互。同样，机器学习也可模拟人类学习的过程，利用已有的知识训练出模型去获取新的知识，并通过不断积累的信息去修正模型，以得到更加准确有用的新模型。不同于被动学习被动地接受知识，主动学习能够有选择性地获取知识。

## 5.2 机器学习基础

扫码观看视频讲解

在人工智能的两个发展阶段中，无论是“推理期”还是“知识期”，都会存在如下两个缺点。

(1) 机器都是按照人类设定的规则和总结的知识运作，永远无法超越其创造者——人类。

(2) 人力成本太高，需要专业人才进行具体实现。

基于上述两个缺点，人工智能技术的发展出现了一个瓶颈期。为了突破这个瓶颈期，一些权威学者就想到：如果机器能够自我学习，问题不就迎刃而解了吗？此时机器学习(Machine Learning)技术便应运而生，人工智能进入“机器学习”时代。本节将简要介绍机器学习的基本知识。

### 5.2.1 机器学习介绍

机器学习(Machine Learning，简称 ML)是一门多领域交叉学科，涉及概率论、统计学、逼近论、凸分析、算法复杂度理论等多门学科。机器学习专门研究计算机怎样模拟或实现人类的学习行为，以获取新的知识或技能，重新组织已有的知识结构，使之不断改善自身的性能。

机器学习是一类算法的总称，这些算法企图从大量历史数据中挖掘出其中隐含的规律，并用于预测或者分类。更具体地说，机器学习可以看作是寻找一个函数，输入是样本数据，输出是期望的结果；只是这个函数过于复杂，以至于不太方便形式化表达。需要注意的是，机器学习的目标是使学到的函数很好地适用于“新样本”，而不仅仅是在训练样本上表现得很好。学到的函数适用于新样本的能力，称为泛化(Generalization)能力。

机器学习有一个显著的特点，也是机器学习最基本的做法，就是使用一个算法从大量的数据中解析且得到有用的信息，并从中学习，然后对之后真实世界中会发生的事情进行预测或做出判断。机器学习需要海量的数据来进行训练，并从这些数据中得到要用的信息，然后反馈到真实世界的用户中。

我们可以用一个简单的例子来说明机器学习。假设在天猫或京东购物的时候，天猫和京东会向我们推送商品信息，这些推荐的商品往往是我们很感兴趣的东西，这个过程是通过机器学习完成的。其实这些推送商品是天猫和京东根据我们以前的购物订单和经常浏览的商品记录而得出的结论，可以从中找出商城中的哪些商品是我们感兴趣的，并且我们会大概率购买的，然后将这些商品定向推送给我们。

## 5.2.2 机器学习的三个发展阶段

机器学习是人工智能的核心，是使计算机具有智能的根本途径，其应用遍及人工智能的各个领域，它主要使用归纳、综合而不是演绎。机器学习的发展分为如下所示的三个阶段。

- 20 世纪 80 年代，连接主义较为流行，代表工作有感知机(Perceptron)和神经网络(Neural Network)。
- 20 世纪 90 年代，统计学习方法开始占据主流舞台，代表性方法有支持向量机(Support Vector Machine)。
- 21 世纪初，深度神经网络技术被提出，连接主义卷土重来，随着数据量和计算能力的不断提升，以深度学习(Deep Learning)为基础的诸多 AI 应用逐渐成熟。

## 5.2.3 机器学习的分类

根据不同的角度，可以将机器学习划分为多种不同的类型。

(1) 按任务类型划分。

机器学习模型可以分为回归模型、分类模型和结构化学习模型，具体说明如下所示。

- 回归模型：又叫预测模型，输出一个不能枚举的数值。
- 分类模型：又分为二分类模型和多分类模型，常见的二分类问题有垃圾邮件过滤，常见的多分类问题有文档自动归类。
- 结构化学习模型：此类型的输出不再是一个固定长度的值，如图片语义分析输出的是图片的文字描述。

(2) 从方法角度划分。

机器学习模型可以分为线性模型和非线性模型，具体说明如下所示。

- 线性模型：虽然比较简单，但是其作用不可忽视。线性模型是非线性模型的基础，很多非线性模型都是在线性模型的基础上变化而来的。
- 非线性模型：又可以分为传统机器学习模型(如 SVM，KNN，决策树等)和深度学习模型。

(3) 按照学习理论划分。

机器学习模型可以分为有监督学习、半监督学习、无监督学习、迁移学习和强化学习，具体说明如下所示。

- 当训练样本带有标记时，是有监督学习。
- 训练样本部分有标记，部分无标记时，是半监督学习。
- 训练样本全部无标记时是无监督学习。
- 迁移学习就是把已经训练好的模型参数迁移到新的模型上，以帮助新模型训练。
- 强化学习是一个学习最优策略(Policy)，可以让本体(Agent)在特定环境(Environment)

中，根据当前状态(State)，做出行动(Action)，从而获得最大回报(Reward)。强化学习和有监督学习最大的不同是，每次的决定没有对与错，而是希望获得最多的累计奖励。

### 5.2.4 深度学习和机器学习的对比

前面介绍的机器学习是一种实现人工智能的方法，深度学习是一种实现机器学习的技术。深度学习本来并不是一种独立的学习方法，其本身也会用到有监督和无监督的学习方法来训练深度神经网络。但由于近几年该领域发展迅猛，一些特有的学习手段相继被提出(如残差网络)，因此越来越多的人将其单独看作一种学习的方法。

假设我们需要识别某个照片是狗还是猫，如果是传统机器学习的方法，会首先定义一些特征，如有没有胡须，耳朵、鼻子、嘴巴的模样，等等。总之，我们首先要确定相应的“面部特征”作为机器学习的特征，以此来对我们的对象进行分类识别。而深度学习的方法则更进一步，它自动找出这个分类问题所需要的重要特征(而传统机器学习则需要我们人工地给出特征)。那么，深度学习是如何做到这一点的呢？继续以猫狗识别的例子进行说明，按照以下步骤进行：

(1) 首先确定出有哪些边和角跟识别出猫狗关系最大。

(2) 然后根据上一步找出的很多小元素(边、角等)构建层级网络，找出它们之间的各种组合。

(3) 在构建层级网络之后，就可以确定哪些组合可以识别出猫和狗。

**注 意**

其实深度学习并不是一个独立的算法，在训练神经网络的时候也通常会用到监督学习和无监督学习。但是由于一些独特的学习方法被提出，我觉得把它看成是单独的一种学习算法应该也没什么问题。深度学习可以大致理解成包含多个隐含层的神经网络结构，深度学习的“深”字指的就是隐藏层的深度。

在机器学习方法中，几乎所有的特征都需要通过行业专家再确定，然后手工对特征进行编码；而深度学习算法会自己从数据中学习特征。这也是深度学习十分引人注目的一点，毕竟特征工程是一项十分烦琐、耗费很多人力物力的工作，深度学习的出现大大减少了发现特征的成本。

在解决问题时，传统机器学习算法通常先把问题分成几块，一个个地解决好之后，再重新组合起来。但深度学习则是一次性地、端到端地解决。

假如存在一个任务：识别出在某图片中有哪些物体，并找出它们的位置。传统机器学习的做法是把问题分为两步：发现物体和识别物体。首先，我们用几个物体边缘的盒型检测算法，把所有可能的物体都框出来。然后，再使用物体识别算法，识别出这些物体中分别是什么。图 5-1 是一个机器学习识别例子。

图5-1　机器学习的识别

但是深度学习不同，它会直接在图片中把对应的物体识别出来，同时还能标明对应物体的名字。这样就可以做到实时的物体识别。

**注　意**

人工智能、机器学习、深度学习三者的关系：机器学习是实现人工智能的方法；深度学习是机器学习算法中的一种算法，是一种实现机器学习的技术和学习方法。

## 5.3　人工智能的研究领域和应用场景

扫码观看视频讲解

本节将对人工智能的研究领域和应用场景进行讲解，为读者步入本书后面知识的学习打下基础。

### 5.3.1　人工智能的研究领域

人工智能的研究领域主要有5层，如图5-2所示。从下往上的具体说明如下所示。

- 第1层：基础设施层，包含大数据和计算能力(硬件配置)两部分，数据越大，人工智能的能力越强。
- 第2层：算法层，例如卷积神经网络、LSTM序列学习、Q-Learning和深度学习等算法都是机器学习的算法。
- 第3层：技术方向层，例如计算机视觉、语音工程和自然语言处理等。另外还有规划决策系统，例如Reinforcement Learning(增强学习)，或类似于大数据分析的统计系统，这些都能在机器学习算法上产生。
- 第4层：具体技术层，例如图像识别、语音识别、语义理解、视频识别、机器翻译等。
- 第5层：行业解决方案层，例如人工智能在金融、医疗、互联网、安防、交通和游戏等领域的应用。

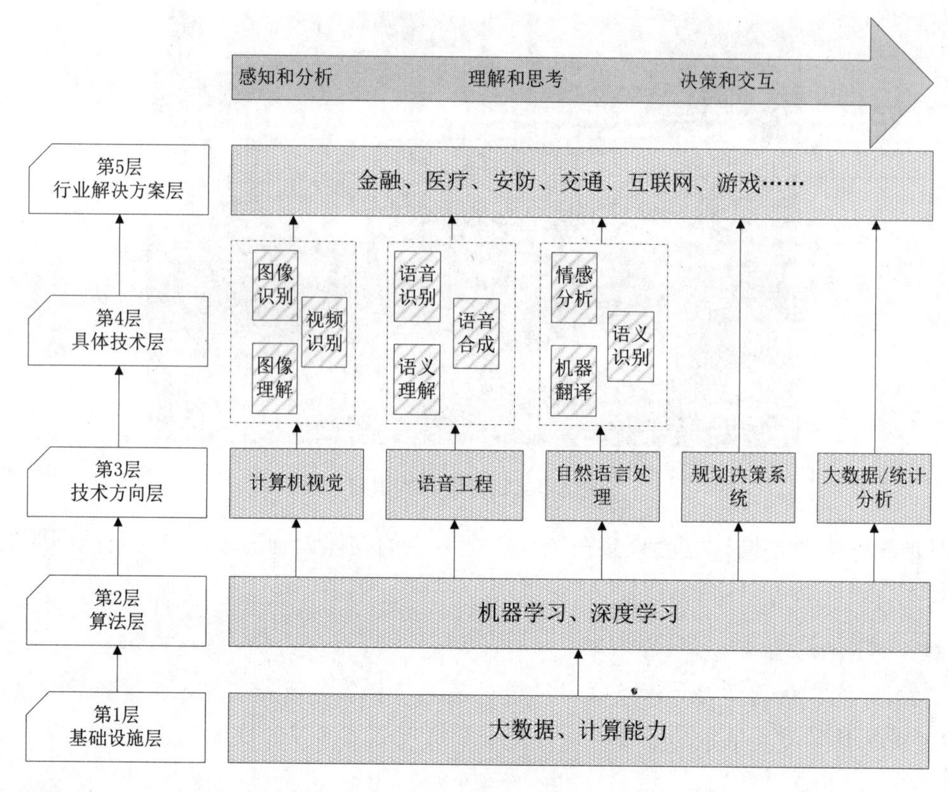


图 5-2 人工智能的研究领域

## 5.3.2 人工智能的应用场景

(1) 计算机视觉。

在 2000 年左右，人们通过机器学习用人工特征实现了较好的计算机视觉系统，如车牌识别、人脸识别等技术。而深度学习则逐渐运用机器代替人工来学习特征，扩大了其应用场景，例如无人驾驶汽车和电商服务等领域。

(2) 语音技术。

在 2010 年后，深度学习的广泛应用使语音识别的准确率大幅提升，像 Siri、Voice Search 和 Echo 等，可以实现不同语言间的交流，用语音说一段话，随之将其翻译为另一种文字。再例如智能助手，我们可以对手机说一段话，它能帮助你完成一些任务。与图像识别相比，自然语言更难、更复杂，不仅需要认知，而且还需要理解。

(3) 自然语言处理。

目前人工智能一个比较重大的突破是机器翻译，这大大提高了原来的机器翻译水平，其中 Google 的 Translation 系统，是人工智能的一个标杆性的事件。在 2010 年左右，IBM 的 Watson 系统在一档综艺节目上和人类冠军进行自然语言的问答并获胜，代表了计算机能力的显著提高。

(4)　决策系统。

决策系统的发展是随着棋类问题的解决而不断提升，从 20 世纪 80 年代西洋跳棋开始，到 20 世纪 90 年代的国际象棋对弈，机器的胜利都标志着科技的进步。决策系统可以在自动化、量化投资等系统上广泛应用。

(5)　大数据应用。

人工智能可以通过你之前看过的文章，理解你所喜欢的内容而进行更精准的推荐；可以分析各个股票的行情，进行量化交易；也可以分析所有客户的喜好而进行精准的营销等。机器能够通过一系列的数据进行判别，找出最适合的一些策略并反馈给我们。

## 5.4　系统需求分析

扫码观看视频讲解

在接下来的内容中，将详细讲解使用人工智能技术实现一个大型人脸识别项目的过程。本节首先讲解项目的功能分析，为步入后面知识的学习打下基础。

### 5.4.1　系统功能分析

本项目是一个人工智能版的人脸识别系统，使用深度学习技术实现。本项目的具体功能模块如下所示。

(1)　采集样本照片。

调用本地电脑摄像头采集照片作为样本，使用快捷键进行采集和取样。一次性可以采集无数个照片，采集的样本照片越多，后面的人脸识别成功率越高。

(2)　图片处理。

处理采集到的原始样本照片，将采集到的原始图像转化为标准数据文件，这样便于被后面的深度学习模块使用。

(3)　深度学习。

使用处理后的图片创建深度学习模型，实现学习训练，将训练结果保存为“.h5”文件。

(4)　人脸识别。

根据训练所得的模型实现人脸识别功能，既可以识别摄像头中的图片，也可以识别 Flask Web 中的上传照片。

### 5.4.2　实现流程分析

实现本项目的具体流程如图 5-3 所示。

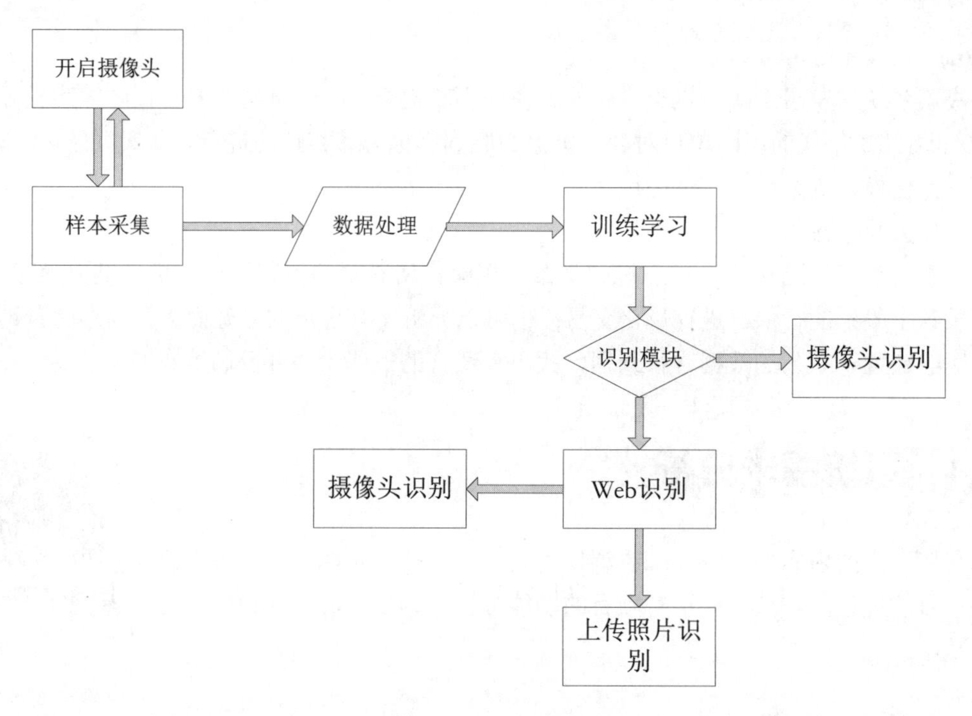


图 5-3　实现流程

## 5.4.3　技术分析

本人脸识别系统是一个综合性的项目，主要用到了如下所示的框架。

(1) Flask：著名的 Python Web 开发框架。

(2) OpenCV-Python：这是著名图像处理框架 OpenCV 的 Python 接口。OpenCV 是一个基于 BSD 许可(开源)发行的跨平台计算机视觉库，可以运行在 Linux、Windows、Android 和 Mac OS 操作系统上。它轻量级而且高效——由一系列 C 函数和少量 C++类构成，同时提供了 Python、Ruby、MATLAB 等语言的接口，实现了图像处理和计算机视觉方面的很多通用算法。OpenCV 用 C++语言编写，它的主要接口也是 C++语言，但是依然保留了大量的 C 语言接口。

可以使用如下命令安装 OpenCV-Python：

```
pip install opencv-python
```

在安装 OpenCV-Python 时，需要安装对应的依赖库，例如常用的 Numpy 等。如果安装 OpenCV-Python 失败，可以下载对应的“.whl”文件，然后通过如下命令进行安装：

```
pip install “.whl”文件
```

(3) Keras：这是一个用 Python 语言编写的高级神经网络 API，它能够以 TensorFlow、CNTK 或者 Theano 作为后端运行。Keras 的开发重点是支持快速的实验学习，能够以最小的时延把你的想法转换为实验结果。可以使用如下命令安装 Keras：

```
pip install keras
```

(4) Sklearn：这是机器学习中常用的第三方模块，对常用的机器学习方法进行了封装，包括回归(Regression)、降维(Dimensionality Reduction)、分类(Classfication)、聚类(Clustering)等方法。当我们面临机器学习问题时，便可选择使用Sklearn中相应的内置模块和方法来实现。在Sklearn中包含了大量的优质数据集，在学习机器学习的过程中，可以通过使用这些数据集实现不同的模型，从而提高我们的动手实践能力。可以使用如下命令安装Sklearn：

```
pip install sklearn
```

**注　意**

在安装Sklearn之前，需要先安装Numpy和Scipy。

## 5.5　照片样本采集

扫码观看视频讲解

在进行人脸识别前，需要先采集一个照片作为样本。本节将详细讲解使用摄像头采集样本照片的过程。编写文件getCameraPics.py，基于摄像头采集视频流中的数据，将截取的人脸图片作为样本照片并存储起来。文件getCameraPics.py的具体实现代码如下所示：

```
import os
import cv2

#python 2运行时加上
# reload(sys)
# sys.setdefaultencoding('utf-8')

def cameraAutoForPictures(saveDir='data/'):
    '''
    调用电脑摄像头来自动获取图片
    '''
    if not os.path.exists(saveDir):
        os.makedirs(saveDir)
    count=1
    cap=cv2.VideoCapture(0)
    width,height,w=640,480,360
    cap.set(cv2.CAP_PROP_FRAME_WIDTH,width)
    cap.set(cv2.CAP_PROP_FRAME_HEIGHT,height)
    crop_w_start=(width-w)//2
    crop_h_start=(height-w)//2
    print('width: ',width)
    print('height: ',height)
    while True:
        ret,frame=cap.read()
        frame=frame[crop_h_start:crop_h_start+w,crop_w_start:crop_w_start+w]
        frame=cv2.flip(frame,1,dst=None)
        cv2.imshow("capture", frame)
        action=cv2.waitKey(1) & 0xFF
        if action==ord('c'):
            saveDir=input(u"请输入新的存储目录：")
            if not os.path.exists(saveDir):
```

```
            os.makedirs(saveDir)
        elif action==ord('p'):
            cv2.imwrite("%s/%d.jpg" % (saveDir,count),cv2.resize(frame, (224,
                        224),interpolation=cv2.INTER_AREA))
            print(u"%s: %d 张图片" % (saveDir,count))
            count+=1
        if action==ord('q'):
            break
    cap.release()
    cv2.destroyAllWindows()


if __name__=='__main__':
    #guanxijing 替换为保存照片的文件名
    cameraAutoForPictures(saveDir='data/guanxijing/')
```

运行上述代码，启动摄像头后需要借助键盘输入操作来完成图片的获取工作，其中键盘按键 c(change)表示设置一个存储样本照片的目录，按键 p(photo)表示执行截图操作，按键 q(quit)表示退出拍摄。运行程序后，会打开本地电脑中的摄像头，如图 5-4 所示。按键盘中的 p 键会截取摄像头中的屏幕照片，并将照片保存起来，上述代码设置的保存路径是 guanxijing。

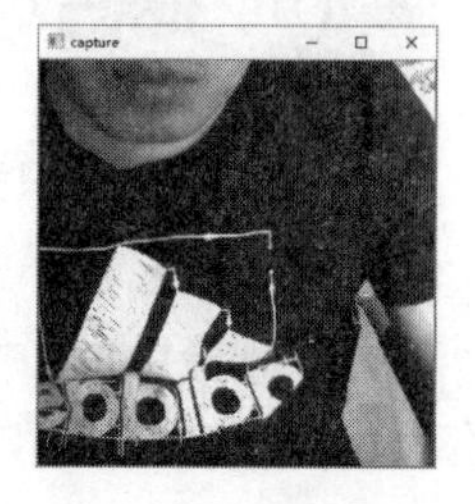


图 5-4　开启摄像头

## 5.6　深度学习和训练

扫码观看视频讲解

在尽可能多地采集样本照片后，对采集到的数据进行分析处理，然后使用人工智能技术实现深度学习训练，将训练结果保存为数据模型文件，根据数据模型文件可以实现人脸识别功能。

### 5.6.1　原始图像预处理

编写文件 dataHelper.py，实现原始图像数据的预处理工作，将原始图像转化为标准数据文件。文件 dataHelper.py 的具体实现代码如下所示：

```
import os
import cv2
import time

def readAllImg(path,*suffix):
    '''
    基于后缀读取文件
    '''
    try:
        s=os.listdir(path)
        resultArray = []
        fileName = os.path.basename(path)
        resultArray.append(fileName)
```

```
        for i in s:
            if endwith(i, suffix):
                document = os.path.join(path, i)
                img = cv2.imread(document)
                resultArray.append(img)
    except IOError:
        print("Error")

    else:
        print("读取成功")
        return resultArray


def endwith(s,*endstring):
    '''
    对字符串的后缀进行匹配
    '''
    resultArray = map(s.endswith,endstring)
    if True in resultArray:
        return True
    else:
        return False

def readPicSaveFace(sourcePath,objectPath,*suffix):
    '''
    图片标准化与存储
    '''
    if not os.path.exists(objectPath):
        os.makedirs(objectPath)
    try:
        resultArray=readAllImg(sourcePath,*suffix)
        count=1
        face_cascade=cv2.CascadeClassifier('config/haarcascade_frontalface_
alt.xml')
        for i in resultArray:
            if type(i)!=str:
              gray=cv2.cvtColor(i, cv2.COLOR_BGR2GRAY)
              faces=face_cascade.detectMultiScale(gray, 1.3, 5)
              for (x,y,w,h) in faces:
                listStr=[str(int(time.time())),str(count)]
                fileName=''.join(listStr)
                f=cv2.resize(gray[y:(y+h),x:(x+w)],(200, 200))
                cv2.imwrite(objectPath+os.sep+'%s.jpg' % fileName, f)
                count+=1
    except Exception as e:
        print("Exception: ",e)
    else:
        print('Read  '+str(count-1)+' Faces to Destination '+objectPath)

if __name__ == '__main__':
    print('dataProcessing!!!')
    readPicSaveFace('data/guanxijing/','dataset/guanxijing/','.jpg','.JPG',
'png','PNG','tiff')

readPicSaveFace('data/KA/','dataset/KA/','.jpg','.JPG','png','PNG','tiff')
```

如果需要处理多人的样本照片，需要在__main__后面添加多个对应的处理目录。运行上述文件后，会在 dataset 目录下得到处理后的照片。

## 5.6.2 构建人脸识别模块

编写文件 faceRegnigtionModel.py，功能是通过深度学习和训练构建人脸识别模块，并将训练后得到的模型保存到本地，默认保存为 face.h5。文件 faceRegnigtionModel.py 的具体实现流程如下所示。

(1) 引入深度学习和机器学习框架，对应的实现代码如下所示：

```
import os
import cv2
import random
import numpy as np
from keras.utils import np_utils
from keras.models import Sequential,load_model
from sklearn.model_selection import train_test_split
from keras.layers import
Dense,Activation,Convolution2D,MaxPooling2D,Flatten,Dropout
```

(2) 编写类 DataSet，功能是保存和读取格式化后的训练数据。对应的实现代码如下所示：

```
class DataSet(object):
   def __init__(self,path):
       '''
       初始化
       '''
       self.num_classes=None
       self.X_train=None
       self.X_test=None
       self.Y_train=None
       self.Y_test=None
       self.img_size=128
       self.extract_data(path)
```

(3) 编写函数 extract_data()抽取数据，使用机器学习 Sklearn 中的函数 train_test_split()将原始数据集按照一定比例划分为训练集和测试集对模型进行训练。通过函数 reshape()将图片转换成指定的尺寸和灰度，通过函数 astype()将图片转换为 float32 数据类型。对应的实现代码如下所示：

```
   def extract_data(self,path):
       imgs,labels,counter=read_file(path)
       X_train,X_test,y_train,y_test=train_test_split(imgs,labels,
test_size=0.2,random_state=random.randint(0, 100))
       X_train=X_train.reshape(X_train.shape[0], 1, self.img_size,
self.img_size)/255.0
       X_test=X_test.reshape(X_test.shape[0], 1, self.img_size,
self.img_size)/255.0
       X_train=X_train.astype('float32')
       X_test=X_test.astype('float32')
```

```
        Y_train=np_utils.to_categorical(y_train, num_classes=counter)
        Y_test=np_utils.to_categorical(y_test, num_classes=counter)
        self.X_train=X_train
        self.X_test=X_test
        self.Y_train=Y_train
        self.Y_test=Y_test
        self.num_classes=counter
```

函数 train_test_split()的原型如下：

```
tain_test_split(trian_data,trian_target,test_size,random_state)
```

函数 train_test_split()各个参数的具体说明如下所示。

- trian_data：表示被划分的样本特征集。
- trian_target：表示划分样本的标记(索引值)。
- test_size：表示将样本按比例划分，返回的第一个参数值为 train_data*test_size。
- random_state：表示随机种子。当为整数的时候，不管循环多少次，X_train 与第一次一样的。其值不能是小数。当 random_state 的值发生改变的时候，其返回值也会发生改变。

(4) 编写函数 check(self)，实现数据校验，打印输出图片的基本信息。对应的实现代码如下所示：

```
    def check(self):
        '''
        校验
        '''
        print('num of dim:', self.X_test.ndim)
        print('shape:', self.X_test.shape)
        print('size:', self.X_test.size)
        print('num of dim:', self.X_train.ndim)
        print('shape:', self.X_train.shape)
        print('size:', self.X_train.size)
```

(5) 编写函数 endwith()，功能是对字符串的后续和标记进行匹配。对应的实现代码如下所示：

```
def endwith(s,*endstring):
    resultArray = map(s.endswith,endstring)
    if True in resultArray:
        return True
    else:
        return False
```

(6) 编写函数 read_file(path)，读取指定路径的图片信息，对应的实现代码如下所示：

```
def read_file(path):
    img_list=[]
    label_list=[]
    dir_counter=0
    IMG_SIZE=128
    for child_dir in os.listdir(path):
        child_path=os.path.join(path, child_dir)
        for dir_image in os.listdir(child_path):
```

```
            if endwith(dir_image,'jpg'):
                img=cv2.imread(os.path.join(child_path, dir_image))
                resized_img=cv2.resize(img, (IMG_SIZE, IMG_SIZE))
                recolored_img=cv2.cvtColor(resized_img,cv2.COLOR_BGR2GRAY)
                img_list.append(recolored_img)
                label_list.append(dir_counter)
        dir_counter+=1
    img_list=np.array(img_list)
    return img_list,label_list,dir_counter
```

(7) 编写函数 read_name_list(path)，读取训练数据集，对应的实现代码如下所示：

```
def read_name_list(path):
    name_list=[]
    for child_dir in os.listdir(path):
        name_list.append(child_dir)
    return name_list
```

(8) 编写类 Model，创建一个基于 CNN 的人脸识别模型，开始构建数据模型并进行训练。对应的实现代码如下所示：

```
class Model(object):
    '''
    人脸识别模型
    '''
    FILE_PATH="face.h5"
    IMAGE_SIZE=128


    def __init__(self):
        self.model=None


    def read_trainData(self,dataset):
        self.dataset=dataset


    def build_model(self):
        self.model = Sequential()
        self.model.add(
            Convolution2D(
                filters=32,
                kernel_size=(5, 5),
                padding='same',
                dim_ordering='th',
                input_shape=self.dataset.X_train.shape[1:]
            )
        )
        self.model.add(Activation('relu'))
        self.model.add(
            MaxPooling2D(
                pool_size=(2, 2),
                strides=(2, 2),
                padding='same'
            )
        )
```

```
        self.model.add(Convolution2D(filters=64, kernel_size=(5,5), padding='same'))
        self.model.add(Activation('relu'))
        self.model.add(MaxPooling2D(pool_size=(2,2), strides=(2,2), padding='same'))
        self.model.add(Flatten())
        self.model.add(Dense(1024))
        self.model.add(Activation('relu'))
        self.model.add(Dense(self.dataset.num_classes))
        self.model.add(Activation('softmax'))
        self.model.summary()


    def train_model(self):
        self.model.compile(
            optimizer='adam',
            loss='categorical_crossentropy',
            metrics=['accuracy'])
        self.model.fit(self.dataset.X_train,self.dataset.Y_train,epochs=10,
batch_size=10)


    def evaluate_model(self):
        print('\nTesting---------------')
        loss, accuracy = self.model.evaluate(self.dataset.X_test,
                                  self.dataset.Y_test)
        print('test loss;', loss)
        print('test accuracy:', accuracy)


    def save(self, file_path=FILE_PATH):
        print('Model Saved Finished!!!')
        self.model.save(file_path)


    def load(self, file_path=FILE_PATH):
        print('Model Loaded Successful!!!')
        self.model = load_model(file_path)


    def predict(self,img):
        img=img.reshape((1, 1, self.IMAGE_SIZE, self.IMAGE_SIZE))
        img=img.astype('float32')
        img=img/255.0
        result=self.model.predict_proba(img)
        max_index=np.argmax(result)
        return max_index,result[0][max_index]
```

(9) 调用上面的函数，打印输出模型训练和评估结果，对应的实现代码如下所示：

```
if __name__ == '__main__':
    dataset=DataSet('dataset/')
    model=Model()
    model.read_trainData(dataset)
    model.build_model()
    model.train_model()
    model.evaluate_model()
    model.save()
```

## 5.7 人脸识别

在使用人工智能技术实现深度学习训练后，可以生成一个数据模型文件，通过调用这个数据模型文件可以实现人脸识别功能。例如在下面的文件 cameraDemo.py 中，通过 OpenCV-Python 直接调用摄像头实现实时人脸识别功能：

```
import os
import cv2
from faceRegnigtionModel import Model

threshold=0.7  #如果模型认为概率高于 70%，则显示为模型中已有的人物

def read_name_list(path):
    '''
    读取训练数据集
    '''
    name_list=[]
    for child_dir in os.listdir(path):
        name_list.append(child_dir)
    return name_list

class Camera_reader(object):
    def __init__(self):
        self.model=Model()
        self.model.load()
        self.img_size=128

    def build_camera(self):
        '''
        调用摄像头来实时人脸识别
        '''
        face_cascade = cv2.CascadeClassifier('config/haarcascade_frontalface_
alt.xml')
        name_list=read_name_list('dataset/')
        cameraCapture=cv2.VideoCapture(0)
        success, frame=cameraCapture.read()
        while success and cv2.waitKey(1)==-1:
            success,frame=cameraCapture.read()
            gray=cv2.cvtColor(frame, cv2.COLOR_BGR2GRAY)
            faces=face_cascade.detectMultiScale(gray, 1.3, 5)
            for (x,y,w,h) in faces:
                ROI=gray[x:x+w,y:y+h]
                ROI=cv2.resize(ROI, (self.img_size, self.img_size),interpolation
                    =cv2.INTER_LINEAR)
                label,prob=self.model.predict(ROI)
                if prob>threshold:
                    show_name=name_list[label]
                else:
                    show_name="Stranger"
                cv2.putText(frame, show_name, (x,y-20),cv2.FONT_HERSHEY_
                        SIMPLEX,1,255,2)
```

```
                frame=cv2.rectangle(frame,(x,y), (x+w,y+h),(255,0,0),2)
            cv2.imshow("Camera", frame)
        else:
            cameraCapture.release()
            cv2.destroyAllWindows()

if __name__ == '__main__':
    camera=Camera_reader()
    camera.build_camera()
```

执行代码后，会开启摄像头并识别摄像头的人物，如图 5-5 所示。

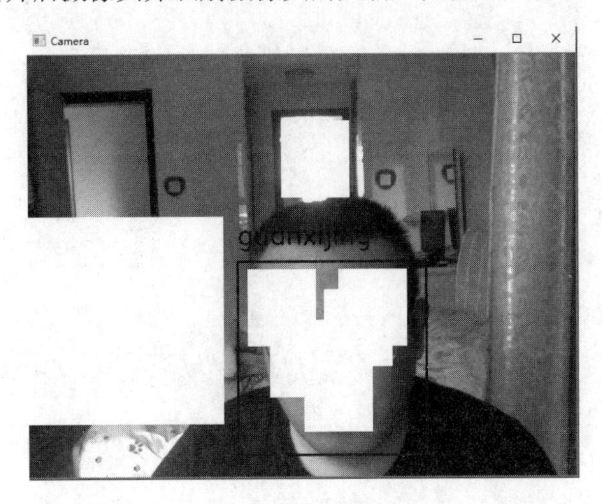


图 5-5　执行效果

## 5.8　Flask Web 人脸识别接口

扫码观看视频讲解

我们可以将前面实现的数据模型和人脸识别功能迁移到 Web 项目中。本节将详细讲解在 Flask Web 中实现人脸识别功能的过程。

### 5.8.1　导入库文件

编写文件 main.py，实现 Flask 项目的主程序功能，首先导入需要的人脸识别库和 Flask 库，具体实现代码如下所示：

```
from flask_uploads import UploadSet, IMAGES, configure_uploads
from flask import redirect, url_for, render_template
import os
import cv2
import time
import numpy as np
from flask import Flask
from flask import request
from faceRegnigtionModel import Model
from cameraDemo import Camera_reader
```

## 5.8.2 识别上传照片

在文件 main.py 中设置 Flask 项目的名字，并设置上传文件的保存目录。通过链接/upload 显示上传表单页面，通过链接/photo/<name>显示上传的照片，并在页面中调用函数 detectOnePicture(path)显示识别结果。具体实现代码如下所示：

```
app = Flask(__name__)
app.config['UPLOADED_PHOTO_DEST'] =
os.path.dirname(os.path.abspath(__file__))
app.config['UPLOADED_PHOTO_ALLOW'] = IMAGES
def dest(name):
    return '{}/{}'.format(app.config.UPLOADED_PHOTO_DEST, name)
photos = UploadSet('PHOTO')


configure_uploads(app, photos)
@app.route('/upload', methods=['POST', 'GET'])
def upload():
    if request.method == 'POST' and 'photo' in request.files:
        filename = photos.save(request.files['photo'])
        return redirect(url_for('show', name=filename))
    return render_template('upload.html')

@app.route('/photo/<name>')
def show(name):
    if name is None:
        print('出错了！')
    url = photos.url(name)

def detectOnePicture(path):
    '''
    单图识别
    '''
    model=Model()
    model.load()
    img=cv2.imread(path)
    img=cv2.resize(img,(128,128))
    img=cv2.cvtColor(img, cv2.COLOR_BGR2GRAY)
    picType,prob=model.predict(img)
    if picType!=-1:
        name_list=read_name_list('dataset/')
        print(name_list[picType],prob)
        res=u"识别为： "+name_list[picType]+u"的概率为： "+str(prob)
    else:
        res=u"抱歉，未识别出该人！请尝试增加数据量来训练模型！"
    return res

    if request.method=="GET":
        picture=name
    start_time=time.time()
    res=detectOnePicture(picture)
    end_time=time.time()
```

```
    execute_time=str(round(end_time-start_time,2))
    tsg=u' 总耗时为: %s 秒' % execute_time
    return render_template('show.html', url=url,
        name=name,xinxi=res,shijian=tsg)

def endwith(s,*endstring):
    '''
    对字符串的后缀进行匹配
    '''
    resultArray=map(s.endswith,endstring)
    if True in resultArray:
        return True
    else:
        return False


def read_file(path):
    '''
    图片读取
    '''
    img_list=[]
    label_list=[]
    dir_counter=0
    IMG_SIZE=128
    for child_dir in os.listdir(path):
        child_path=os.path.join(path, child_dir)
        for dir_image in os.listdir(child_path):
            if endwith(dir_image,'jpg'):
                img=cv2.imread(os.path.join(child_path, dir_image))
                resized_img=cv2.resize(img, (IMG_SIZE, IMG_SIZE))
                recolored_img=cv2.cvtColor(resized_img,cv2.COLOR_BGR2GRAY)
                img_list.append(recolored_img)
                label_list.append(dir_counter)
        dir_counter+=1
    img_list=np.array(img_list)
    return img_list,label_list,dir_counter


def read_name_list(path):
    '''
    读取训练数据集
    '''
    name_list=[]
    for child_dir in os.listdir(path):
        name_list.append(child_dir)
    return name_list
```

## 5.8.3 在线识别

设置 Web 首页显示一个“打开摄像头识别”链接，单击链接后调用摄像头实现在线识别功能。具体实现代码如下所示。

```
@app.route("/")
def init():
      return render_template("index.html",title = 'Home')

@app.route("/she/")
def she():
      camera = Camera_reader()
      camera.build_camera()
      return render_template("index.html", title='Home')

if __name__ == "__main__":
   print('faceRegnitionDemo')
   app.run(debug=True)
```

执行代码后，将在主页显示“打开摄像头识别”链接，单击链接后会实现在线人脸识别功能，如图 5-6 所示。

127.0.0.1:5000

打开摄像头识别

(a) 系统主页

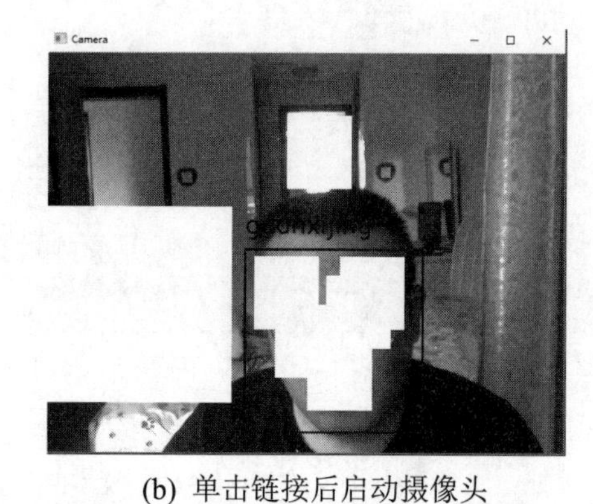


(b) 单击链接后启动摄像头

图 5-6　在线人脸识别

输入 http://127.0.0.1:5000/upload 后，显示图片上传页面，上传照片并单击“提交”按钮后，显示上传照片并和识别结果，如图 5-7 所示。

(a) 上传图片表单页面

(b) 显示识别结果

图 5-7　识别上传照片

# 第 6 章

## 在线生鲜商城系统

## （Django+Vue+新浪微博账号登录+支付宝支付）

近几年生鲜市场一直是创投者们格外关注的焦点，而应用市场上也涌现了许多生鲜配送 App 和在线生鲜商城系统。据不完全统计，生鲜市场仍以每年 50%的幅度增长，由此可见线上生鲜配送商城的前景大有可为。本章将通过一个综合实例的实现过程，详细讲解使用 Django 开发一个在线生鲜商城系统的方法。

## 6.1 系统背景介绍

扫码观看视频讲解

生鲜电商是指借助电子商务平台直接进行新鲜水果、蔬菜、肉类等生鲜农产品交易，并以自营物流或第三方冷链物流完成生鲜产品配送的一系列商业活动。生鲜电商的出现，在很大程度上顺应了消费者个性化、多元化的需求，也是体验经济以及共享经济发展的趋势。过去的生鲜电商受到技术的限制，发展有限，毕竟生鲜产品需要全程冷链配送，周期 3 天左右，时间长、成本高是生鲜电商难以发展的问题之一。而随着互联网和我国科技的不断发展，以上问题有了一定的解决方案。生鲜商城使用“线下门店+线上网店+配送”的模式，用户下单之后，就能在短时间内收到生鲜产品，在保证商品质量的情况下为用户提供良好的配送服务。

在过去的几年中，生鲜行业是发展较好的行业，而依靠生鲜电商获得融资的企业纷纷涌现。另外，阿里巴巴旗下的盒马鲜生的出现，让生鲜电商正式走入大众的视野。虽然电商 App 在这几年迅猛发展，但生鲜电商市场份额占比较少，所以在线生鲜商城系统的前景大有可为。

中国生鲜电商市场发展迅速，2019 年我国生鲜电商交易规模为 2554.5 亿元，较 2018 年 1950 亿元，同比增长 31%。2019 年我国生鲜电商行业渗透率达 4.67%，较 2018 年的 3.8%，同比增长 22.89%。

中国生鲜电商行业经历了多年的高速发展，原本已进入发展的平台期，2019 年还遭遇了生鲜电商的发展“寒冬”。但 2020 年初，受疫情影响，线上买菜刺激了生鲜电商的发展，居民消费习惯逐步改变，生鲜市场的新一轮混战已然开始。数据显示，2020 年 3 月，中国生鲜电商平台月活排名中，多点月活达 1026.4 万人，排名第一；盒马鲜生和每日优鲜排名二、三位，月活分别为 892.7 万人和 735.7 万人。

近年来，国家出台多项政策，鼓励通过“互联网+”推动农业发展，同时在农产品流通、技术发展等方面出台利好政策和规范；“一带一路”政策和自贸区的建立促进了跨境生鲜电商业务。国家政策的支持，冷链物流、信息技术的发展，大资本的介入，新零售的布局及大数据的发展，都为生鲜电商注入了新的动力，生鲜电商发展已进入快车道。另外，网购生鲜的消费习惯逐渐形成，刚需高频商品特性及消费需求升级，直接拉动生鲜电商行业发展。

## 6.2 功能需求分析

扫码观看视频讲解

作为一个在线生鲜商城系统，必须具备如下所示的功能模块。

(1) 会员系统。

包括会员注册、登录验证、个人信息管理子模块，并且可以使用新浪微博账号和手机登录系统。

(2) 热门生鲜商品。

在商城首页为用户展示各种类型的热门生鲜商品，也为用户提供行业资讯，方便用户在线浏览相关内容，让用户深入了解生鲜市场。

(3) 生鲜商品分类。

为了方便用户在短时间内找到合适的生鲜商品，在商城系统中会对生鲜商品进行分类。用户可以选择相应的生鲜商品板块来搜索商品，也能直接在线查找生鲜产品信息，获得良好的购买体验。

(4) 生鲜商品介绍。

通过精美图片来展示生鲜商品是店铺吸引用户消费的手段，用户在浏览商品信息的同时，也能查看生鲜商品图片，给用户选购提供一定的便利。

(5) 购物车。

网上商城系统中的购物车是对现实购物车拟化，买家可以像在超市里购物一样，随意添加、删除商品，选购完毕后，统一下单。

(6) 在线支付功能。

当用户选购好相关商品之后，能在线支付购买费用，平台支持用户使用微信或是支付宝进行在线支付。

(7) 订单管理。

在购物过程中提交订单后，需要填写收货地址信息和联系电话信息，在付款成功后是一次完整的购物过程。购物者可以随时管理自己的订单信息，在商城后台也可以管理订单信息。

(8) 后台管理。

网站后台管理系统主要用于对网站前台的信息进行管理，如文字、图片、影音和其他日常使用文件的发布、更新、删除等操作，同时也包括会员信息、订单信息、访客信息的统计和管理。

上述各个构成模块的具体说明如图 6-1 所示。

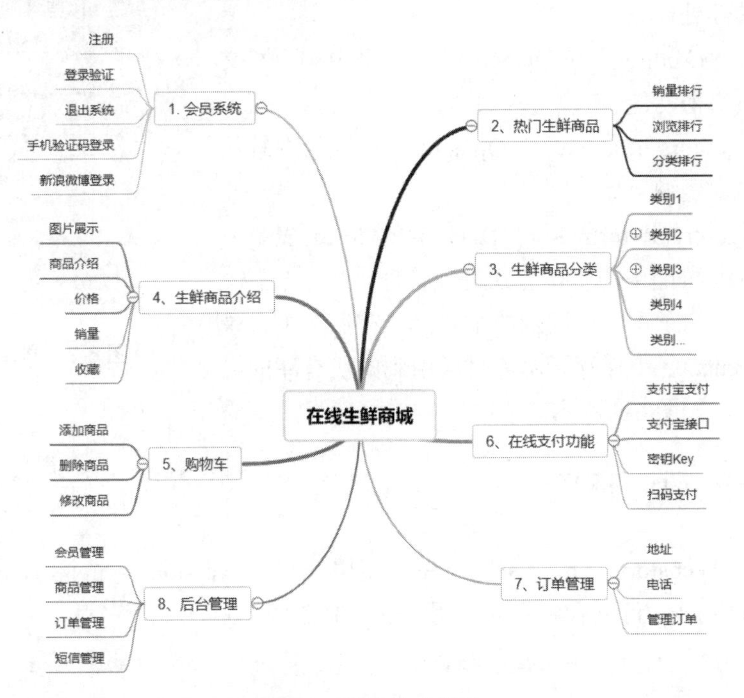


图 6-1　系统构成模块

## 6.3 准备工作

在做好系统需求分析之后，在开发软件项目之前，还需要根据项目需求做好准备工作。因为Python的最大优势是使用框架来提高开发效率，所以在编码之前需要选择用到的框架。

### 6.3.1 用到的库

在本项目中，主要用到了如下所示的库。

- Django：著名的企业级Web开发库，本项目的后端主要是基于Django实现的。
- Vue：这是一套构建用户界面的渐进式库，本项目的前端主要是基于Vue实现的。
- Django Rest Framework：这是基于Django实现的一个RESTful风格API库，能够帮助我们快速开发RESTful风格的API。本项目将使用Django Rest Framework实现前后端的分离功能。
- drf-extensions：用于处理Django Rest Framework的缓存。
- social-auth-app-django：可以实现基于QQ、微信和微博的第三方账号登录。
- django-redis：使用redis在Django Web项目中实现缓存处理。
- django-ckeditor：在Django Web项目中实现富文本编辑器功能。
- django-cors-headers：在Django Web项目中解决跨域问题。
- django-crispy-forms：对Django的form表单在HTML页面中的呈现方式进行管理。

上面只是介绍了本项目用到的主要库，在后端主目录文件requirements.txt中保存了本项目所用到的所有库的名称和版本信息，如图6-2所示。

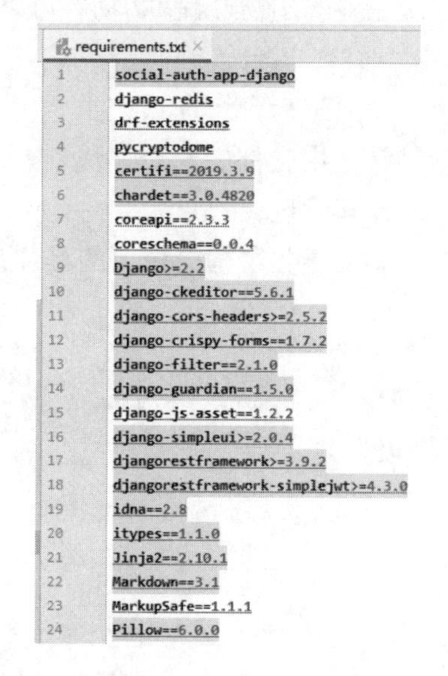

```
requirements.txt
1  social-auth-app-django
2  django-redis
3  drf-extensions
4  pycryptodome
5  certifi==2019.3.9
6  chardet==3.0.4820
7  coreapi==2.3.3
8  coreschema==0.0.4
9  Django>=2.2
10 django-ckeditor==5.6.1
11 django-cors-headers>=2.5.2
12 django-crispy-forms==1.7.2
13 django-filter==2.1.0
14 django-guardian==1.5.0
15 django-js-asset==1.2.2
16 django-simpleui>=2.0.4
17 djangorestframework>=3.9.2
18 djangorestframework-simplejwt>=4.3.0
19 idna==2.8
20 itypes==1.1.0
21 Jinja2==2.10.1
22 Markdown==3.1
23 MarkupSafe==1.1.1
24 Pillow==6.0.0
```

图6-2 文件requirements.txt

### 6.3.2 准备Vue环境

下载安装Webstorm、nodejs(安装完后使用node --version测试)和cnpm。用户访问https://npm.taobao.org/可以看到安装说明。安装命令如下：

```
npm install -g cnpm --registry=https://registry.npm.taobao.org
```

然后使用如下命令安装依赖包，安装后的效果如图6-3所示：

```
cnpm install
```

| 名称 | 修改日期 | 类型 |
|---|---|---|
| dist | 2020/6/15 21:05 | 文件夹 |
| mock | 2020/6/15 21:05 | 文件夹 |
| node_modules | 2020/7/23 13:37 | 文件夹 |
| src | 2020/6/15 21:05 | 文件夹 |
| .babelrc | 2018/11/26 17:01 | BABELRC 文件 |
| .editorconfig | 2018/11/26 17:01 | EDITORCONFIG ... |
| .gitignore | 2018/11/26 17:01 | 文本文档 |
| package.json | 2019/8/8 22:33 | JSON File |
| package-lock.json | 2018/11/26 17:01 | JSON File |
| postcss.config.js | 2018/11/26 17:01 | JavaScript 文件 |
| proxy.js | 2019/8/10 20:45 | JavaScript 文件 |
| README.md | 2018/11/26 17:01 | MD 文件 |
| run server.bat | 2019/7/15 21:14 | Windows 批处理... |
| server.js | 2019/4/19 8:34 | JavaScript 文件 |
| template.html | 2018/11/26 17:01 | HTML 文件 |
| webpack.config.js | 2018/11/26 17:01 | JavaScript 文件 |
| webpack.prod.js | 2019/8/8 22:51 | JavaScript 文件 |

图 6-3　安装完成

最后使用如下命令启动 Vue：

```
cnpm run dev
```

## 6.3.3　创建应用

本在线生鲜商城系统的功能比较强大，规模也比较庞大。为了便于系统的设计、实现和后期维护，将整个功能通过几个模块应用来实现。在 Django Web 项目中，不同的模块应用被称为 App。分别用如下命令创建 users、goods、trade 和 user_operation 共计 4 个 App，如图 6-4 所示。

```
startapp users
startapp goods
startapp trade
startapp user_operation
```

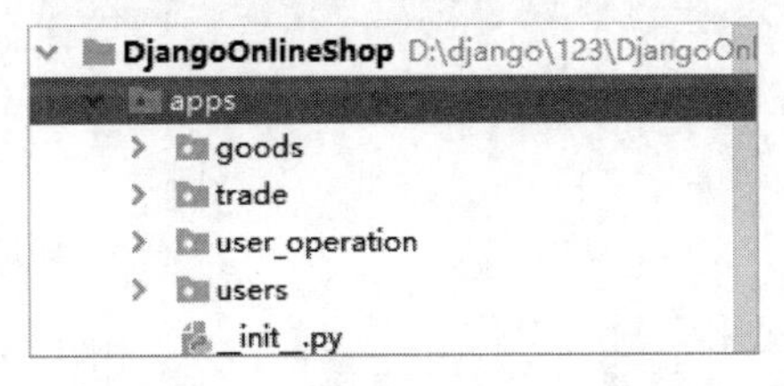


图 6-4　创建的 4 个 App

在文件 settings.py 中，需要将上面的 4 个 App 添加到 Django Web 中：

```
INSTALLED_APPS = [
    ......
    'users.apps.UsersConfig',
    'goods.apps.GoodsConfig',
    'trade.apps.TradeConfig',
    'user_operation.apps.UserOperationConfig',
]
```

## 6.3.4 系统配置

在文件settings.py中配置Django项目，具体实现流程如下。

(1) 设置后端认证方式，本项目不但支持使用自己的注册验证系统，还可以使用微博认证、QQ认证和微信认证等方式。代码如下：

```
AUTHENTICATION_BACKENDS = (
    'users.views.CustomBackend',  # 自定义认证后端
    'social_core.backends.weibo.WeiboOAuth2',  # 微博认证后端
    'social_core.backends.qq.QQOAuth2',  # QQ认证后端
    'social_core.backends.weixin.WeixinOAuth2',  # 微信认证后端
    'django.contrib.auth.backends.ModelBackend',
     # 使用了`django.contrib.auth应用程序，支持账密认证
)  # 指定认证后台
```

(2) 在INSTALLED_APPS中，添加在本项目中用到是App，主要包括drf应用(rest_framework.authtoken)、social_django跨域、admin后台管理和ckeditor富文本编辑器等。代码如下：

```
INSTALLED_APPS = [
    'django.contrib.admin',
    'django.contrib.auth',
    'django.contrib.contenttypes',
    'django.contrib.sessions',
    'django.contrib.messages',
    'django.contrib.staticfiles',
    # 添加drf应用
    'rest_framework',
    'rest_framework.authtoken',
    'django_filters',
    # 添加Django联合登录
    'social_django',
    # Django跨域解决
    'corsheaders',
    # 注册富文本编辑器ckeditor
    'ckeditor',
    # 注册富文本上传图片ckeditor_uploader
    'ckeditor_uploader',
    'users.apps.UsersConfig',
    'goods.apps.GoodsConfig',
    'trade.apps.TradeConfig',
    'user_operation.apps.UserOperationConfig',
]
```

(3) 在本项目默认使用SQLite3数据库，也可以用注释部分的MySQL数据库。代码如下：

```
DATABASES = {
    'default': {
        'ENGINE': 'django.db.backends.sqlite3',
```

```
        'NAME': os.path.join(BASE_DIR, 'db.sqlite3'),
        # 'ENGINE': 'django.db.backends.mysql',
        # 'NAME': 'online_shop',
        # 'USER': 'root',
        # 'PASSWORD': '123456',
        # 'HOST': 'localhost',
        # 'POST': 3306
    }
}
```

(4) 设置保存媒体文件和上传文件的路径，代码如下：

```
# 配置媒体文件
MEDIA_URL = '/media/'
MEDIA_ROOT = os.path.join(BASE_DIR, 'media')

# 配置富文本上传路径
CKEDITOR_UPLOAD_PATH = 'upload/'
```

(5) 使用 DRF Token 认证功能，此身份验证方案使用一个简单的基于令牌的 HTTP 身份验证方案。令牌身份验证适用于客户机/服务器模式，如本机桌面和移动客户机。要使用 TokenAuthentication 方案，可以使用 DEFAULT_AUTHENTICATION_CLASSES 设置全局缺省身份验证方案。代码如下：

```
# DRF 配置
REST_FRAMEWORK = {
    # 'DEFAULT_PAGINATION_CLASS':
'rest_framework.pagination.PageNumberPagination',
    # 'PAGE_SIZE': 5,
    'DEFAULT_AUTHENTICATION_CLASSES': (
        'rest_framework.authentication.BasicAuthentication',
        'rest_framework.authentication.SessionAuthentication',
        # 上面两个用于 DRF 基本验证
        # 'rest_framework.authentication.TokenAuthentication',
        # TokenAuthentication，取消全局 token，放在视图中进行
        # 'rest_framework_simplejwt.authentication.JWTAuthentication',
        # djangorestframework_simplejwt JWT 认证
    ),
    # throttle 对接口访问限速
    'DEFAULT_THROTTLE_CLASSES': [
        # 'rest_framework.throttling.AnonRateThrottle',
        # 用户未登录请求限速，通过 IP 地址判断
        # 'rest_framework.throttling.UserRateThrottle'
        # 用户登录后请求限速，通过 token 判断
        'rest_framework.throttling.ScopedRateThrottle',
        # 限制用户对于每个视图的访问频次，使用 ip 或 user id
    ],
    'DEFAULT_THROTTLE_RATES': {
        # 'anon': '60/minute',
        # 限制所有匿名未认证用户，使用 IP 区分用户。使用 DEFAULT_THROTTLE_RATES['anon']
来设置频次
        # 'user': '200/minute'
        # 限制认证用户，使用 User id 来区分。使用 DEFAULT_THROTTLE_RATES['user'] 来设置频次
        'goods_list': '600/minute'
```

```
    }
}
```

(6) 设置跨域请求，在 Django 设置中配置中间件的行为。必须将允许执行跨站点请求的主机添加到 CORS_ORIGIN_WHITELIST，或者将 CORS_ORIGIN_ALLOW_ALL 设置为 True 以允许所有主机。

① CORS_ORIGIN_ALLOW_ALL：如果是 True，将不使用白名单，所有的连接将被接受。默认值为 False。

② CORS_ORIGIN_WHITELIST：授权发出跨站点 HTTP 请求的源主机名列表。值 null 也可以出现在这个列表中，并将与在浏览器隐私敏感上下文中使用的 Origin: null 头匹配，例如当客户机从 file:// domain 运行时。默认为[]。

在本项目中实现 CORS 配置的代码如下：

```
# 跨域 CORS 设置
# CORS_ORIGIN_ALLOW_ALL = False  # 默认为 False，如果为 True 则允许所有连接
CORS_ORIGIN_WHITELIST = (  # 配置允许访问的白名单
    'http://localhost:8080',
    'http://localhost:8000',
    'http://127.0.0.1:8080',
    'http://127.0.0.1:8000',
    'http://127.0.0.1:8081',
)
```

(7) 本项目使用 JWT 在用户和服务器之间传递安全可靠的信息，下面是自定义配置 JWT 的代码：

```
from datetime import timedelta

SIMPLE_JWT = {
    'ACCESS_TOKEN_LIFETIME': timedelta(days=7),  # 配置过期时间
    'REFRESH_TOKEN_LIFETIME': timedelta(days=15),
}
```

(8) 下面是支付宝的相关配置信息，app_id、app_private_key 和 alipay_public_key 等信息需要去支付宝开发者中心申请：

```
app_id = ""
alipay_debug = True
app_private_key_path = os.path.join(BASE_DIR,
'apps/trade/keys/private_key_2048.txt')
alipay_public_key_path = os.path.join(BASE_DIR,
"apps/trade/keys/alipay_key_2048.txt")

# drf-extensions 配置
REST_FRAMEWORK_EXTENSIONS = {
    'DEFAULT_CACHE_RESPONSE_TIMEOUT': 60 * 10
    # 缓存全局过期时间(60 * 10 表示 10 分钟)
}
```

(9) 使用 social_django 配置认证密钥，将本项目上传到网络服务器后，将涉密信息保存在配置文件中。分别设置自己的微博账号登录信息，包括 weibo_key 和 weibo_secret。代

码如下：

```
import configparser
config = configparser.ConfigParser()
config.read(os.path.join(BASE_DIR, 'ProjectConfig.ini'))
weibo_key = ''
weibo_secret = '3'
SOCIAL_AUTH_WEIBO_KEY = weibo_key
SOCIAL_AUTH_WEIBO_SECRET = weibo_secret

SOCIAL_AUTH_LOGIN_REDIRECT_URL = '/index/'  # 登录成功后跳转，一般为项目首页
```

## 6.4 设计数据库

扫码观看视频讲解

数据库技术是动态 Web 的根本，在线生鲜商城系统中的数据信息都被保存在数据库中。在 Django Web 项目中，可以使用 migrate 和 Model 模型方便地实现数据库的设计和创建工作。

### 6.4.1 为 users 应用创建 Model 模型

在 users 应用的数据库模型中，主要用于保存系统用户信息，包括会员和管理员。

(1) 首先在文件 settings.py 中设置认证模型，添加如下代码：

```
AUTH_USER_MODEL = 'users.UserProfile'  # 使用自定义的 model 做认证
```

(2) 在 users 目录下编写文件 models.py，分别创建模型类 UserProfile 和 VerifyCode，对应代码如下所示：

```
class UserProfile(AbstractUser):
    """
    扩展用户，需要在 settings 设置认证 model
    """
    name = models.CharField(max_length=30, blank=True, null=True,
    verbose_name='姓名', help_text='姓名')
    birthday = models.DateField(null=True, blank=True,
    verbose_name='出生年月', help_text='出生年月')
    mobile = models.CharField(max_length=11, blank=True, null=True,
    verbose_name='电话', help_text='电话')
    gender = models.CharField(max_length=6, choices=(('male', '男'), ('female',
    '女')), default='male', verbose_name='性别', help_text='性别')

    class Meta:
        verbose_name_plural = verbose_name = '用户'

    def __str__(self):
        # 要判断 name 是否有值，如果没有，则返回 username，
        #否则使用 createsuperuser 创建用户访问与用户关联的模型会报错，
        # 页面(A server error occurred. Please contact the administrator.)
```

```
        # 后台(UnicodeDecodeError: 'gbk' codec can't decode byte 0xa6 in position
        # 9737: illegal multibyte sequence)
        if self.name:
            return self.name
        else:
            return self.username


class VerifyCode(models.Model):
    """
    短信验证码，可以保存在 redis 中
    """
    code = models.CharField(max_length=20, verbose_name='验证码',
        help_text='验证码')
    mobile = models.CharField(max_length=11, verbose_name='电话',
        help_text='电话')
    add_time = models.DateTimeField(auto_now_add=True, verbose_name='添加时间')

    class Meta:
        verbose_name_plural = verbose_name = '短信验证码'

    def __str__(self):
        return self.code
```

(3) 在 users 目录下编写文件 apps.py，设置在后台将应用名显示为中文，对应代码如下所示：

```
from django.apps import AppConfig
class UsersConfig(AppConfig):
    name = 'users'
    verbose_name = '用户'
```

(4) 在 users 目录下编写文件 admin.py，功能是采用批量注册方式将应用 users 关联到 admin 后台。对应代码如下所示：

```
from django.contrib import admin
from django.apps import apps

all_models = apps.get_app_config('user_operation').get_models()
for model in all_models:
    try:
        admin.site.register(model)
    except:
        pass
```

### 6.4.2 为 goods 应用创建 Model 模型

在 goods 应用的数据库模型中，主要用于保存和商品有关的信息，包括类别、品牌、商品详情、图片、首页轮播图和首页广告等信息。

(1) 在向数据库中添加商品信息时用到了富文本编辑器，所以首先在文件 settings.py 中添加富文本编辑器应用，添加如下代码：

```
    # 注册富文本编辑器 ckeditor
    'ckeditor',
    # 注册富文本上传图片 ckeditor_uploader
    'ckeditor_uploader',
```

并在文件 settings.py 中设置文件上传路径：

```
# 配置媒体文件
MEDIA_URL = '/media/'
MEDIA_ROOT = os.path.join(BASE_DIR, 'media')

# 配置富文本上传路径
CKEDITOR_UPLOAD_PATH = 'upload/'
```

(2) 在路径导航文件 urls.py 中添加富文本编辑器的路由，代码如下：

```
from django.contrib import admin
from django.urls import path, include
from django.conf.urls.static import static
# 上传的文件能直接通过 url 打开，以及 settings 中设置
from django.conf import settings

urlpatterns = [
    path('admin/', admin.site.urls),
    path('api-auth/', include('rest_framework.urls')),
    path('ckeditor/', include('ckeditor_uploader.urls')),  # 配置富文本编辑器 url
]

# 上传的文件能直接通过 url 打开
if settings.DEBUG:
    urlpatterns += static(settings.MEDIA_URL,
                          document_root=settings.MEDIA_ROOT)
```

(3) 在 goods 目录下编写文件 models.py，分别创建模型类 GoodsCategory、GoodsCategoryBrand、Goods、GoodsImage、Banner 和 IndexCategoryAd，对应代码如下所示：

```
class GoodsCategory(models.Model):
    """
    商品类别
    """
    CATEGORY_TYPE = (
        (1, '一级类目'),
        (2, '二级类目'),
        (3, '三级类目'),
    )
    name = models.CharField(max_length=30, default='',
     verbose_name='类别名称', help_text='商品类别名称')
    # help_text 说明，生成文档很有用
    code = models.CharField(max_length=30, default='',
    verbose_name='类别编码', help_text='商品类别编码')
    desc = models.TextField(default='', verbose_name='类别描述',
    help_text='类别描述')
    category_type = models.SmallIntegerField(choices=CATEGORY_TYPE, default=1,
verbose_name='类目级别', help_text='商品类目的级别')
```

```
    is_tab = models.BooleanField(default=False, verbose_name='是否导航',
help_text='类别是否导航')
    add_time = models.DateTimeField(auto_now_add=True, verbose_name='添加时间')
    parent_category = models.ForeignKey('self', null=True, blank=True,
                        verbose_name='父级目录', help_text='父级目录', on_delete=
                        models.CASCADE, related_name='sub_category')

    class Meta:
        verbose_name_plural = verbose_name = '商品类别'

    def __str__(self):
        return self.name


class GoodsCategoryBrand(models.Model):
    """
    品牌
    """
    category = models.ForeignKey(GoodsCategory, null=True, blank=True,
        on_delete=models.CASCADE, verbose_name='商品类别', help_text='商品类别',
        related_name='brands')
    name = models.CharField(max_length=30, default='',
        verbose_name='品牌名称', help_text='品牌名称')
    desc = models.TextField(default='', max_length=200,
        verbose_name='品牌描述', help_text='品牌描述')
    image = models.ImageField(max_length=200, upload_to='brand/images/',
        verbose_name='品牌图片', help_text='品牌图片')
    add_time = models.DateTimeField(auto_now_add=True, verbose_name='添加时间')

    class Meta:
        verbose_name_plural = verbose_name = '品牌'

    def __str__(self):
        return self.name


class Goods(models.Model):
    """
    商品
    """
    category = models.ForeignKey(GoodsCategory, on_delete=models.CASCADE,
            verbose_name='商品类别', help_text='商品类别', related_name='goods')
    goods_sn = models.CharField(max_length=100, default='',
            verbose_name='商品编码', help_text='商品唯一货号')
    name = models.CharField(max_length=300, verbose_name='商品名称',
        help_text='商品名称')
    click_num = models.IntegerField(default=0, verbose_name='点击数',
            help_text='点击数')
    sold_num = models.IntegerField(default=0, verbose_name='销售量',
            help_text='销售量')
    fav_num = models.IntegerField(default=0, verbose_name='收藏数',
            help_text='收藏数')
    goods_num = models.IntegerField(default=0, verbose_name='库存量',
            help_text='库存量')
```

```
    market_price = models.FloatField(default=0, verbose_name='市场价格',
                   help_text='市场价格')
    shop_price = models.FloatField(default=0, verbose_name='本店价格',
                 help_text='本店价格')
    goods_brief = models.TextField(max_length=500, verbose_name='简短描述',
                 help_text='商品简短描述')
    goods_desc = RichTextUploadingField(verbose_name='详情描述',
                 help_text='详情描述')
    ship_free = models.BooleanField(default=True, verbose_name='是否免运费',
                 help_text='是否免运费')
    goods_front_image = models.ImageField(upload_to='goods/front/', null=True,
                        blank=True, verbose_name='封面图', help_text='封面图')
    is_new = models.BooleanField(default=False, verbose_name='是否新品',
             help_text='是否新品')
    is_hot = models.BooleanField(default=False, verbose_name='是否热销',
             help_text='是否热销')
    add_time = models.DateTimeField(auto_now_add=True, verbose_name='添加时间')

    class Meta:
        verbose_name_plural = verbose_name = '商品'

    def __str__(self):
        return self.name


class GoodsImage(models.Model):
    """
    商品图片
    """
    goods = models.ForeignKey(Goods, verbose_name='商品', help_text='商品',
            on_delete=models.CASCADE, related_name='images')
    image = models.ImageField(upload_to='goods/images/', verbose_name='图片',
            help_text='图片')
    add_time = models.DateTimeField(auto_now_add=True, verbose_name='添加时间')

    class Meta:
        verbose_name_plural = verbose_name = '商品图片'

    def __str__(self):
        return self.goods.name


class Banner(models.Model):
    """
    首页轮播图
    """
    goods = models.ForeignKey(Goods, verbose_name='商品', help_text='商品',
            on_delete=models.CASCADE, related_name='banners')
    image = models.ImageField(upload_to='goods/banners/', verbose_name='图片',
           help_text='图片')
    index = models.IntegerField(default=0, verbose_name='轮播顺序', help_text='
           轮播顺序')
    add_time = models.DateTimeField(auto_now_add=True, verbose_name='添加时间')
```

```
    class Meta:
        verbose_name_plural = verbose_name = '首页轮播图'

    def __str__(self):
        return self.goods.name


class IndexCategoryAd(models.Model):
    """
    首页广告
    """
    category = models.ForeignKey(GoodsCategory, null=True, blank=True,
               on_delete=models.CASCADE, verbose_name='商品类别', help_text='
               商品类别', related_name='ads')
    goods = models.ForeignKey(Goods, verbose_name='商品', help_text='商品',
            on_delete=models.CASCADE, related_name='ads')
    add_time = models.DateTimeField(auto_now_add=True, verbose_name='添加时间')

    class Meta:
        verbose_name_plural = verbose_name = '首页类别广告'

    def __str__(self):
        return '{}: {}'.format(self.category.name, self.goods.name)
```

(4) 在 goods 目录下编写文件 apps.py，设置在后台将应用名显示为中文，对应代码如下所示：

```
from django.apps import AppConfig

class GoodsConfig(AppConfig):
    name = 'goods'
    verbose_name = '商品'
```

(5) 在 goods 目录下编写文件 admin.py，功能是采用批量注册方式将应用 goods 关联到 admin 后台。对应代码如下所示：

```
@admin.register(GoodsCategory)
class GoodsCategoryAdmin(admin.ModelAdmin):
    list_display = ['name', 'category_type', 'is_tab', 'parent_category']
    # 列表页显示
    list_display_links = ('name', 'parent_category',)
    # 列表页外键链接，字段需在 list_display 中
    list_editable = ('is_tab',)  # 列表页可编辑
    list_filter = ('category_type',)  # 列表页可筛选
    search_fields = ('name', 'desc')  # 列表页可搜索


class GoodsImageInline(admin.TabularInline):
    model = GoodsImage


@admin.register(Goods)
class GoodsAdmin(admin.ModelAdmin):
```

```
    list_display = ['name']
    inlines = [
        GoodsImageInline
    ]


@admin.register(IndexCategoryAd)
class IndexCategoryAdAdmin(admin.ModelAdmin):
    list_display = ['category', 'goods']

    def formfield_for_foreignkey(self, db_field, request, **kwargs):
        if db_field.name == 'category':
            # 外键下拉框添加过滤
            kwargs['queryset'] = GoodsCategory.objects.filter(category_type=1)
        return super(IndexCategoryAdAdmin, self).formfield_for_foreignkey
                    (db_field, request, **kwargs)


all_models = apps.get_app_config('goods').get_models()
for model in all_models:
    try:
        admin.site.register(model)
    except:
        pass
```

## 6.4.3　为 trade 应用创建 Model 模型

在 trade 应用的数据库模型中，主要用于保存系统交易信息，包括购物车、订单和订单商品等信息。

(1)　在 trade 目录下编写文件 models.py，分别创建模型类 ShoppingCart、OrderInfo 和 OrderGoods，对应代码如下所示：

```
from django.db import models
from goods.models import Goods
from django.contrib.auth import get_user_model

User = get_user_model()
class ShoppingCart(models.Model):
    """
    购物车
    """
    user = models.ForeignKey(User, verbose_name='用户', help_text='用户',
           on_delete=models.CASCADE, related_name='shopping_carts')
    goods = models.ForeignKey(Goods, verbose_name='商品', help_text='商品',
            on_delete=models.CASCADE)
    nums = models.IntegerField(default=0, verbose_name='购买数量', help_text='
        购买数量')
    add_time = models.DateTimeField(auto_now_add=True, verbose_name='添加时间')

    class Meta:
        verbose_name_plural = verbose_name = '购物车'
```

```
    def __str__(self):
        return "{}({})".format(self.goods.name, self.nums)


class OrderInfo(models.Model):
    """
    订单
    """
    ORDER_STATUS = (
        ('success', '成功'),
        ('cancel', '取消'),
        ('topaid', '待支付')
    )
    user = models.ForeignKey(User, verbose_name='用户', help_text='用户',
           on_delete=models.CASCADE, related_name='order_infos')
    order_sn = models.CharField(max_length=30, unique=True,
               verbose_name='订单号', help_text='订单号')
    trade_no = models.CharField(max_length=100, unique=True, null=True,
               blank=True, verbose_name='支付')
    pay_status = models.CharField(choices=ORDER_STATUS, max_length=20,
                 verbose_name='订单状态', help_text='订单状态')
    post_script = models.CharField(max_length=50, blank=True, null=True,
                  verbose_name='订单留言', help_text='订单留言')
    order_amount = models.FloatField(default=0.0, verbose_name='订单金额',
                   help_text='订单金额')
    pay_time = models.DateTimeField(null=True, blank=True,
               verbose_name='支付时间', help_text='支付时间')
    # 用户信息
    address = models.CharField(max_length=200, default='',
              verbose_name='收货地址', help_text='收货地址')
    signer_name = models.CharField(max_length=20, default='',
                  verbose_name='签收人', help_text='签收人')
    signer_mobile = models.CharField(max_length=11, verbose_name='联系电话',
                    help_text='联系电话')

    add_time = models.DateTimeField(auto_now_add=True, verbose_name='添加时间')

    class Meta:
        verbose_name_plural = verbose_name = '订单'

    def __str__(self):
        return "{}".format(self.order_sn)


class OrderGoods(models.Model):
    """
    订单商品详情
    """
    order = models.ForeignKey(OrderInfo, on_delete=models.CASCADE, verbose_name
          ='订单信息', help_text='订单信息', related_name='order_goods')
    goods = models.ForeignKey(Goods, verbose_name='商品', help_text='商品',
            blank=True, null=True, on_delete=models.SET_NULL)
    goods_nums = models.IntegerField(default=0, verbose_name='购买数量',
                 help_text='购买数量')
```

```
    add_time = models.DateTimeField(auto_now_add=True, verbose_name='添加时间')

    class Meta:
        verbose_name_plural = verbose_name = '订单商品'

    def __str__(self):
        return str(self.order.order_sn)
```

(2) 在 trade 目录下编写文件 apps.py，设置在后台将应用名显示为中文，对应代码如下所示：

```
from django.apps import AppConfig
class TradeConfig(AppConfig):
    name = 'trade'
    verbose_name = '交易'
```

(3) 在“trade”目录下编写文件 admin.py，功能是采用批量注册方式将 trade 应用关联到 admin 后台。对应代码如下所示：

```
all_models = apps.get_app_config('trade').get_models()
for model in all_models:
    try:
        admin.site.register(model)
    except:
        pass
```

### 6.4.4 为 user_operation 应用创建 Model 模型

在 user_operation 应用的数据库模型中，主要用于保存会员用户的资料信息，包括收藏、留言和收货地址等。

(1) 在 user_operation 目录下编写文件 models.py，分别创建模型类 ShoppingCart、OrderInfo 和 OrderGoods，对应代码如下所示：

```
from django.db import models
from goods.models import Goods
from django.contrib.auth import get_user_model

User = get_user_model()

class UserFav(models.Model):
    """
    用户收藏
    """
    user = models.ForeignKey(User, verbose_name='用户', help_text='用户', on_delete
        =models.CASCADE, related_name='favs')
    goods = models.ForeignKey(Goods, on_delete=models.CASCADE,
            verbose_name='商品', help_text='商品')
    add_time = models.DateTimeField(auto_now_add=True, verbose_name='添加时间')
    class Meta:
        verbose_name_plural = verbose_name = '用户收藏'

    def __str__(self):
```

```
        return self.user.name

class UserLeavingMessage(models.Model):
    """
    用户留言
    """
    MESSAGE_TYPE = (
        (1, '留言'),
        (2, '投诉'),
        (3, '询问'),
        (4, '售后'),
        (5, '求购')
    )
    user = models.ForeignKey(User, verbose_name='用户', help_text='用户', on_
            delete=models.CASCADE, related_name='leaving_msgs')
    message_type = models.IntegerField(default=1, choices=MESSAGE_TYPE, verbose_
    name='留言类型', help_text='留言类型: 1-留言, 2-投诉, 3-询问, 4-售后, 5-求购')
    subject = models.CharField(max_length=100, default='', verbose_name='主题',
            help_text='主题')
    message = models.TextField(default='', verbose_name='留言内容', help_text='
            留言内容')
    file = models.FileField(upload_to='upload/leaving_msg/', blank=True, null
        =True, verbose_name='上传文件', help_text='上传文件')
    add_time = models.DateTimeField(auto_now_add=True, verbose_name='添加时间')

    class Meta:
        verbose_name_plural = verbose_name = '用户留言'
    def __str__(self):
        return '{} {}:{}'.format(self.user.name, self.get_message_type_
            display(), self.subject)

class UserAddress(models.Model):
    """
    用户收货地址
    """
    user = models.ForeignKey(User, verbose_name='用户', help_text='用户',
            on_delete=models.CASCADE, related_name='addresses')
    district = models.CharField(max_length=100, default='',
            verbose_name='区域', help_text='区域')
    address = models.CharField(max_length=200, default='',
            verbose_name='收货地址', help_text='收货地址')
    signer_name = models.CharField(max_length=20, default='',
            verbose_name='签收人', help_text='签收人')
    signer_mobile = models.CharField(max_length=11, verbose_name='联系电话',
            help_text='联系电话')
    add_time = models.DateTimeField(auto_now_add=True, verbose_name='添加时间')

    class Meta:
        verbose_name_plural = verbose_name = '收货地址'
    def __str__(self):
        return self.address
```

(2) 在 user_operation 目录下编写文件 apps.py，设置在后台将应用名显示为中文，对应代码如下所示：

```
from django.apps import AppConfig
class UserOperationConfig(AppConfig):
    name = 'user_operation'
    verbose_name = '操作'
```

(3) 在 user_operation 目录下编写文件 admin.py，功能是采用批量注册方式将 user_operation 应用关联到 admin 后台。对应代码如下所示：

```
all_models = apps.get_app_config('user_operation').get_models()
for model in all_models:
    try:
        admin.site.register(model)
    except:
        pass
```

## 6.4.5　生成数据库表

通过如下命令在数据库中生成数据库表：

```
manage.py DjangoOnlineFreshSupermarket > makemigrations
manage.py DjangoOnlineFreshSupermarket > migrate
```

使用创建的管理员账号登录后台系统后，会显示后台管理页面，在页面中显示有上面创建的数据库表信息，如图 6-5 所示。

交易
订单 +增加 修改
订单商品 +增加 修改
购物车 +增加 修改
商品
品牌 +增加 修改
商品 +增加 修改
商品图片 +增加 修改
商品类别 +增加 修改
首页轮播图 +增加 修改
操作
收货地址 +增加 修改
用户收藏 +增加 修改
用户留言 +增加 修改
用户
用户 +增加 修改
短信验证码 +增加 修改
最近动作
我的动作
无可用的

图 6-5　后台首页

## 6.5 使用 Restful API

扫码观看视频讲解

为了便于系统开发和维护，实现前端资源和后端资源的分离，本项目使用 Restful API 实现后台 View 视图和前台 Vue 的关联。Restful API 是当今被公认的实现 Django 前后端分离的最佳工具库，在 Django Web 中使用 Restful API 后，相当于为 Django Web 设计了一套 API 标准。当一个 Web 在使用 Restful API 后，这个 Web 可以直接通过 http 协议拥有 post、get、put、delete 等操作方法，而不需要额外的协议。

### 6.5.1 商品列表序列化

商品列表页面是 http://127.0.0.1:8000/goods/，此 URL 对应的 View 视图文件是 goods 应用的文件 views_base.py 和 views.py。

(1) 在文件 views_base.py 中通过 Django 的 View 获取商品列表页，代码如下：

```
class GoodsListView(View):
    def get(self, request):
        """
        通过Django的View获取商品列表页
        :param request:
        :return:
        """
        from django.core import serializers
        json_data = serializers.serialize('json', all_goods)  # 序列化

        from django.http import HttpResponse, JsonResponse
        import json
        # return HttpResponse(json_data, content_type='application/json')
        json_data = json.loads(json_data)  # 转换为数组
        return JsonResponse(json_data, safe=False)
```

(2) 在文件 views.py 中使用 DRF 实现商品视图功能，Django+DRF 将后端变成一种声明式的工作流，只要按照 Models→Serializer→Views→urls 的流程去实现一个个 Python 文件，即可生成一个很全面的通用的后端。文件 views.py 的具体实现流程如下。

▶ 通过 GoodsPagination 实现自定义分页功能，代码如下：

```
class GoodsPagination(PageNumberPagination):
    page_size = 12  # 每一页个数，由于前段
    page_query_description = _('使用分页后的页码')  # 分页文档中文描述
    page_size_query_param = 'page_size'
    page_size_query_description = _('每页返回的结果数')
    page_query_param = 'page'  # 参数?p=xx，将其修改为page，适应前端，也方便识别
    max_page_size = 36  #指定每页最大个数
```

Django 的分页 API 可以支持以下两种方式：

① 作为响应内容的一部分提供的分页链接。

② 包含在响应头中的分页链接，如内容范围或链接。

只有在使用 Generic Views(根据 URL 中传递的参数从数据库中获取数据，加载模板并且返回显示的模板。因为这很常见，Django 提供了一个快捷方式，称为 Generic Views 系统)或 Viewsets 时才会自动执行分页。如果使用常规的 API View，则需要自己调用分页 API，以确保返回分页响应。Restful API 的任何全局设置都被保存在一个名为 REST_FRAMEWORK 的配置字典中，分页样式可以使用 DEFAULT_PAGINATION_CLASS 和 PAGE_SIZE 设置，这两个值默认都是 None。例如，要使用内置的"限制/偏移"分页，在文件 settings.py 中添加以下代码：

```
# DRF 配置
REST_FRAMEWORK = {
    'DEFAULT_PAGINATION_CLASS':
'rest_framework.pagination.PageNumberPagination',
    'PAGE_SIZE': 5
}
```

在本实例中使用了单独设置分页的方式。在 GenericAPIView 子类上，还可以设置 pagination_class 属性，以根据每个视图选择 PageNumberPagination。如果希望修改分页样式的特定方面，则需要覆盖分页类中的一个，并设置要更改的属性。

类 PageNumberPagination 包含许多属性，覆盖这些属性可以修改分页样式。要想设置这些属性，应该覆盖类 PageNumberPagination，然后像上面那样启用自定义分页类。类 PageNumberPagination 的常用属性如下所示。

① django_paginator_class：要使用的 Django Paginator 类。默认值为 django.core.paginator.Paginator，对于大多数用例来说都可用。

② page_size：设置每个分页的大小。如果设置，则会覆盖 PAGE_SIZE 设置。默认值为与 settings.py 中 PAGE_SIZE 相同的值。

③ page_query_param：一个字符串值，指示分页控件使用的查询参数的名称。

④ page_size_query_param：这是一个字符串值，指示查询参数的名称，允许客户端根据每个请求设置页面大小。默认值为 None，表示客户机可能无法控制请求的页面大小。

⑤ max_page_size：这是一个数字值，指示允许的最大页面大小。只有在设置 page_size_query_param 时，此属性才有效。

⑥ last_page_strings：字符串值的列表或元组，指示可以与 page_query_param 一起使用的值，用于请求集合中的最终页面。

⑦ template：在可浏览 API 中显示分页控件时使用的模板名称。可以重写以修改显示样式，也可以将其设置为 None 以完全禁用 HTML 分页控件。

在上述代码的类 GoodsPagination 中，我们使用 REST_FRAMEWORK 中的模块 PageNumberPagination 实现了分页功能。在使用上述自定义分页功能后，需要取消文件 settings.py 中的默认分页，防止影响之后商品分类的结果。代码如下：

```
# DRF 配置
REST_FRAMEWORK = {
    'DEFAULT_PAGINATION_CLASS':
'rest_framework.pagination.PageNumberPagination',
```

```
    # 'PAGE_SIZE': 5
}
```

▶ 定义类 GoodsListView，使用分页样式显示商品信息。代码如下：

```
class GoodsListView(generics.ListAPIView):
    """
    显示所有的商品列表
    """
    queryset = Goods.objects.all()
    serializer_class = GoodsSerializer
    pagination_class = GoodsPagination
```

▶ 通过多个 ViewSet 类显示商品列表信息。ViewSet 类几乎与视图类相同，只是它提供了 read 或 update 之类的操作，而不是 get 或 put 之类的方法处理程序。ViewSet 类只在最后时刻绑定到一组方法处理程序，当它被实例化为一组视图时，通常通过使用一个 Router 类来定义 URL Conf 的复杂性。代码如下：

```
class GoodsListViewSet(CacheResponseMixin, mixins.ListModelMixin,
mixins.RetrieveModelMixin, viewsets.GenericViewSet):
    """
    list:
        显示商品列表，分页、过滤、搜索、排序

    retrieve:
        显示商品详情
    """
    queryset = Goods.objects.all()  # 使用get_queryset函数，依赖queryset的值
    serializer_class = GoodsSerializer
    pagination_class = GoodsPagination
    filter_backends = (DjangoFilterBackend, filters.SearchFilter,
        filters.OrderingFilter,)  # 将过滤器后端添加到单个视图或视图集
    filterset_class = GoodsFilter
    # authentication_classes = (TokenAuthentication, )  # 只在本视图中验证Token
    search_fields = ('name', 'goods_desc', 'category__name')  # 搜索字段
    ordering_fields = ('click_num', 'sold_num', 'shop_price')  # 排序
    # throttle_classes = [UserRateThrottle, AnonRateThrottle]
    # DRF默认限速类，可以仿写自己的限速类
    throttle_scope = 'goods_list'

    def retrieve(self, request, *args, **kwargs):
        # 增加点击数
        instance = self.get_object()
        instance.click_num += 1
        instance.save()
        serializer = self.get_serializer(instance)
        return Response(serializer.data)

    def get_queryset(self):
        keyword = self.request.query_params.get('search')
        if keyword:
            from utils.hotsearch import HotSearch
            hot_search = HotSearch()
            hot_search.save_keyword(keyword)
```

```
        return self.queryset

class CategoryViewSet(mixins.ListModelMixin, mixins.RetrieveModelMixin,
                      viewsets.GenericViewSet):
    # 注释很有用，在 drf 文档中
    """
    list:
        商品分类列表

    retrieve:
        商品分类详情
    """
    # queryset = GoodsCategory.objects.all()
    # 取出所有分类，没必要分页，因为分类数据量不大
    queryset = GoodsCategory.objects.filter(category_type=1)
    # 只获取一级分类数据
    serializer_class = CategorySerializer
    # 使用商品类别序列化类，写商品的分类外键已有，直接调用


class ParentCategoryViewSet(mixins.ListModelMixin,
                       mixins.RetrieveModelMixin, viewsets.GenericViewSet):
    """
    list:
        根据子类别查询父类别

    retrieve:
        根据子类别查询父类别详情
    """
    queryset = GoodsCategory.objects.all()
    serializer_class = ParentCategorySerializer


class BannerViewSet(mixins.ListModelMixin, viewsets.GenericViewSet):
    """
    list:
        获取轮播图列表
    """
    queryset = Banner.objects.all()
    serializer_class = BannerSerializer


class IndexCategoryGoodsViewSet(mixins.ListModelMixin, viewsets.GenericViewSet):
    """
    list:
        首页分类、商品数据
    """
    queryset = GoodsCategory.objects.filter(category_type=1)
    serializer_class = IndexCategoryGoodsSerializer

    def get_queryset(self):
        # 随机取出几个分类
        import random
        category_id_list = self.queryset.values_list('id', flat=True)
        selected_ids = random.sample(list(category_id_list), 3)
```

```
        qs = self.queryset.filter(id__in=selected_ids)
        return qs
```

在默认情况下，GenericViewSet类不提供任何操作，但它包含了基本的通用视图行为集，例如get_object 和 get_queryset方法，也就是之前继承的viewsets.GenericViewSet没有定义get、post方法，处理程序方法只在定义URL Conf时绑定到操作。因为使用的是ViewSet类而不是View类，所以实际上不需要自己设计URL。可以使用Router类自动处理将资源连接到视图和URL的约定，需要做的就是用路由器注册适当的视图集，然后让它完成剩下的工作。所以在urls.py文件中，API url现在由路由器自动确定：

```
urlpatterns = [
    path('admin/', admin.site.urls),
    path('api-auth/', include('rest_framework.urls')),  # drf 认证url
    path('ckeditor/',include('ckeditor_uploader.urls')),  # 配置富文本编辑器url

    path('', include(router.urls)),  # API url现在由路由器自动确定

    # DRF 文档
    path('docs/', include_docs_urls(title='DRF文档')),
]
```

(3) 在上面曾经提到过，Django+DRF 开发只要按照 Models→Serializer→Views→urls的流程去实现一个个Python文件即可。在上述文件views.py中，基于Serializer实现了DRF，通过代码GoodsSerializer(serializers.Serializer)实现了Serializer功能。编写文件serializers.py实现GoodsSerializer，通过使用DRF的Serializer可以将数据保存到数据库中。Serializer的功能可以相当于Django的Form功能，也可以完成序列化为JSON的功能。文件serializers.py的具体实现流程如下所示。

- 通过CategorySerializer3获取所有三级分类的所有信息，代码如下：

```
class CategorySerializer3(serializers.ModelSerializer):
    class Meta:
        model = GoodsCategory
        fields = '__all__'
```

- 通过CategorySerializer2获取所有二级分类的信息，代码如下：

```
class CategorySerializer2(serializers.ModelSerializer):
    sub_category = CategorySerializer3(many=True)  # 通过二级分类获取三级分类
    class Meta:
        model = GoodsCategory
        fields = '__all__'
```

- 通过CategorySerializer获取所有一级分类的所有信息，代码如下：

```
class CategorySerializer(serializers.ModelSerializer):
    sub_category = CategorySerializer2(many=True)
    # 通过一级分类获取到二级分类，由于一级分类下有多个二级分类，需要设置many=True

    class Meta:
        model = GoodsCategory
        fields = '__all__'
```

▶ 通过 GoodsImageSerializer 获取商品图片表中 image 字段的信息，代码如下：

```
# 商品图片序列化
class GoodsImageSerializer(serializers.ModelSerializer):
    class Meta:
        model = GoodsImage
        fields = ['image']  # 需要的字段只有 image
```

▶ 通过 GoodsSerializer 获取所有的商品信息，代码如下：

```
class GoodsSerializer(serializers.ModelSerializer):
    category = CategorySerializer()  # 自定义字段覆盖原有的字段，实例化
    images = GoodsImageSerializer(many=True)
    # 字段名和外键名称一样，商品轮播图需要加 many=True，因为一个商品有多个图片
    class Meta:
        model = Goods
        fields = '__all__'
```

▶ 通过 ParentCategorySerializer3 获取 CategorySerializer3 父级分类的所有信息，代码如下：

```
# 获取父级分类
class ParentCategorySerializer3(serializers.ModelSerializer):
    class Meta:
        model = GoodsCategory
        fields = '__all__'
```

▶ 通过 ParentCategorySerializer2 获取 CategorySerializer2 父级分类的所有信息，代码如下：

```
class ParentCategorySerializer2(serializers.ModelSerializer):
    parent_category = ParentCategorySerializer3()

    class Meta:
        model = GoodsCategory
        fields = '__all__'
```

▶ 通过 ParentCategorySerializer 获取 CategorySerializer 父级分类的所有信息，代码如下：

```
class ParentCategorySerializer(serializers.ModelSerializer):
    parent_category = ParentCategorySerializer2()

    class Meta:
        model = GoodsCategory
        fields = '__all__'
```

▶ 通过 BannerSerializer 获取 Banner 的所有信息，代码如下：

```
class BannerSerializer(serializers.ModelSerializer):
    class Meta:
        model = Banner
        fields = "__all__"
```

▶ 通过 BrandsSerializer 获取 GoodsCategoryBrand 的所有信息，代码如下：

```
class BrandsSerializer(serializers.ModelSerializer):
    class Meta:
        model = GoodsCategoryBrand
        fields = "__all__"
```

▶ 通过 IndexCategoryGoodsSerializer 序列化首页中的分类商品信息，代码如下：

```
class IndexCategoryGoodsSerializer(serializers.ModelSerializer):
    brands = BrandsSerializer(many=True)  # 分类下的品牌图片
    # goods = GoodsSerializer(many=True)
    # 不能这样用，因为现在需要的是一级分类，而大多数商品是放在三级分类中的，所以很多商品是
    # 取不到的，所以到一级分类子类别下查询所有商品
    goods = serializers.SerializerMethodField()
    sub_category = CategorySerializer2(many=True)  # 序列化二级分类
    ad_goods = serializers.SerializerMethodField()
    # 广告商品可能加了很多，取每个分类第一个

    def get_ad_goods(self, obj):
        all_ads = obj.ads.all()
        if all_ads:
            ad = all_ads.first().goods  # 获取到商品分类对应的商品
            ad_serializer = GoodsSerializer(ad, context={'request': self.context
['request']})
            # 序列化该广告商品，嵌套的序列化类中添加 context 参数，可在序列化时添加域名
            return ad_serializer.data
        else:
            # 在该分类没有广告商品时，必须要返回空字典，否则 Vue 中取 obj.id 会报错
            return {}

    def get_goods(self, obj):
        # 查询每级分类下的所有商品
        all_goods = Goods.objects.filter(Q(category_id=obj.id) | Q(category_
                    _parent_category_id=obj.id) | Q(category__parent_category
                    __parent_category_id=obj.id))
        # 将查询的商品集进行序列化
        goods_serializer = GoodsSerializer(all_goods, many=True, context=
                           {'request': self.context['request']})
        # 返回 json 对象
        return goods_serializer.data

    class Meta:
        model = GoodsCategory
        fields = '__all__'
```

如果此时设置 goods 应用在文件 urls.py 中的 url 为：

```
urlpatterns = [
    path('admin/', admin.site.urls),
    path('api-auth/', include('rest_framework.urls')),  # drf 认证 url
    path('ckeditor/', include('ckeditor_uploader.urls')),  # 配置富文本编辑器 url
    path('goods/', GoodsListView.as_view(), name='goods_list'),

    # DRF 文档
    path('docs/', include_docs_urls(title='DRF 文档')),
]
```

例如访问 http://127.0.0.1:8000/goods/，会显示 DRF 格式的商品列表，如图 6-6 所示。

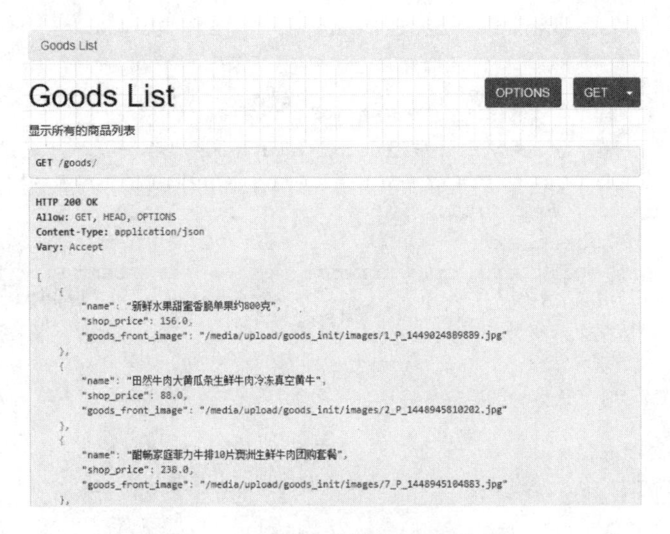


图 6-6　DRF 格式的商品列表页面

如果访问 http://127.0.0.1:8000/goods/?format=json，就可以看到 JSON 的内容，如图 6-7 所示。

```
[
  - {
        name: "新鲜水果甜蜜香脆单果约800克",
        shop_price: 156,
        goods_front_image: "/media/upload/goods_init/images/1_P_1449024889889.jpg"
    },
  - {
        name: "田然牛肉大黄瓜条生鲜牛肉冷冻真空黄牛",
        shop_price: 88,
        goods_front_image: "/media/upload/goods_init/images/2_P_1448945810202.jpg"
    },
  - {
        name: "酣畅家庭菲力牛排10片澳洲生鲜牛肉团购套餐",
        shop_price: 238,
        goods_front_image: "/media/upload/goods_init/images/7_P_1448945104883.jpg"
    },
  - {
        name: "日本蒜蓉粉丝扇贝270克6只装",
```

图 6-7　JSON 的内容

(4) 编写文件 filters.py，功能是使用库 django-filter 实现商品过滤功能。库 django-filter 包含一个 DjangoFilterBackend 类，它支持 REST 框架的高度可定制字段过滤。要想使用 DjangoFilterBackend，首先要安装 django-filter，然后将 django_filter 添加到 Django 的 INSTALLED_APPS 中，再在文件 views.py 的类 GoodsListViewSet 中通过如下代码增加商品列表的过滤器：

```
# 将过滤器后端添加到单个视图或视图集
filter_backends = (DjangoFilterBackend, filters.SearchFilter,
                   filters.OrderingFilter,)
```

文件 filters.py 的具体实现代码如下所示：

```
class GoodsFilter(filters.FilterSet):
    """
```

```
    商品的过滤类
    """
    name = filters.CharFilter(field_name='name', lookup_expr='contains',
                                 help_text='分类名模糊匹配')  # 包含关系，模糊匹配
    goods_desc = filters.CharFilter(field_name='name', lookup_expr='contains',
                 help_text='商品描述模糊匹配')
    min_price = filters.NumberFilter(field_name="shop_price",
                 lookup_expr='gte', help_text='最低价格')  # 自定义字段
    max_price = filters.NumberFilter(field_name="shop_price",
                 lookup_expr='lte', help_text='最高价格')
    top_category = filters.NumberFilter(method='top_category_filter',
field_name='category_id', lookup_expr='=', help_text='自定义过滤某个一级分类')
# 自定义过滤，过滤某个一级分类

    def top_category_filter(self, queryset, field_name, value):
        """
        自定义过滤内容
        这儿是传递一个分类的 id，在已有商品查询集基础上获取分类 id，一级一级往上找，直到将三
级类别找完
        :param queryset:
        :param field_name:
        :param value: 需要过滤的值
        :return:
        """
        queryset = queryset.filter(Q(category_id=value) |
                   Q(category__parent_category_id=value) |
                   Q(category__parent_category__parent_category_id=value))
        return queryset

    class Meta:
        model = Goods
        fields = ['name', 'goods_desc', 'min_price', 'max_price', 'is_hot', 'is_new']
```

### 6.5.2 在前端展示左侧分类、排序、商品列表和分页

在前端通过顶部导航显示的左侧分类，和通过搜索显示的左侧分类的格式是相同的，所以在本项目中它们用的是同一个 Vue 组件。尽管格式相同，但它们还是有一定区别的。

- 点击顶部导航，左侧的分类显示该一级分类下的所有子分类(包括所有二级分类和二级分类对应的三级分类)。
- 搜索显示的分类，只显示二级分类和一级分类，也就是说，搜索某一个商品，可以获取该商品对应的分类和对应的子分类。

(1) 获取所有数据。

在文件 list/list.vue 中通过函数 getAllData()获取所有的数据，此函数使用 if 语句判断用户是点击分类还是搜索进入。代码如下：

```
methods: {
    getAllData() {
        console.log('list/list.vue 获取左侧菜单');
        console.log(this.$route.params);
        var curloc_id = '';  //当前点击分类的 id
```

```
      if (this.$route.params.id) {
         //如果能获取到 id 值，那么就是点击分类
         this.top_category = this.$route.params.id;
         this.pageType = 'list';
         this.getMenu(this.top_category); // 获取左侧菜单列表
         curloc_id = this.$route.params.id;
      } else {
         this.getMenu(null); // 获取左侧菜单列表
         this.pageType = 'search';
         this.searchWord = this.$route.params.keyword;
         curloc_id = ''
      }

      this.getCurLoc(curloc_id); // 获取当前位置
      this.getListData(); //获取产品列表
      this.getPriceRange(); // 获取价格区间
   },
```

(2) 获取菜单。

在文件 list/list.vue 中通过函数 getMenu(id)获取菜单信息，能够传递某个一级分类 id 参数，然后通过函数 getCategory()获取这个 id 分类法详情(二、三级分类)信息并显示出来。代码如下：

```
         getMenu(id) {
            if (id != null) {
               getCategory({
                  id: this.$route.params.id
               }).then((response) => {
                  this.cateMenu = response.data.sub_category;
                  console.log('list/list.vue 获取分类数据：');
                  console.log(response.data);
                  this.currentCategoryName = response.data.name;
                   //获取当前分类的名称
                  this.currentCategoryID = response.data.id;
                  //获取请求的一级分类的 ID
                  this.isObject = true
                  //console.log(response.data)
               }).catch(function (error) {
                  console.log(error);
               });
            } else {
               getCategory({}).then((response) => {
                  this.cateMenu = response.data;
                  console.log('list/list.vue 获取分类数据：');
                  console.log(response.data);
                  this.isObject = false
               }).catch(function (error) {
                  console.log(error);
               });
            }
         },
```

(3) Nav 功能。

在文件 list/list.vue 中引入 list/listNav.vue 组件功能，listNav.vue 的代码如下：

```
        <div class="cate-menu" id="cate-menu">
            <!--<h3 v-if="isObject"><a
href=""><strong>{{currentCategoryName}}</strong><i id="total_count">商品共
{{proNum}}件</i></a></h3>-->
            <h3 v-if="isObject">
                <a
@click="changeMenu(currentCategoryID)"><strong>{{currentCategoryName}}</str
ong><i id="total_count">商品共{{proNum}}件</i></a>
            </h3>
            <h3 v-else><a @click="changeMenu('')"><strong>全部分类</strong><i
id="total_count">商品共{{proNum}}件</i></a></h3>
            <dl>
                <template v-for="item in cateMenu">
                    <dt><a @click="changeMenu(item.id)" style="color:
#888">{{ item.name}}</a></dt>
                    <dd v-for="subItem in item.sub_category">
                        <a @click="changeMenu(subItem.id)">{{ subItem.name}}</a>
                    </dd>
                </template>
            </dl>
        </div>
```

运行程序，在前端通过导航点击进入，此时链接为 http://127.0.0.1:8080/#/app/home/list/121，效果如图 6-8 所示。

图 6-8 “精品肉类”分类

在后端只需请求某个分类的 API 即可获取这个分类的父级分类信息，例如获取“牛肉”分类的父类信息的链接是 http://127.0.0.1:8000/parent_categories/126/，效果如图 6-9 所示。

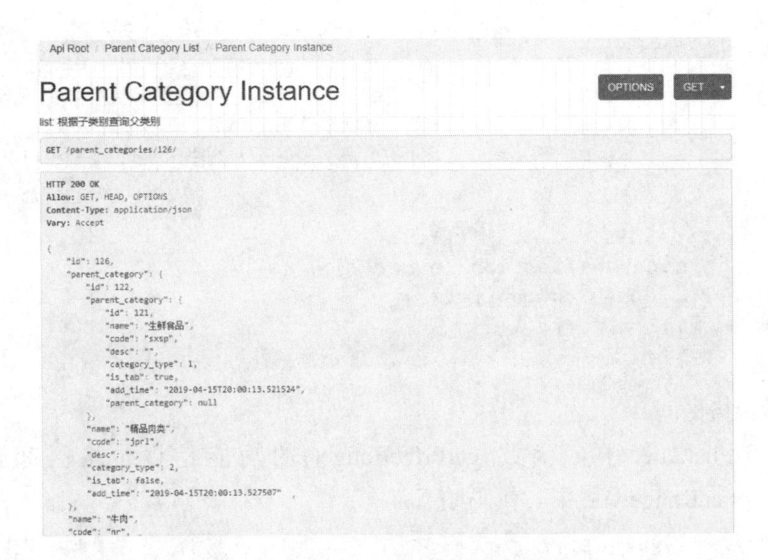


图 6-9　获取“牛肉”分类的父类信息

(4) 获取当前请求位置。

在文件 list/list.vue 中，函数 getCurLoc(id)用于提供点击的分类 id，例如代码 this.getCurLoc(curloc_id)能够获取当前位置，在获取请求后会组合分类面包屑。面包屑是我们在项目中经常使用的一个功能，一般情况下用来表示当前所处的站点位置，也可以帮助我们能够更快地回到上个层级。函数 getCurLoc(id)的代码如下：

```
//获取当前位置
getCurLoc(id) { // 当前位置
    // this.$http.post('/currentLoc', {
    //     params: {
    //         proType: this.type, //商品类型
    //     }
    // }).then((response) => {
    //     //console.log(response.data);
    //     this.curLoc = response.data;
    // }).catch(function (error) {
    //     console.log(error);
    // });
    getParentCategory({
        id: id //传递指定分类的 id
    }).then((response) => {
        console.log('list/list.vue 获取当前位置: ');
        console.log(response.data);
        var dt = response.data;
        var curLoc;
        //组合类别
        var index_p = {'id': 0, 'name': '首页'};
        var first = {'id': dt.id, 'name': dt.name};
        curLoc = [index_p, first];
        if (dt.parent_category != null) {
            var second = {'id': dt.parent_category.id, 'name':
                        dt.parent_category.name};
            curLoc = [index_p, second, first];
```

```
                if (dt.parent_category.parent_category != null) {
                    var third = {'id': dt.parent_category.parent_category.id,
                                'name': dt.parent_category.parent_
                                category.name};
                    curLoc = [index_p, third, second, first];
                }
            }
            this.curLoc = curLoc
        }).catch(function (error) {
            console.log(error);
        });
    },
```

(5) 显示价格区间。

在文件 list/list.vue 中，函数 getPriceRange()的功能是将价格区间数据填充到 list/price-range/priceRange.vue 中。代码如下：

```
    getPriceRange() {
        //价格区间显示，不使用上方mock.js中的内容
        this.priceRange = [
            {
                min: 1,
                max: 30,
            },
            {
                min: 31,
                max: 80,
            },
            {
                min: 81,
                max: 150,
            },
            {
                min: 151,
                max: 300,
            },
        ]
    },
```

## 6.6 登录认证

扫码观看视频讲解

在购买商品时，需要用户登录后才能操作，所以就必须增加会员登录验证和注册功能。由于在 Django 中配置了 path('api-auth/', include('rest_framework.urls'))，所以在 DRF 中可以直接使用认证功能。在本项目中，将使用 DRF Token 认证登录机制。

### 6.6.1 使用 DRF Token 认证

DRF Token 使用一个简单的基于令牌的 HTTP 身份验证方案，令牌身份验证适用于客

户机/服务器模式，例如本机桌面和移动客户机。要使用 TokenAuthentication 方案，需要将身份验证类配置为包含 TokenAuthentication；另外，在 INSTALLED_APPS 设置中还包括 rest_framework.authtoken。可以使用 DEFAULT_AUTHENTICATION_CLASSES 设置全局缺省身份验证方案，也就是默认为：

```
REST_FRAMEWORK = {
    'DEFAULT_AUTHENTICATION_CLASSES': (
        'rest_framework.authentication.BasicAuthentication',
        'rest_framework.authentication.SessionAuthentication',
    )
}
```

要想使用 TokenAuthentication 认证，还需要增加 rest_framework.authentication.SessionAuthentication 设置。

(1) 配置全局 TokenAuthentication。

在本项目中配置为：

```
# DRF 配置
REST_FRAMEWORK = {
    'DEFAULT_PAGINATION_CLASS':
'rest_framework.pagination.PageNumberPagination',
    # 'PAGE_SIZE': 5,
    'DEFAULT_AUTHENTICATION_CLASSES': (
        'rest_framework.authentication.BasicAuthentication',
        'rest_framework.authentication.SessionAuthentication',
        # 上面两个用于 DRF 基本验证
        'rest_framework.authentication.TokenAuthentication',
        # TokenAuthentication
    )
}
```

然后将 rest_framework.authtoken 添加到 INSTALLED_APPS 中：

```
INSTALLED_APPS = [
    # 添加 drf 应用
    'rest_framework',
    'rest_framework.authtoken',
]
```

最后执行下面的命令实现数据库迁移：

```
manage.py DjangoOnlineFreshSupermarket > makemigrations
manage.py DjangoOnlineFreshSupermarket > migrate
```

执行上述命令后，会发现在数据库中增加了一个名为 authtoken_token 的表，如图 6-10 所示。其中 user_id 是一个外键，指向现有的用户。如果之前已经创建了用户，在这个 token 表中是没有记录的。这个表和用户表是一一对应的。创建用户表后，需要手动创建 token 表，也就是当用户注册时，需要调用 token = Token.objects.create(user=...)来生成 token。

对于要进行身份验证的客户机，令牌密钥应该包含在授权 HTTP 头中。键应该以字符串 Token 作为前缀，并用空格分隔两个字符串。例如：

```
Authorization: Token 9944b09199c62bcf9418ad846dd0e4bbdfc6ee4b
```

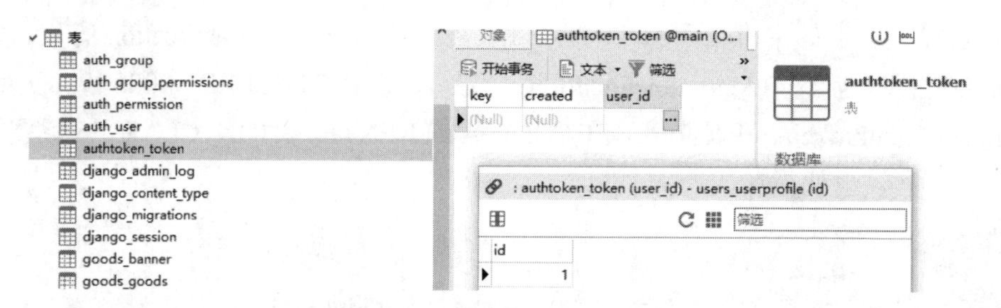

图 6-10　表 authtoken_token

(2)　添加 api-token-auth 的 URL。

在使用 TokenAuthentication 时，可能希望为客户端提供一种机制，以获得给定用户名和密码的令牌。REST 框架提供了一个内置视图来满足这个要求，但需要将视图 obtain_auth_token 添加到 URLconf:

```
path('api-token-auth/', views.obtain_auth_token),  # drf token 获取的 url
```

## 6.6.2　使用 JWT 认证

JSON Web Toke 是一个相当新的标准，可以用于基于令牌的身份验证。与内置的 TokenAuthentication 方案不同，JWT 身份验证不需要使用数据库来验证 Token。用于 JWT 身份验证的包是 djangorestframework-simplejwt，它提供了一些特性以及一个可扩展的 token 黑名单应用程序。因为 django-rest-framework-jwt 不支持新版的 Django 和 Django Restful Framework，所以可以使用 djangorestframework_simplejwt 来代替。

(1)　安装配置 djangorestframework_simplejwt。

使用如下命令安装包:

```
pip install djangorestframework_simplejwt
```

(2)　配置 JWT 认证类。

将 Django 项目配置为使用该库。在配置文件 settings.py 中，将 rest_framework_simplejwt.authentication.JWTAuthentication 添加到认证类中。

```
'rest_framework_simplejwt.authentication.JWTAuthentication',
# djangorestframework_simplejwt  JWT 认证
```

(3)　添加 JWT 认证 URL。

然后在主路径导航文件 url.py (或任何其他 url 配置)中，添加包括 Simple JWT 的 TokenObtainPairView 和 TokenRefreshView 视图的路由:

```
   path('api/token/', simplejwt_views.TokenObtainPairView.as_view(),
        name='token_obtain_pair'),  # simplejwt 认证接口
   path('api/token/refresh/', simplejwt_views.TokenRefreshView.as_view(),
        name='token_refresh'),  # simplejwt 认证接口
```

(4)　使用 JWT。

获取 JWT Access，以验证 JWT 接口是否生效。请求 http://127.0.0.1:8000/ api/token/，会

得到如图 6-11 所示的结果。

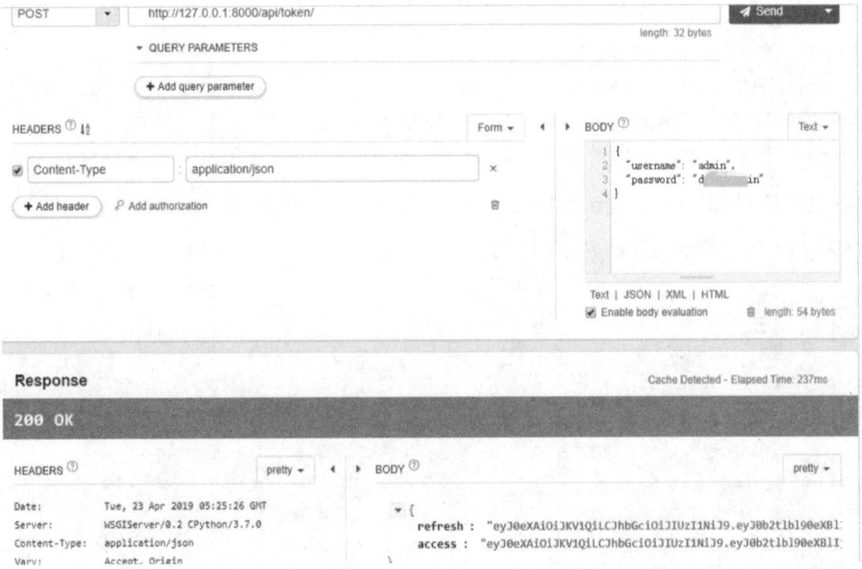

图 6-11　获取 JWT Access

图中的 refresh 和 access 是获取的结果，我们可以使用返回的 accesstoken 来验证受保护视图的身份验证。格式为：

```
Authorization: Bearer [access 对应的值]
```

请注意，在文件 src/axios/index.js 的 http request 拦截器中，一定要将 JWT 修改为 Bearer，也就是 Bearer ${store.state.userInfo.token}，否则之后获取个人信息类的页面肯定会出错：

```
// http request 拦截器
axios.interceptors.request.use(
  config => {
    if (store.state.userInfo.token) {
    // 判断是否存在 token，如果存在的话，则每个 http header 都加上 token
      config.headers.Authorization = 'Bearer ${store.state.userInfo.token}';
    }
    return config;
  },
  err => {
    return Promise.reject(err);
  });
```

当这个临时的 accesstoken 过期时，可以使用较长时间的 refreshtoken 获得另一个 accesstoken。如果中间等待时间过长，导致 access 过期，这时访问 http://127.0.0.1:8000/goods/ 就会出现 401 错误，这样做是为了提高安全性。

(5)　使用 refresh 获取新的 accesstoken。

当 accesstoken 过期后，可以使用 refresh 来获取新的 accesstoken，http://127.0.0.1:8000/api/token/refresh/需要将 refresh 的值传入后 POST(提交)得到新的 accesstoken。

(6) Vue 登录和 JWT 接口调试。

在文件 src/api/api.js 中，前端的登录接口是：

```
//登录
export const login = params => {
   return axios.post('${local_host}/login/', params)
};
```

而现在 DRF 登录 url 是 api/token/。修改为 JWT 方式的方法有两种：一种是修改 Vue 中的登录接口，二是修改 Django 的 URL。采用修改后台的方式修改主 urls.py：

```
urlpatterns = [
   path('admin/', admin.site.urls),
   path('api-auth/', include('rest_framework.urls')),  # drf 认证url
   path('api-token-auth/', views.obtain_auth_token),  # drf token获取的url
   # path('api/token/', simplejwt_views.TokenObtainPairView.as_view(),
         name='token_obtain_pair'),  # simplejwt认证接口
   path('login/', simplejwt_views.TokenObtainPairView.as_view(),
        name='token_obtain_pair'),  # 登录一般是login
   path('api/token/refresh/', simplejwt_views.TokenRefreshView.as_view(),
        name='token_refresh'),  # simplejwt认证接口
   path('ckeditor/', include('ckeditor_uploader.urls')),  # 配置富文本编辑器url

   path('', include(router.urls)),  # API url现在由路由器自动确定

   # DRF文档
   path('docs/', include_docs_urls(title='DRF文档')),
]
```

此时访问前端登录页面 http://127.0.0.1:8080/#/app/login，效果如图 6-12 所示。输入用户名、密码并单击“立即登录”按钮后，可以登录商城前端。

图 6-12　登录界面

在前端文件 src/views/login/login.vue 中，登录验证的逻辑代码如下：

```
      methods: {
         login() {
            // if(this.userName==''||this.parseWord==''){
            //   this.error = true;
            //   return
            // }
```

```
            var that = this;
            login({
                username: this.userName, //当前用户名
                password: this.parseWord
            }).then((response) => {
                //console.log(response);
                //本地存储用户信息
                console.log('用户登录信息: ');
                console.log(response.data);
                cookie.setCookie('name', this.userName, 7); // 设置过期时间为7天
                // cookie.setCookie('token', response.data.token, 7);
                cookie.setCookie('token', response.data.access, 7);
                //存储在 store
                // 更新 store 数据
                that.$store.dispatch('setInfo');
                //跳转到首页页面
                this.$router.push({name: 'index'})
```

### 6.6.3　增加用户名和手机号短信验证登录功能

(1) 在后端文件 users/views.py 中增加自定义的后台认证类，代码如下：

```
from django.shortcuts import render
from django.contrib.auth import get_user_model
from django.db.models import Q
from django.contrib.auth.backends import ModelBackend

User = get_user_model()

class CustomBackend(ModelBackend):
    """
    自定义用户登录，可以使用用户名和手机登录，重写 authenticate 方法
    """
    def authenticate(self, request, username=None, password=None, **kwargs):
        try:
            user = User.objects.get(Q(username=username) | Q(mobile=username))
            if user.check_password(password):
                return user
        except Exception as e:
            return None
```

(2) 在配置文件 settings.py 中添加设置，使用认证后台的代码：

```
AUTHENTICATION_BACKENDS = ('users.views.CustomBackend',)  # 指定认证后台
```

(3) 手机登录需要发送短信验证码，需要后端提供一个发送短信的接口。本项目使用云片公司的第三方短信发送服务：

```
https://www.yunpian.com/doc/zh_CN/introduction/demos/python.html
```

在项目下创建新包 utils，然后在里面创建文件 user_op.py，用于创建发送短信的方法。代码如下：

```
def send_sms(mobile, code):
```

```
    """
    调用短信服务商 API 发送短信逻辑
    :param mobile:
    :param code:
    :return:
    """
    print('\n\n【生鲜电商】你的验证码为：{}\n\n'.format(code))
    return {'status_code': 0, 'msg': '短信发送成功'}
```

(4) 创建序列化类。在 users 应用下创建文件 serializers.py，在里面创建 3 个序列化类：验证手机号的序列化类 VerifyCodeSerializer，用户测试序列化类 UserSerializer，用户详情序列化类 UserDetailSerializer。创建序列化类的具体原理和 Django 中的 Form 几乎完全一样。文件 serializers.py 的代码如下：

```
class VerifyCodeSerializer(serializers.Serializer):
    """"
    不用 ModelSerializer 原因：发送验证码只需要提交手机号码
    """
    mobile = serializers.CharField(max_length=11, help_text='手机号码',
    label='手机号码')

    def validate_mobile(self, mobile):
        """
        验证手机号码
        :param mobile:
        :return:
        """
        # 是否已注册
        if User.objects.filter(mobile=mobile):
            raise serializers.ValidationError('用户已存在')

        # 正则验证手机号码
        regexp = "^[1][3,4,5,7,8][0-9]{9}$"
        if not re.match(regexp, mobile):
            raise serializers.ValidationError('手机号码不正确')

        # 验证发送频率
        one_minute_ago = now() - timedelta(hours=0, minutes=1, seconds=0)
        # 获取一分钟以前的时间
        # print(one_minute_ago)
        if VerifyCode.objects.filter(add_time__gt=one_minute_ago, mobile=mobile):
            # 如果添加时间大于一分钟以前的时间，则在这一分钟内已经发过短信，不允许再次发送
            raise serializers.ValidationError('距离上次发送未超过 60s')

        return mobile

class UserSerializer(serializers.ModelSerializer):
   code = serializers.CharField(required=True,
                        min_length=4,
                        max_length=4,
                        help_text='验证码',
                        label='验证码',
```

```
                            write_only=True,
                            # 更新或创建实例时可以使用该字段，但序列化时不包含该字段
                            error_messages={
                                'blank': '请输入验证码',
                                'required': '该字段必填项',
                                'min_length': '验证码格式不正确',
                                'max_length': '验证码格式不正确',
                            })
username = serializers.CharField(required=True,
                                allow_blank=False,
                                help_text='用户名',
                                label='用户名',
                                validators=[UniqueValidator(queryset=User.
                                        objects.all(), message='用户已存在')])
password = serializers.CharField(required=True,
                                help_text='密码',
                                label='密码',
                                write_only=True,
                                style={'input_type': 'password'})

def validate_code(self, code):
    # 验证 code
    # self.initial_data 为用户前端传过来的所有值
    verify_codes = VerifyCode.objects.filter(mobile=self.initial_data
                ['username']).order_by('-add_time')
    if verify_codes:
        last_record = verify_codes[0]

        # 发送验证码如果超过某个时间就提示过期
        three_minute_ago = now() - timedelta(hours=0, minutes=3, seconds=0)
        # 获取三分钟以前的时间
        if last_record.add_time < three_minute_ago:
            raise serializers.ValidationError('验证码已过期')

        # 比较传入的验证码
        if last_record.code != code:
            raise serializers.ValidationError('验证码输入错误')
        # return code
        # 没必要 return，因为 code 这个字段只是用来验证的，不是用来保存到数据库中的

    else:
        # 没有查到该手机号对应的验证码
        raise serializers.ValidationError('验证码错误')

def validate(self, attrs):
    """
    code 这个字段是不需要保存到数据库中的，不需要改字段
    validate 这个函数作用于所有的字段之上
    :param attrs: 每个字段 validate 之后返回的一个总的 dict
    :return:
    """
```

```
        attrs['mobile'] = attrs['username']
        # mobile 不需要前端传过来，就直接后台取 username 中的值填充
        del attrs['code']  # 删除不需要的 code 字段
        return attrs

    class Meta:
        model = User
        fields = ('username', 'mobile', 'code', 'password')
        # username 是 Django 自带的字段，与 mobile 的值保持一致

class UserDetailSerializer(serializers.ModelSerializer):
    """
    用户详情序列化类
    """

    class Meta:
        model = User
        fields = ('username', 'name', 'email', 'birthday', 'mobile', 'gender')
```

(5) 在视图文件 users/views.py 中创建发送验证码视图，代码如下：

```
class SendSmsCodeViewSet(mixins.CreateModelMixin, viewsets.GenericViewSet):
    """
    发送短信验证码
    """
    serializer_class = VerifyCodeSerializer

    def generate_code(self):
        # 定义一个种子，从这里面随机拿出一个值，可以是字母
        seeds = "1234567890"
        # 定义一个空列表，每次循环，将拿到的值，加入列表
        random_str = []
        # choice 函数：每次从 seeds 拿一个值，加入列表
        for i in range(4):
            # 将列表里的值，变成四位字符串
            random_str.append(choice(seeds))
        return ''.join(random_str)

    # 直接复制 CreateModelMixin 中的 create 方法进行重写
    def create(self, request, *args, **kwargs):
        serializer = self.get_serializer(data=request.data)
        serializer.is_valid(raise_exception=True)
        # raise_exception=True 表示 is_valid 验证失败，就直接抛出异常，
        # 被 drf 捕捉到，直接会返回 400 错误，不会往下执行

        mobile = serializer.validated_data['mobile']
       #直接取 mobile，上方无异常，那么 mobile 字段肯定是有的

        # 生成验证码
        code = self.generate_code()
        sendsms = send_sms(mobile=mobile, code=code)  # 模拟发送短信

        if sendsms.get('status_code') != 0:
            return Response({
                'mobile': sendsms['msg']
```

```
            }, status=status.HTTP_400_BAD_REQUEST)
        else:
            # 在短信发送成功之后保存验证码
            code_record = VerifyCode(mobile=mobile, code=code)
            code_record.save()

            return Response({
                'mobile': mobile
            }, status=status.HTTP_201_CREATED)  # 可以创建成功代码为 201

        # 以下就不需要了
        # self.perform_create(serializer)
        # headers = self.get_success_headers(serializer.data)
        # return Response(serializer.data, status=status.HTTP_201_CREATED,
                    headers=headers)
```

在网页中访问 http://127.0.0.1:8000/code/，测试发送验证码接口，如图 6-13 所示。

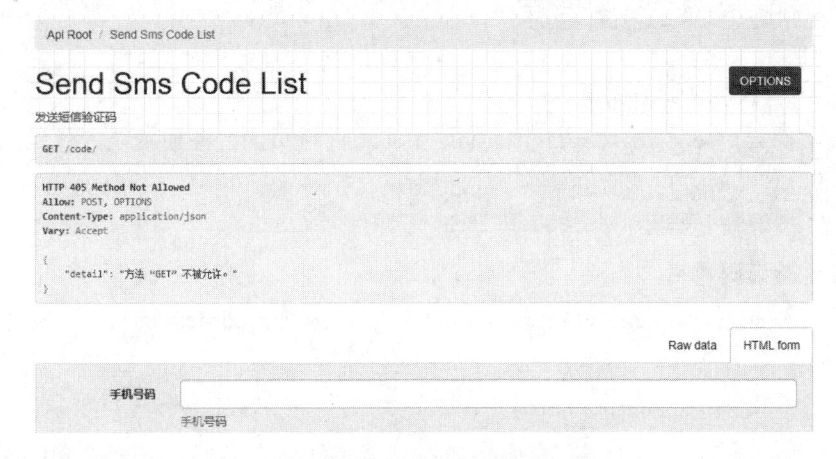


图 6-13　发送手机短信验证码

如果输入了一个不正确的手机号码，则会返回错误提示，如图 6-14 所示。

Api Root / Send Sms Code List

Send Sms Code List

OPTIONS

发送短信验证码

```
POST /code/

HTTP 400 Bad Request
Allow: POST, OPTIONS
Content-Type: application/json
Vary: Accept

{
    "mobile": [
        "手机号码不正确"
    ]
}
```

Raw data　HTML form

手机号码　123

手机号码不正确

手机号码

图 6-14　输入了一个不正确的手机号码

(6) 密码加密保存。在序列化文件 serializers.py 中，通过重载 create()方法实现密码加密功能，代码如下：

```
def create(self, validated_data):
    user = super(UserSerializer, self).create(validated_data=validated_data)
    # user 对象是 Django 中继承的 AbstractUser
    # UserProfile-->AbstractUser-->AbstractBase
      User 中有个 set_password(self,
     raw_password) 方法
    user.set_password(validated_data['password'])
    # 取出 password 密码，进行加密后保存
    user.save()
    # ModelSerializer 有一个 save() 方法，save() 里面会调用 create() 函数，
    # 这儿重载了 create() 函数，加入加密的逻辑
    return user
```

## 6.6.4 注册会员和退出登录

(1) 在 Vue 中获取验证码逻辑。

在前端文件 src/views/register/register.vue 中实现前端页面的注册功能，代码如下：

```
<input class="verify-code-btn sendcode" type="button" id="jsSendCode"
@click="seedMessage" :value="getMessageText">
```

(2) 发送验证码逻辑。

在前端文件 src/views/register/register.vue 中，通过函数 seedMessage()实现发送验证码功能，代码如下：

```
        seedMessage() {
            var that = this;
            //开启倒计时
            var countdown = 60;
            settime();

            function settime() {
                if (countdown === 0) {
                    that.getMessageText = "免费获取验证码";
                    countdown = 60;
                    return;
                } else {
                    that.getMessageText = "重新发送(" + countdown + ")";
                    countdown--;
                }
                setTimeout(function () {
                    settime()
                }, 1000)
            }

            getMessage({
                mobile: that.mobile
            }).then((response) => {
                console.log();
```

```
        })
            .catch(function (error) {
                that.error.mobile = error.mobile ? error.mobile[0] : '';
            });
    }
```

(3) Vue 的注册逻辑。

在前端文件 src/views/register/register.vue 中，实现注册功能的代码如下：

```
<input class="btn btn-green" id="jsMobileRegBtn" @click="isRegister"
type="button" value="注册并登录">
```

在前端文件 src/views/register/register.vue 中，通过函数 isRegister()实现会员注册功能，代码如下：

```
        methods: {
            isRegister() {
                var that = this;
                register({
                    password: that.password,
                    username: that.mobile,
                    code: that.code,
                }).then((response) => {
                    cookie.setCookie('name', response.data.username, 7);
                    cookie.setCookie('token', response.data.access, 7);
                    //存储在 store 中
                    // 更新 store 数据
                    that.$store.dispatch('setInfo');
                    //跳转到首页页面
                    this.$router.push({name: 'index'})

                })
                    .catch(function (error) {
                        that.error.mobile = error.username ? error.username[0] : '';
                        that.error.password = error.password ? error.password[0] : '';
                        that.error.username = error.mobile ? error.mobile[0] : '';
                        that.error.code = error.code ? error.code[0] : '';
                    });
            },
            seedMessage() {
                var that = this;
                //开启倒计时
                var countdown = 60;
                settime();

                function settime() {
                    if (countdown === 0) {
                        that.getMessageText = "免费获取验证码";
                        countdown = 60;
                        return;
                    } else {
                        that.getMessageText = "重新发送(" + countdown + ")";
                        countdown--;
```

```
            }
            setTimeout(function () {
               settime()
            }, 1000)
         }

         getMessage({
            mobile: that.mobile
         }).then((response) => {
            console.log();

         })
            .catch(function (error) {
               that.error.mobile = error.mobile ? error.mobile[0] : '';
            });

      }
   }
```

在注册过程中，传递的参数有password、username和code。有如下两种注册模式。

① 第一种：注册完成，自己跳转到登录页面登录。此时可以注释掉上方的保存cookie的逻辑。这种登录很简单，代码如下：

```
                //cookie.setCookie('name', response.data.username, 7);
                //cookie.setCookie('token', response.data.access, 7);
                //存储在store中
                //更新store数据
                //that.$store.dispatch('setInfo');
```

这样就直接跳回首页，或者也可以跳回登录页面。

② 第二种：注册完成后就直接登录。假如注册完成后，需要自己手动登录，此时需要后端返回一个Token。配置注册的url，按住Ctrl键，单击注册就可以跳转到src/api/api.js，修改为后端注册的url即可：

```
//注册
export const register = parmas => {
   return axios.post('${local_host}/users/', parmas)
};
```

如果使用第二种方法，但是在后端并没有写返回Token的接口，就需要在注册视图UserViewSet(mixins.CreateModelMixin, viewsets.GenericViewSet)中重载mixins.CreateModelMixin的create(self, request, *args, **kwargs)函数。直接将create()函数拷贝到注册视图中，其他逻辑基本不变，代码如下：

```
class UserViewSet(mixins.CreateModelMixin, viewsets.GenericViewSet):
   """
   创建用户
   """
   serializer_class = UserSerializer
   queryset = User.objects.all()

   def create(self, request, *args, **kwargs):
```

```
        serializer = self.get_serializer(data=request.data)
        serializer.is_valid(raise_exception=True)
        self.perform_create(serializer)
        # 添加自己的逻辑，通过 user，生成 token 并返回
        headers = self.get_success_headers(serializer.data)
        return Response(serializer.data, status=status.HTTP_201_CREATED,
                        headers=headers)
```

在生成 Token 之前，self.perform_create(serializer)函数只是调用了 serializer.save()，因为要生成用户的 Token 之前，必须要拿到 user，所以这个函数也需要被重载。原函数中只是调用.save()，并没有返回值。实际上 serializer.save()中的 serializer 就是 UserSerializer (serializers.ModelSerializer)中 model 的对象。重载它并把它返回，才能得到 user 对象。代码如下：

```
class UserViewSet(mixins.CreateModelMixin, viewsets.GenericViewSet):
    """
    创建用户
    """
    serializer_class = UserSerializer
    queryset = User.objects.all()

    def create(self, request, *args, **kwargs):
        serializer = self.get_serializer(data=request.data)
        serializer.is_valid(raise_exception=True)
        user = self.perform_create(serializer)

        # 添加自己的逻辑，生成 token 并返回

        headers = self.get_success_headers(serializer.data)
        return Response(serializer.data, status=status.HTTP_201_CREATED,
                        headers=headers)

    def perform_create(self, serializer):
        return serializer.save()
```

例如手机注册界面的效果如图 6-15 所示。

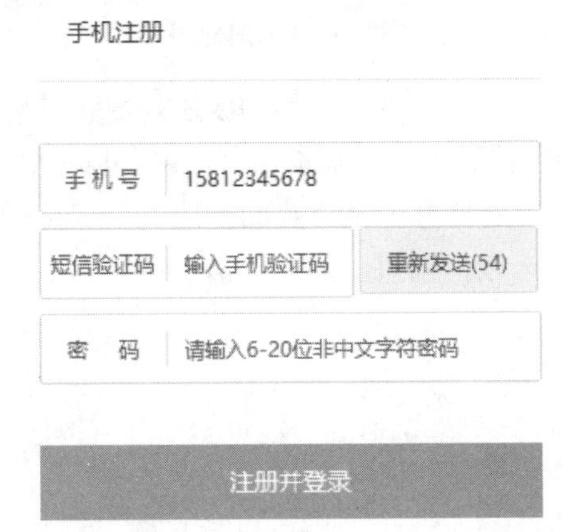


图 6-15　手机注册界面的效果

(4) Vue 中 logout 功能逻辑。

JWT 的 Token 并不是保存在服务器端的，而是保存在客户端。在前端文件 src/views/head/shophead.vue 中，实现退出功能的代码如下：

```
<a @click="loginOut">退出</a>
```

在前端文件 src/views/head/shophead.vue 中，通过函数 loginOut()实现会员退出功能，代码如下：

```
    methods: {
        loginOut() {
            cookie.delCookie('token');
            cookie.delCookie('name');
            //重新触发 store
            //更新 store 数据
            this.$store.dispatch('setInfo');
            //跳转到登录
            this.$router.push({name: 'login'})
        },
    }
}
```

## 6.6.5 微博账户登录

(1) 搜索微博开放平台，或者直接登录 https://open.weibo.com/connect，注册个人信息。

(2) 在注册微博开放平台后创建应用，如图 6-16 所示。

图 6-16 创建应用

(3) 在“我的应用”中可以看到刚刚创建的应用，测试的时候不用通过审核也可以运行。在“我的应用”→“应用信息”→“基本信息”中可以看到 App Key、App Secret 等信息，这是在我们程序中要使用的信息。在“我的应用”→“应用信息”→“高级信息”里面填写授权回调页的 URL，可以根据自己的实际情况填写，如图 6-17 所示。

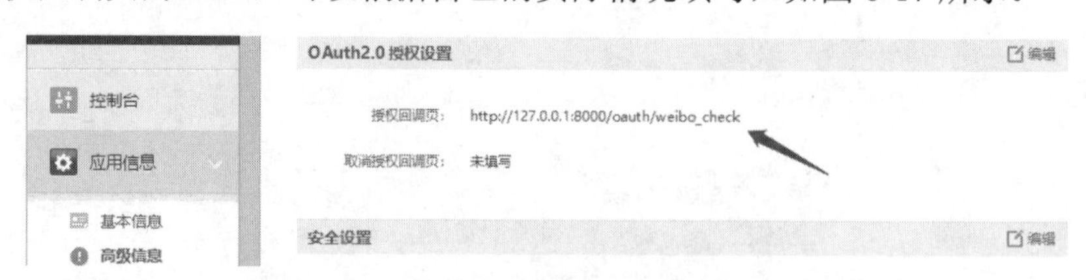


图 6-17 填写授权回调页的 URL

(4) 在“我的应用”→“应用信息”→“测试信息”中将自己的账号添加到测试账号中，未审核的应用只能通过测试账号登录。

(5) 微博授权登录。可以访问如下地址查看官方的文档：

```
https://open.weibo.com/wiki/授权机制
```

常用接口文档的具体说明如下。

- OAuth2/authorize：请求用户授权 Token。
- OAuth2/access_token：获取授权过的 Access Token。
- OAuth2/get_token_info：授权信息查询接口。
- OAuth2/revokeoauth2：授权回收接口。
- OAuth2/get_oauth2_token：OAuth 1.0 的 Access Token 更换至 OAuth 2.0 的 Access Token。

官方给出的各个请求参数的具体说明如表 6-1 所示。

表 6-1　请求参数的具体说明

| 参　数 | 必选 | 类型及范围 | 说　明 |
| --- | --- | --- | --- |
| client_id | true | string | 申请应用时分配的 AppKey |
| redirect_uri | true | string | 授权回调地址，站外应用需与设置的回调地址一致，站内应用需填写 canvas page 的地址 |
| scope | false | string | 申请 scope 权限所需参数，可一次申请多个 scope 权限，用逗号分隔 |
| state | false | string | 用于保持请求和回调的状态，在回调时，会在 Query Parameter 中回传该参数。开发者可以用这个参数验证请求有效性，也可以记录用户请求授权页前的位置。这个参数可用于防止跨站请求伪造(CSRF)攻击 |

返回参数的具体说明如表 6-2 所示。

表 6-2　返回参数的具体说明

| 返回值字段 | 字段类型 | 字段说明 |
| --- | --- | --- |
| code | string | 用于第二步调用 oauth2/access_token 接口，获取授权后的 access token |
| state | string | 如果传递参数，会回传该参数 |

例如下面是请求示例：

```
//请求
https://api.weibo.com/oauth2/authorize?client\_id=123050457758183&redirect\
_uri=http://www.example.com/response&response_type=code

//同意授权后会重定向
http://www.example.com/response&code=CODE
```

在 apps/users 中创建文件 oauth_weibo.py 用于测试微博登录功能，主要实现代码如下所示：

```
class OAuth_Weibo(object):
    def __init__(self, client_id, client_secret, redirect_uri, state):
        self.client_id = client_id  # 申请应用时分配的AppKey
        self.client_secret = client_secret  # 申请的密钥
        self.redirect_uri = redirect_uri
        # 授权回调地址，站外应用需与设置的回调地址一致，站内应用需填写canvas page的地址。
        self.state = state  # 防跨域攻击，随机码
        self.access_token = ''  # 获取到的token，初始化为空
        self.uid = ''  # 记录用户id

    def get_auth_url(self):
        """
        登录时，获取认证的url，跳转到该url进行github认证
        :return: https://api.weibo.com/oauth2/authorize?client_id=********
&redirect_uri=http://1270.0.1:8000/oauth/weibo_check&state=******
        """
        auth_url = "https://api.weibo.com/oauth2/authorize"

        params = {
            'client_id': self.client_id,  # 申请应用时分配的AppKey
            'redirect_uri': self.redirect_uri,
            # 授权回调地址，站外应用需与设置的回调地址一致，站内应用需填写canvas page的地址
            'state': self.state  # 不可猜测的随机字符串。用于防止跨站点请求伪造攻击
        }

        url = "{}?{}".format(auth_url, urlencode(params))
        # urlencode将字典拼接成url参数
        # print(url)
        return url

    def get_access_token(self, code):
        """
        认证通过后，生成code，放在url中，视图中获取这个code，调用该函数，post提交请求
token，最终得到token
        :param code: get_auth_url这一步中认证通过后，跳转回来的url中的code，10分钟
过期。
        :return:
        """
        access_token_url = 'https://api.weibo.com/oauth2/access_token'

        data = {
            'client_id': self.client_id,  # 申请应用时分配的AppKey
            'client_secret': self.client_secret,  # 申请应用时分配的AppSecret
            'grant_type': 'authorization_code',
            # 请求的类型，填写authorization_code
            'code': code,
            # 调用authorize获得的code值，请求get_auth_url的地址返回的值
            'redirect_uri': self.redirect_uri,
            # 授回调地址，需与注册应用里的回调地址一致
        }

        r = requests.post(access_token_url, data=data)
        res = json.loads(r.text)
```

```
        if 'error' in res:
            # token 错误
            print(res['error_description'])
        else:
            self.access_token = res['access_token']
            self.uid = res['uid']

        return self.access_token

    def get_user_info(self):
        """
        根据 token 和 uid 获取用户信息
        :return:
        """
        user_info_url = 'https://api.weibo.com/2/users/show.json'

        params = {'access_token': self.access_token, 'uid': self.uid}
        # 根据 token 获取用户信息

        r = requests.get(user_info_url, params=params)
        # print(r.json())
        return r.json()


if __name__ == '__main__':
    app_key = 'xxxxxx'          # 申请的 key
    app_secret = 'xxxxx'        # 申请的密钥
    redirect_uri = 'http://127.0.0.1:8000/oauth/weibo_check'
    state = 'hj*&(hkjhfs76^hJHKULKG89798we'

    oauth = OAuth_Weibo(client_id=app_key, client_secret=app_secret,
                redirect_uri=redirect_uri, state=state)
    auth_url = oauth.get_auth_url()
    print(auth_url)
    # 访问该页面进行授权认证
    ''' 完成后跳回本地 URL: http://127.0.0.1:8000/oauth/weibo_check?state=hj%2A%
26%28hkjhfs76%5EhJHKULKG89798we&code=3b108579d3f025811ca22e79d4b62bed'''

    # 本地获取到 url 的参数值，判断 return_state 是否和以前的 state 相等
    return_state = 'hj%2A%26%28hkjhfs76%5EhJHKULKG89798we'
    return_code = '3b108579d3f025811ca22e79d4b62bed'
    # 使用 code 来获取 token，code 只能使用一次，使用后失效

    access_token = oauth.get_access_token(code=return_code)
    # 将 token 和 uid 保存在类中
    print(access_token)

    user_info = oauth.get_user_info()
    # 根据 token 和 uid 获取用户信息，最终用户把这些信息保存在 session 中
    print(user_info)
    # 通过 user_info['name'] 获取该用户昵称
```

## 6.6.6 social-app-django 集成第三方登录

Python Social Auth 是一种易于设置的社交认证/注册机制，这时 python-social-auth 生态系统的 Django 组件，实现了在基于 Django 的项目中集成 social-auth-core 所需的功能。django-social-auth 本身是来自 django-twitter-oauth 和 django-openid-auth 项目的修改代码的产物。

(1) 本项目使用 social-auth-app-django 集成了第三方登录功能，首先需要使用下面的命令安装 social-auth-app-django：

```
pip install social-auth-app-django
```

(2) 然后将 social_django 添加到 INSTALLED_APPS 中：

```
INSTALLED_APPS = [
    # 添加 Django 联合登录
    'social_django',
]
```

(3) 将 social_django 添加到已安装的应用程序后，需要同步数据库以创建所需的模型：

```
python manage.py migrate
```

执行上述命令后会在数据库中生成新表，如图 6-18 所示。

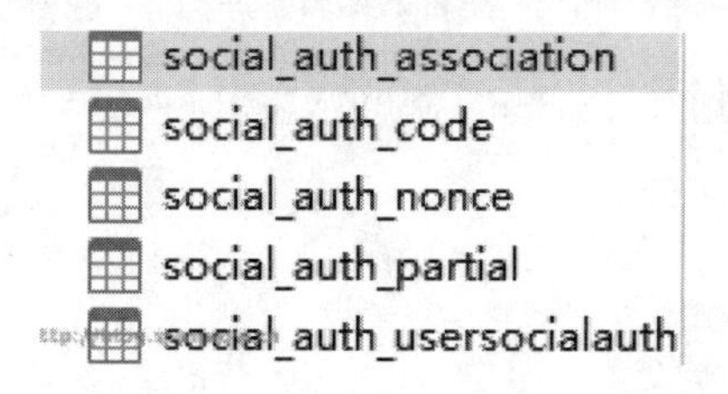


图 6-18 生成的新表

(4) 如果使用 django.contrib.auth 应用程序，不要忘记添加 django.contrib.auth.backends.ModelBackend，否则用户将无法使用“用户名/密码”方法登录。在上述代码中，支持的认证类有 OpenIdAuth、GoogleOAuth2 等，想要查看其他支持的类型，可以访问安装库的位置 *PATH*/site-packages/social_core/backends。例如要使用微博登录本系统，打开文件 apps_extend\social_core\backends\weibo.py，它的源代码如下：

```
from .oauth import BaseOAuth2

class WeiboOAuth2(BaseOAuth2):
    """Weibo (of sina) OAuth authentication backend"""
    name = 'weibo'
    ID_KEY = 'uid'
    AUTHORIZATION_URL = 'https://api.weibo.com/oauth2/authorize'
    REQUEST_TOKEN_URL = 'https://api.weibo.com/oauth2/request_token'
    ACCESS_TOKEN_URL = 'https://api.weibo.com/oauth2/access_token'
    ACCESS_TOKEN_METHOD = 'POST'
    REDIRECT_STATE = False
```

```
    EXTRA_DATA = [
        ('id', 'id'),
        ('name', 'username'),
        ('profile_image_url', 'profile_image_url'),
        ('gender', 'gender')
    ]

    def get_user_details(self, response):
        """Return user details from Weibo. API URL is:
        https://api.weibo.com/2/users/show.json/?uid=<UID>&access_token=<TOKEN>
        """
        if self.setting('DOMAIN_AS_USERNAME'):
            username = response.get('domain', '')
        else:
            username = response.get('name', '')
        fullname, first_name, last_name = self.get_user_names(
            first_name=response.get('screen_name', '')
        )
        return {'username': username,
                'fullname': fullname,
                'first_name': first_name,
                'last_name': last_name}

    def get_uid(self, access_token):
        """Return uid by access_token"""
        data = self.get_json(
            'https://api.weibo.com/oauth2/get_token_info',
            method='POST',
            params={'access_token': access_token}
        )
        return data['uid']

    def user_data(self, access_token, response=None, *args, **kwargs):
        """Return user data"""
        # If user id was not retrieved in the response, then get it directly
        # from weibo get_token_info endpoint
        uid = response and response.get('uid') or self.get_uid(access_token)
        user_data = self.get_json(
            'https://api.weibo.com/2/users/show.json',
            params={'access_token': access_token, 'uid': uid}
        )
        user_data['uid'] = uid
        return user_data
```

打开 Django 项目的配置文件 settings.py，将所需的身份验证后端信息添加到 AUTHENTICATION_BACKENDS 中：

```
AUTHENTICATION_BACKENDS = (
    'users.views.CustomBackend',  # 自定义认证后端
    'social_core.backends.weibo.WeiboOAuth2',  # 微博认证后端
    'social_core.backends.qq.QQOAuth2',  # QQ 认证后端
    'social_core.backends.weixin.WeixinOAuth2',  # 微信认证后端
    'django.contrib.auth.backends.ModelBackend',
    # 使用了`django.contrib.auth 应用程序，支持账密认证
)  # 设置认证后台
```

(5) 添加认证 URL。修改文件 DjangoOnlineFreshSupermarket/urls.py 中的 urlpatterns，添加以下路由：

```
urlpatterns = [
    # ...省略其他 path

    # social_django 认证登录
    path('', include('social_django.urls', namespace='social'))
]
```

如果需要自定义命名空间，还需要在文件 settings.py 中增加以下设置：

```
SOCIAL_AUTH_URL_NAMESPACE = 'social'
```

(6) 使用 Key 和 Secret 配置 OAuth。在 social_django 的文件 PATH/site-packages/social_django/urls.py 中，回调 url 是：

```
    url(r'^complete/(?P<backend>[^/]+){0}$'.format(extra), views.complete,
        name='complete'),
```

所以需要将微博认证的授权回调页设置为 http://127.0.0.1:8000/complete/weibo/，和本机的 IP 地址保持一致。如果是部署到网络服务器，就需要用服务器的 IP，如图 6-19 所示。

图 6-19　授权回调页

(7) 填写密钥。在后端配置文件 settings.py 中填写密钥信息，代码如下：

```
# social_django 配置 OAuth keys，项目上传完，将涉密信息保存在配置文件中
import configparser
config = configparser.ConfigParser()
config.read(os.path.join(BASE_DIR, 'ProjectConfig.ini'))
weibo_key = config['DjangoOnlineFreshSupermarket']['weibo_key']
weibo_secret = config['DjangoOnlineFreshSupermarket']['weibo_secret']
SOCIAL_AUTH_WEIBO_KEY = weibo_key
SOCIAL_AUTH_WEIBO_SECRET = weibo_secret

SOCIAL_AUTH_LOGIN_REDIRECT_URL = '/index/'  # 登录成功后跳转，一般为项目首页
```

## 6.7　支付宝支付

扫码观看视频讲解

对于一个在线商城系统来说，在线支付功能十分重要。从当前技术条件和使用频率来看，常用的在线支付手段有支付宝、微信和网银等。本节将详细讲解在本系统中添加支付宝支付功能的方法。

## 6.7.1　配置支付宝的沙箱环境

(1) 访问支付宝开放平台主页 https://open.alipay.com/platform/home.htm，登录后进入“账户中心”页面。第一次登录需要填写个人信息，如图 6-20 所示。

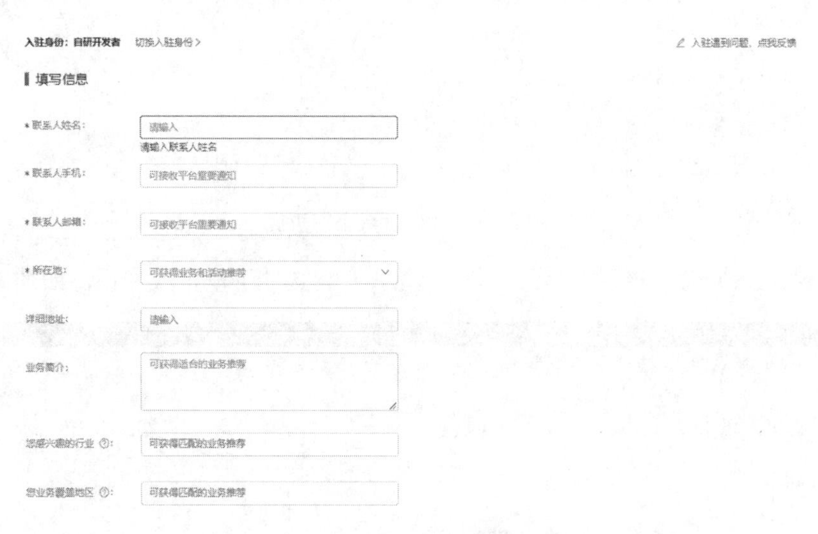


图 6-20　填写开发者信息

(2) 只有企业支付宝账户才能创建应用程序，个人账户则不行。但是个人账户可以在沙箱中进行调试，这个应用是自动生成的。填写基本信息完成后，会在“账户中心”显示自己的信息，如图 6-21 所示。

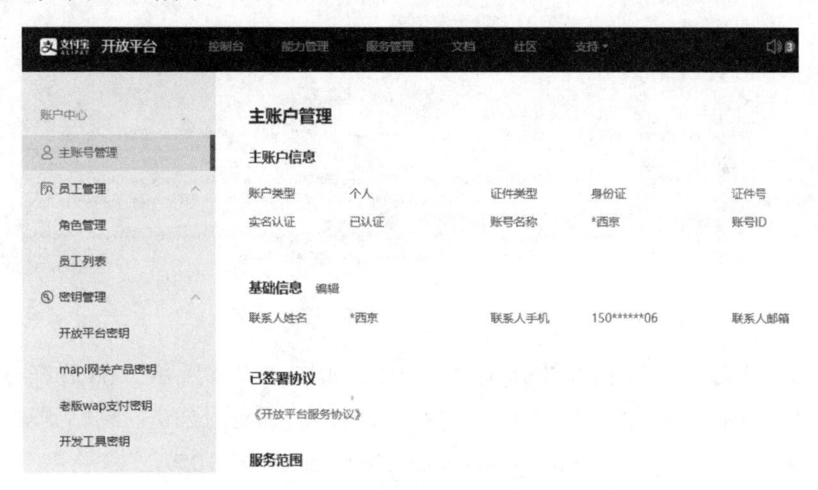


图 6-21　在“账户中心”显示自己的信息

(3) 登录“研发服务”页面 https://openhome.alipay.com/platform/developerIndex.htm，如图 6-22 所示。在这个页面中，可以设置密钥和公钥。记住，这是沙箱配置页面，是为不是商户的开发者模拟商户身份的，这样可以实现模拟收款功能。

图 6-22 “研发服务”页面

(4) 下载 RSA 签名验签工具，运行里面的 .bat 文件可以生成密钥，如图 6-23 所示。

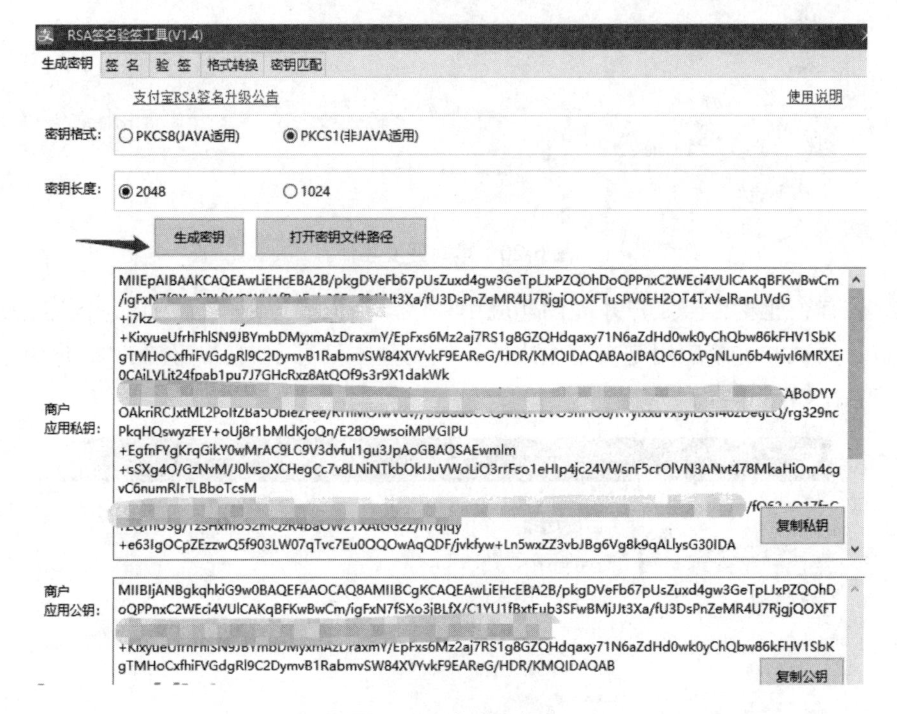


图 6-23 生成密钥

生成的密钥文件被保存在本地计算机中，如图 6-24 所示。然后在沙箱页面中“RSA2(SHA256)密钥(推荐)”栏单击“设置”按钮，复制应用公钥，粘贴到输入框中，具体方法请参看最新的官方文档 https://docs.open.alipay.com/291/105972/。

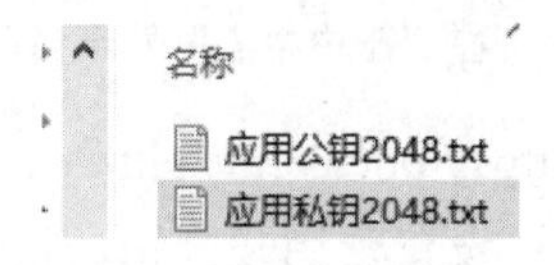


图 6-24 生成的密钥文件

**注 意**

生成的私钥必须妥善保管，避免遗失，不要泄露。在项目中需要将私钥填写到代码中，供签名时使用。公钥需提供给支付宝账号管理者并上传到支付宝开放平台。

(5) 将上面生成的密钥保存到项目目录中，在本项目中保存在文件 apps/trade/keys 下的两个记事本文件中，如图 6-25 所示。应用私钥和支付宝公钥特别重要，应用公钥的用处不大。请求支付宝接口时，用应用私钥来进行加密，支付宝用我们上传的应用公钥对签名做验证；支付宝公钥用来查询订单状态。

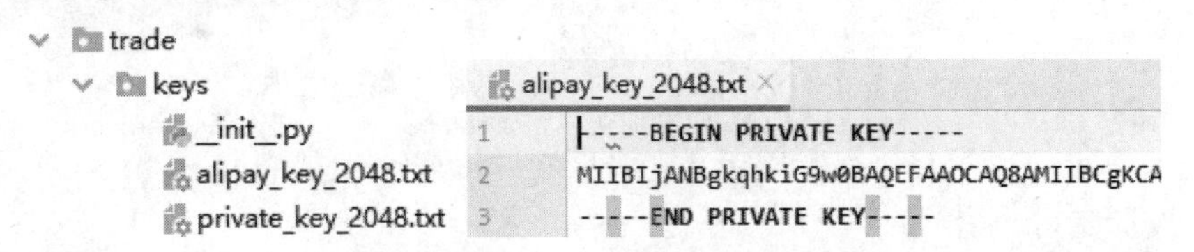


图 6-25　将密钥文件保存到项目目录中

## 6.7.2　编写程序

(1) 支付接口类。

编写文件 utils/alipay.py，实现一个支付接口类，在此文件中用到了上面申请的密钥。文件 alipay.py 的主要实现代码如下所示：

```
class AliPay(object):
    """
    支付宝支付接口
    """

    def __init__(self, app_id, notify_url, app_private_key_path, alipay_public
                _key_path, return_url, debug=True):
        self.app_id = app_id  # 支付宝分配的应用 ID
        self.notify_url = notify_url
        # 支付宝服务器主动通知商户服务器里指定的页面 http/https 路径；用户一旦支付，会向该
          url 发一个异步的请求给自己服务器，这一定需要公网可访问
        self.app_private_key_path = app_private_key_path  # 个人私钥路径
        self.app_private_key = None  # 个人私钥内容
        self.return_url = return_url  # 网页上支付完成后跳转回自己服务器的 url
        with open(self.app_private_key_path) as fp:
            # 读取个人私钥文件，提取到私钥内容
            self.app_private_key = RSA.importKey(fp.read())

        self.alipay_public_key_path = alipay_public_key_path
        with open(self.alipay_public_key_path) as fp:
            # 读取支付宝公钥文件，提取公钥内容，支付宝公钥在代码中验签使用
            self.alipay_public_key = RSA.import_key(fp.read())

        if debug is True:
            # 使用沙箱的网关
            self.__gateway = "https://openapi.alipaydev.com/gateway.do"
```

```
        else:
            self.__gateway = "https://openapi.alipay.com/gateway.do"

    def direct_pay(self, subject, out_trade_no, total_amount, return_url=None,
                   **kwargs):
        biz_content = {  # 请求参数的集合
            "subject": subject,  # 订单标题
            "out_trade_no": out_trade_no,  # 商户订单号,
            "total_amount": total_amount,  # 订单总金额
            "product_code": "FAST_INSTANT_TRADE_PAY",  # 销售产品码，默认
            # "qr_pay_mode":4
        }

        biz_content.update(kwargs)  # 合并其他请求参数字典
        data = self.build_body("alipay.trade.page.pay", biz_content, return_url)
        # 将请求参数合并到公共参数字典的键 biz_content 中
        return self.sign_data(data)

    def build_body(self, method, biz_content, return_url=None):
        """
        组合所有的请求参数到一个字典中
        :param method:
        :param biz_content:
        :param return_url:
        :return:
        """
        data = {
            "app_id": self.app_id,
            "method": method,
            "charset": "utf-8",
            "sign_type": "RSA2",
            "timestamp": datetime.now().strftime("%Y-%m-%d %H:%M:%S"),
            "version": "1.0",
            "biz_content": biz_content
        }

        if return_url is None:
            data["notify_url"] = self.notify_url
            data["return_url"] = self.return_url

        return data

    def ordered_data(self, data):
        """
        按照第一个字符的键值 ASCII 码递增排序(字母升序排序)，如果遇到相同字符则按照第二个字
符的键值 ASCII 码递增排序，以此类推
        :param data:
        :return: 返回的是数组列表，按照数据中的 k 进行排序的
        """
        complex_keys = []
        for key, value in data.items():
            if isinstance(value, dict):
                complex_keys.append(key)

        # 将字典类型的数据 dump 出来
```

```
        for key in complex_keys:
            data[key] = json.dumps(data[key], separators=(',', ':'))

        return sorted([(k, v) for k, v in data.items()])

    def sign(self, unsigned_string):
        """
        使用各自语言对应的 SHA256WithRSA(对应 sign_type 为 RSA2)或 SHA1WithRSA(对应
sign_type 为 RSA)签名函数利用商户私钥对待签名字符串进行签名，并进行 Base64 编码
        :param unsigned_string:
        :return:
        """
        # 开始计算签名
        key = self.app_private_key
        signer = PKCS1_v1_5.new(key)
        signature = signer.sign(SHA256.new(unsigned_string))
        # base64 编码，转换为 unicode 表示并移除回车
        sign = encodebytes(signature).decode("utf8").replace("\n", "")
        return sign

    def sign_data(self, data):
        """
        获取所有请求参数，不包括字节类型参数，如文件、字节流，剔除 sign 字段，剔除值为空的参数。
        进行排序。
        将排序后的参数与其对应值，组合成“参数=参数值”的格式，并且把这些参数用&字符连接起来，
此时生成的字符串为待签名字符串。
        然后对该字符串进行签名。
        把生成的签名赋值给 sign 参数，拼接到请求参数中。
        :param data:
        :return:
        """
        data.pop("sign", None)
        # 排序后的字符串
        ordered_items = self.ordered_data(data)  # 数组列表，进行遍历拼接
        unsigned_string = "&".join("{0}={1}".format(k, v) for k, v in
ordered_items)  # 使用“参数=值”的格式，在各个“参数=值”之间用&连接

        sign = self.sign(unsigned_string.encode("utf-8"))  # 得到签名后的字符串
        quoted_string = "&".join("{0}={1}".format(k, quote_plus(v)) for k, v in
            ordered_items)  # quote_plus 给 url 进行预处理，特殊字符串在 url 中会有问题

        # 获得最终的订单信息字符串
        signed_string = quoted_string + "&sign=" + quote_plus(sign)
        return signed_string

    def _verify(self, raw_content, signature):
        # 开始计算签名
        key = self.alipay_public_key
        signer = PKCS1_v1_5.new(key)
        digest = SHA256.new()
        digest.update(raw_content.encode("utf8"))
        if signer.verify(digest, decodebytes(signature.encode("utf8"))):
            return True
        return False
```

```
    def verify(self, data, signature):
        if "sign_type" in data:
            sign_type = data.pop("sign_type")
        # 排序后的字符串
        unsigned_items = self.ordered_data(data)
        message = "&".join(u"{}={}".format(k, v) for k, v in unsigned_items)
        return self._verify(message, signature)
```

(2) 配置项目的服务器 IP。

在本项目下创建配置文件 ProjectConfig.ini，主要用来设置在项目中用到的涉密信息，比如连接数据库账密，服务器信息等。配置结构如下：

```
 [DjangoOnlineFreshSupermarket]
server_ip=xx.ip.ip.xx        # 服务器的 IP 地址
```

然后在接口类文件 utils/alipay.py 中创建函数 get_server_ip()，用于获取这个服务器 IP 信息。代码如下：

```
def get_server_ip():
    """
    在项目根目录下创建 ProjectConfig.ini 配置文件，读取其中配置的 IP 地址
    :return:
    """
    import configparser
    import os
    import sys

    # 获取当前文件的路径(运行脚本)
    pwd = os.path.dirname(os.path.realpath(__file__))

    # 获取项目的根目录
    sys.path.append(pwd + "../")

    # 要想单独使用 django 的 model，必须指定一个环境变量，去 settings 配置中查找
    # 参照 manage.py
    os.environ.setdefault('DJANGO_SETTINGS_MODULE',
'DjangoOnlineFreshSupermarket.settings')

    import django
    django.setup()
    from django.conf import settings

    config = configparser.ConfigParser()
    config.read(os.path.join(settings.BASE_DIR, 'ProjectConfig.ini'))
    server_ip = config['DjangoOnlineFreshSupermarket']['server_ip']
    return server_ip
```

为了保证能够成功获取 IP，在文件 utils/alipay.py 中添加函数 main()进行测试：

```
if __name__ == "__main__":
    print(get_server_ip())
    server_ip = get_server_ip()
    # 得到自己服务器的 IP 地址测试订单创建支付
```

在函数 main()中添加订单支付测试的代码如下：

```
if __name__ == "__main__":
    print(get_server_ip())
    server_ip = get_server_ip()  # 得到自己服务器的 IP 地址
    alipay = AliPay(
        app_id="2016100900646609",  # 自己支付宝沙箱 APP ID
        notify_url="http://{}:8000/".format(server_ip),
        app_private_key_path="../apps/trade/keys/private_key_2048.txt",
        # 可以使用相对路径
        alipay_public_key_path="../apps/trade/keys/alipay_key_2048.txt",
        # 支付宝的公钥，验证支付宝回传消息使用，不是你自己的公钥
        debug=True,  # 默认 False
        return_url="http://{}:8000/".format(server_ip)
    )

    # 创建订单
    url = alipay.direct_pay(
        subject="测试订单",
        out_trade_no="2019080716060001",
        total_amount=0.01
    )
    re_url = "https://openapi.alipaydev.com/gateway.do?{data}".format(data=url)
    print(re_url)
```

直接运行文件 utils/alipay.py，这时候会得到一个支付链接：

```
https://openapi.alipaydev.com/gateway.do?app_id=2016100900646609&biz_conten
t=%7B%22subject%22%3A%22%5Cu6d4b%5Cu8bd5%5Cu8ba2%5Cu5355%22%2C%22out_trade_
no%22%3A%222019080716060001%22%2C%22total_amount%22%3A0.01%2C%22product_cod
e%22%3A%22FAST_INSTANT_TRADE_PAY%22%7D&charset=utf-8&method=alipay.trade.pa
ge.pay&notify_url=http%3A%2F%2Fxx.ip.ip.xx%3A8000%2F&return_url=http%3A%2F%
2Fxx.ip.ip.xx%3A8000%2F&sign_type=RSA2&timestamp=2019-08-07+16%3A09%3A25&ve
rsion=1.0&sign=crNYPmSRAccnEb%2BnvnYqgG6qpp4n5NrOHP4sBLyjNBWws6RWS5JrGntGX%
2FG2SGqf21dIwvUtt5sV5XY%2Bol1dId%2Bn%2BVBykzJShjB4Y0mt%2Bgm498Tv5ecUCUFvOFX
Y%2BpWRu3HiuuiJXxCHHzEZ795sw1x8xSQaKZCTEHCBZsfwexKwE1UKsCWLv1cfgjO3O8rCMziS
ASTMta%2BlfmPcZTdO9tTI9qTXE%2Bq2TMQpZWqBZvN1LPKHdzv1TZL3efjI64qEKglYK6KUCtU
gNJoUBJrmYj4Ao3XZMro06Lu73MTPpheg8v56yBXGe4FyMdpxOvOS2t%2FnRZtyM2cx5io6lUoS
cQ%3D%3D
```

使用浏览器访问这个链接后，可以看到沙箱支付页面，如图 6-26 所示。这就说明我们前面配置的密钥信息和支付接口信息完全正确。

我们可以使用登录账户付款(在“开放平台”→“沙箱环境”→“沙箱账号”中)，也可以下载支付宝沙箱测试应用进行扫码支付。付款成功后的效果如图 6-27 所示。

(3) 修改 Vue 指向线上地址。

由于需要调试支付宝的相关功能，将 Vue 项目指向线上的接口，需要修改 src/api/api.js 中的 local_host 值：

```
//let local_host = 'http://localhost:8000';
let local_host = 'http://xx.ip.ip.xx:8000';  #服务器的 IP 地址
```

启动后端 Django 的线上服务，而不是在本地运行；然后重启 Vue 服务器，再访问 http://127.0.0.1:8080 检查是否能够正常加载接口的数据。

图 6-26　支付宝沙箱生成的订单支付页面

图 6-27　付款成功页面

当用户在购物车中单击“去结算”按钮，会进入 OrderInfoViewSet 逻辑，此时会创建一个订单，然后将当前用户购物车中的商品取出来，放在该订单对应的商品列表中。

(4) 使用 OrderInfoSerializer 序列化生成支付 url。

SerializerMethodField 是一个只读字段，能够通过序列化类中的方法获取其值，可以用来向序列化对象中添加任何类型的数据。序列化器方法应该接收单个参数(除了 self)，该参数是被序列化的对象，能够返回想要包含在序列化对象中的任何内容。

在本项目的序列化文件 apps/trade/serializers.py 中添加一个 alipay_url 字段，代码如下：

```
from utils.alipay import AliPay, get_server_ip
from DjangoOnlineFreshSupermarket.settings import app_id, alipay_debug,
alipay_public_key_path, app_private_key_path


class OrderInfoSerializer(serializers.ModelSerializer):
    user = serializers.HiddenField(
        default=serializers.CurrentUserDefault()
        # 表示 user 为隐藏字段，默认为获取当前登录用户
    )
    order_sn = serializers.CharField(read_only=True)
    # 只能读，不能显示给用户修改，只能后台修改
    trade_no = serializers.CharField(read_only=True)  # 只读
    pay_status = serializers.CharField(read_only=True)  # 只读
    pay_time = serializers.DateTimeField(read_only=True)  # 只读
```

```
alipay_url = serializers.SerializerMethodField()  # 生成支付宝url

def generate_order_sn(self):
    # 当前时间+userid+随机数
    import time
    from random import randint
    order_sn = '{time_str}{user_id}{random_str}'.format(time_str=
              time.strftime ('%Y%m%d%H%M%S'), user_id=self.context
              ['request'].user.id, random_str=randint(10, 99))
    return order_sn

def validate(self, attrs):
    # 数据验证成功后，生成一个订单号
    attrs['order_sn'] = self.generate_order_sn()
    return attrs

def get_alipay_url(self, obj):
    # 方法命名规则为：get_<field_name>
    server_ip = get_server_ip()
    alipay = AliPay(
        app_id=app_id,  # 自己支付宝沙箱 APP ID
        notify_url="http://{}:8000/alipay/return/".format(server_ip),
        app_private_key_path=app_private_key_path,  # 可以使用相对路径
        alipay_public_key_path=alipay_public_key_path,
        # 支付宝的公钥，验证支付宝回传消息使用，不是你自己的公钥，
        debug=alipay_debug,  # 默认False,
        return_url="http://{}:8000/alipay/return/".format(server_ip)
    )
    # 创建订单
    order_sn = obj.order_sn
    order_amount = obj.order_amount
    url = alipay.direct_pay(
        subject="生鲜超市-{}".format(order_sn),
        out_trade_no=order_sn,
        total_amount=order_amount,
        # return_url="http://{}:8000/alipay/return/".format(server_ip)
        # 支付完成后自动跳回该url，可以不填，因为初始化已经加上了
    )
    re_url = "https://openapi.alipaydev.com/gateway.do?{data}".format(data=url)
    return re_url

class Meta:
    model = OrderInfo
    fields = "__all__"
```

上述 alipay_url 字段用于在订单创建时生成支付宝的支付 URL，然后在文件 alipay.py 中编写如下代码设置返回支付 URL。

```
if return_url is None:
    data["notify_url"] = self.notify_url
    data["return_url"] = self.return_url
```

当没有指定 return_url 时，直接使用初始化的 return_url。将上述代码上传到服务器。为了避免出错，在每次修改代码后都需要将新的代码上传到服务器，然后重启 Debug。接下来

访问 http://xx.ip.ip.xx:8000/，验证查看 API 是否正常，然后再访问 http://xx.ip.ip.xx:8000/orderinfo/ 测试创建新订单，如图 6-28 所示。

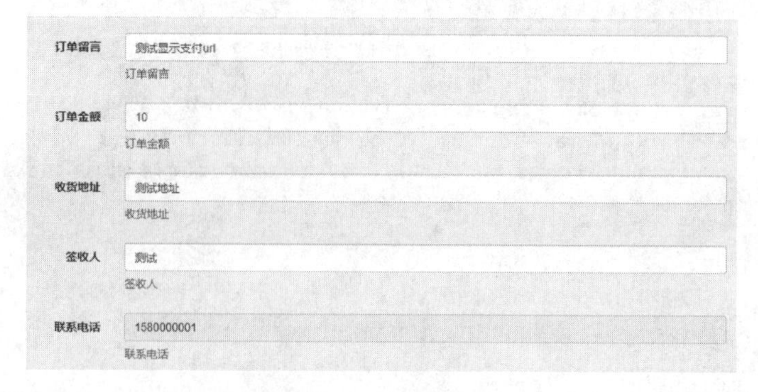


图 6-28　创建测试订单

创建上述测试订单，可以看到返回的值，如图 6-29 所示。

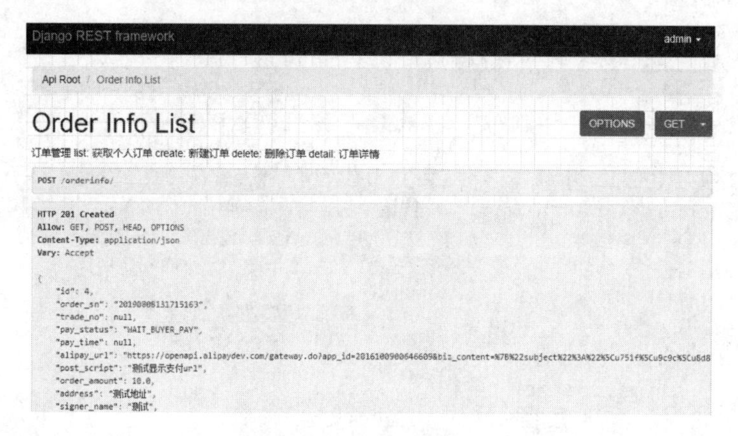


图 6-29　返回的值

在返回值中提取 alipay_url 的内容，然后在浏览器中访问时，会看到刚刚创建的订单，如图 6-30 所示。

图 6-30　刚刚创建的订单

## 6.8 测试程序

(1) 运行如下命令，启动后端 Django Web 模块，效果如图 6-31 所示。

```
python manage.py runserver
```

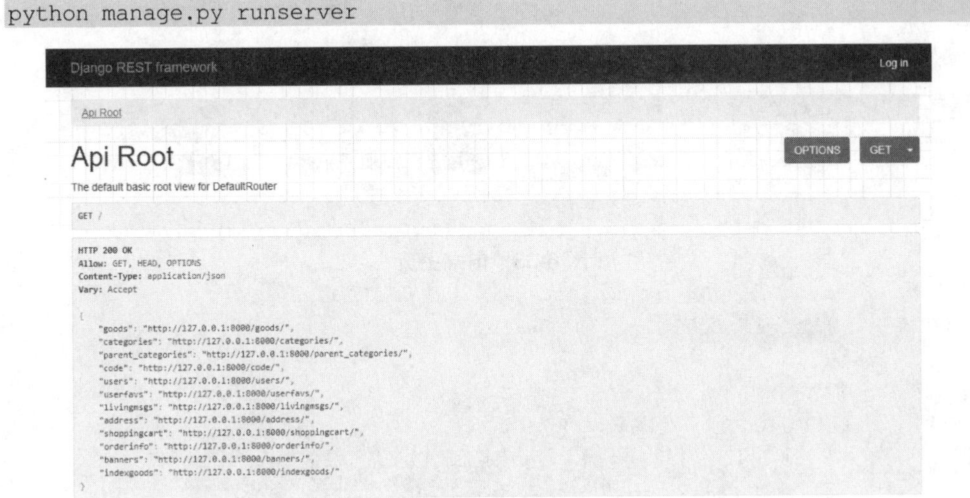


图 6-31　后端主页

(2) 运行如下命令，启动前端 Vue 模块，效果如图 6-32 所示。

```
npm install -g cnpm --registry=https://registry.npm.taobao.org
cnpm install
cnpm run dev
```

(3) 在浏览器中输入 http://127.0.0.1:8081/#/app/home/index 后显示前端主页，如图 6-33 所示。

(4) 订单结算页面的效果如图 6-34 所示。

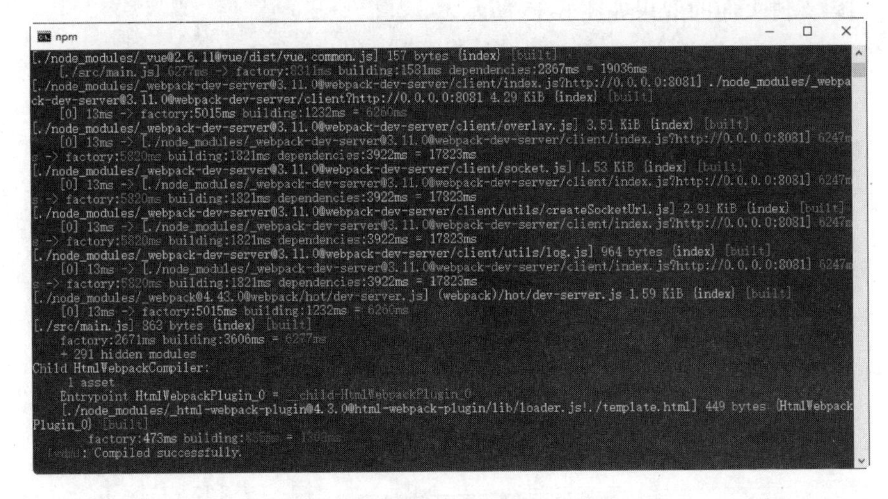


图 6-32　启动 Vue

图 6-33　前端主页

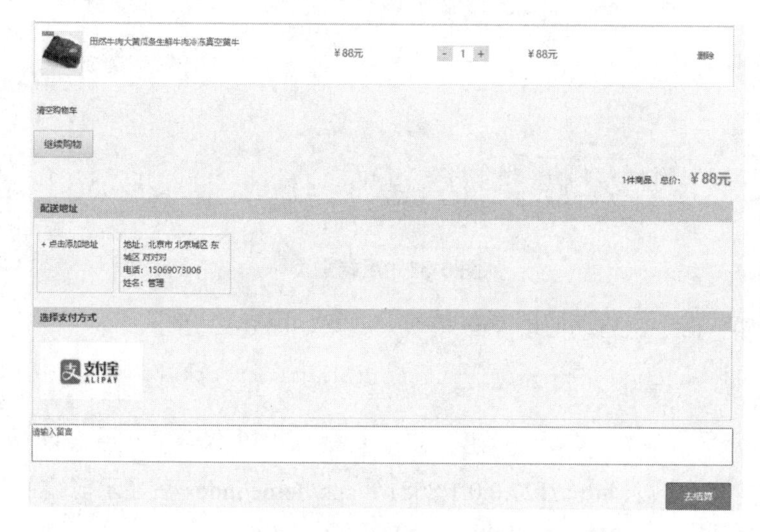


图 6-34　订单结算页面

(5)　扫码支付页面的效果如图 6-35 所示。

图 6-35　扫码支付页面

# 第 7 章

# 民宿信息可视化分析系统

## （网络爬虫+Django+Echarts 可视化）

很多工作累了想出去放松放松的游客，不想住酒店，而是想通过民宿这个平台，感受一下当地的风土人情，体验一下不同的生活方式。基于越来越多的人喜欢住民宿的市场需求，分析民宿市场的发展和市场定位变得越发重要。本章将通过一个综合实例的实现过程，详细讲解爬虫抓取民宿信息的方法，并讲解可视化分析民宿信息的过程。

扫码观看视频讲解

## 7.1 系统背景介绍

绿水青山就是金山银山。近年来，随着国家建设美丽乡村政策的实施，各地纷纷加大特色小镇的建设力度，相继出台对民宿的补贴扶持方案。之所以选择从事民宿行业，大部分人是因为自己喜欢旅行，也有“隐于野”的诗意情结，他们或是放弃了稳定的工作，或是逃离了大都市的生活，希望能通过民宿传递自己的生活理念。

在新的消费观念下，越来越多的客人厌倦了千篇一律的酒店住宿形态。在旅游过程中，游客更希望体验多样化的住宿形态，深入到当地的特色文化中。大众化景点路线会越来越被轻视，个性化住宿、个性化旅游线路的选择会带给客栈民宿更多的发展机遇。

作为一种新兴的非标准住宿业态，民宿对传统标准酒店住宿业起到明显的补充作用。目前美团民宿交易额占美团酒店交易额的比例约为 4.8%，且整体呈现上升趋势。从各省份民宿交易额看，广东省占据首位，交易额占全国市场的 11.6%。交易额排前 10 位的省市依次为广东省、北京市、四川省、江苏省、山东省、陕西省、重庆市、上海市、浙江省、湖北省，上述十省市交易额占全国民宿市场交易额比例超过 65%。

数据显示，2017 年民宿预订以女性消费者为主，占比 55.7%。从民宿产品用户年龄层分布来看，40 岁以下人群占整体消费者比例达到 86.2%，可见国内民宿产品受众偏向年轻化。其中，90 后是民宿消费的主力军，90 后消费者的订单量占比约 58.9%，80 后占比约 27.3%。从消费品类偏好看，用户在住民宿期间，同时消费餐饮品类的比例约占 30.8%，同时消费非餐饮品类的比例约占 28.2%。这说明民宿消费对其他品类的消费具有一定的带动作用。

在民宿市场大发展的前提下，可视化分析民宿市场的发展现状对商家来说具有重要的意义。另外，对于消费者来说，也可以通过可视化系统及时了解民宿行情，帮助自己取得更加物美价廉的服务。

扫码观看视频讲解

## 7.2 爬虫抓取信息

本项目将使用 Scrapy 作为爬虫框架，使用代理 IP 爬取业内知名民宿网中的数据信息，然后将爬取的信息保存到 MySQL 数据库中。最后使用 Django 可视化展示在数据库中保存的民宿数据信息。本节将首先讲解爬虫功能的具体实现过程。

### 7.2.1 系统配置

在 Django 模块中设置整个项目的配置信息，在文件 settings.py 中设置数据库和缓存等配置信息，主要代码如下所示。

```
# 配置mysql数据库
DATABASES = {
    'default': {
```

```
        'ENGINE': 'django.db.backends.mysql',
        'NAME': "scrapy_django",
        'USER': 'root',
        'PASSWORD': '66688888',
        'OPTIONS': {
                'charset':'utf8mb4', # 值编码方式，避免 emoji 无法存储
                },
        'HOST': "127.0.0.1",  # IP 地址
        'PORT': '3306'        # 端口
    }
}
CACHES = {  # redis 做缓存
    'default': {
        'BACKEND': 'django_redis.cache.RedisCache',
        "LOCATION": "redis://127.0.0.1:6379/3",  # 本机 django 的 redis 缓存路径
        # 'LOCATION':"redis://127.0.0.1:6378/3",
        'OPTIONS':{
            "CLIENT_CLASS":"django_redis.client.DefaultClient",
        }
    }
}
```

## 7.2.2　Item 处理

Scrapy 提供了 Item 类，利用这些 Item 类可以让我们自己指定字段。比方说在某个 Scrapy 爬虫项目中定义了一个 Item 类，在这个 Item 里边包含了 title、release_date、url 等。这样，通过各种爬取方法爬取过来的字段，再通过 Item 类进行实例化，就不容易出错了，因为我们在一个地方统一定义过了字段，而且这个字段具有唯一性。在本项目实例文件 items.py 中设置了 4 个 ORM 对象，这 4 个对象和本项目数据库字段是一一对应的。文件 items.py 的具体实现代码如下所示：

```
class HouseItem(DjangoItem):
    django_model = House
    jsonString = scrapy.Field()   #增加临时字段，用来一次性传递多个其他对象的属性

class HostItem(DjangoItem):
    django_model = Host

class LabelsItem(DjangoItem):
    django_model = Facility

class FacilityItem(DjangoItem):
    django_model = Labels

class CityItem(DjangoItem):
    django_model = City

class urlItem(scrapy.Item):  # master 专用 item
    # define the fields for your item here like:
    url = scrapy.Field()
```

## 7.2.3 具体爬虫

编写文件 hotel.py，功能是实现具体的网络爬虫功能，具体实现流程如下所示。

(1) 创建类 HotwordspiderSpider，设置爬虫项目的名字是 hotel，然后分别设置爬虫的并发请求数、延时、最大的并发请求数量、保存数据管道和使用代理等信息。

(2) 设置爬虫网页 HTTP 请求协议的请求报文(Request Headers)。

(3) 编写函数 regexMaxNum()，功能是使用正则表达式返回匹配到的最大的数字(就是页数)：

```
def regexMaxNum(self,reg,text):
    temp = re.findall(reg,text)
    return max([int(num) for num in temp if num != ""])
```

(4) 编写函数 start_requests()，功能是设置爬虫启动时要爬取的城市列表。

(5) 编写函数 getRSXFPrice()，功能是提取爬虫数据中的价格信息。

(6) 编写函数 detail()，功能是获取每个民宿的详细信息，包括面积、标签、标题、地址、房型、位置、城市、留言数量等信息。在函数 detail()中，关于价格的计算比较麻烦，因为民宿网的价格进行了数据加密，所以需要用专门的逻辑来破解这个反扒机制。

## 7.2.4 破解反扒字体加密

网站中的价格信息是加密的，为了获取每个民宿的价格信息，需要对“.woff”格式的加密字体进行破解。编写文件 parseTool.py，功能是破解“.woff”格式的价格信息，主要实现代码如下所示：

```
# 获得 j-gallery 这段的字符串
def getFontUrl(UserJson):
    j_gallery_text = UserJson  # 准备过滤
    UserJson = j_gallery_text.replace("'",'"').replace("\\","")   # 过滤斜杠
    test = re.findall('(?<=cssPath\\"\\:\\").*?(?=\\}\\,)',UserJson)[0]

    print()
    wofflist = re.findall('(?<=\\(\\").*?(?=\\)\\;)',test) # j-gallery 字段处理
#     print(wofflist)
    print()
    font_url = ''
    for woffurl in wofflist:
        if woffurl.find("woff")!=-1:
            tempwoff = re.findall('(?<=\\").*?(?=\\")',woffurl)
    #         print(tempwoff)
            for j in tempwoff:
                if j.find("woff")!=-1:
                    print("https:"+j)
                    font_url = "https:"+ j

    # 提取字体成功
#     print(font_url)
```

```
    return font_url

def download_font(img_url,imgName,path=None):
    headers = {'User-Agent':"Mozilla/5.0 (Windows NT 6.1; WOW64)
AppleWebKit/537.1 (KHTML, like Gecko) Chrome/22.0.1207.1 Safari/537.1",
            }  ##浏览器请求头
    try:
        img = requests.get(img_url, headers=headers)
        dPath = os.path.join("woff",imgName)  # 传递 imgName
        # print(dPath)
        print("字体的文件名 "+dPath)
        f = open(dPath, 'ab')
        f.write(img.content)
        f.close()
        print("下载成功")
        return dPath
    except Exception as e:
        print(e)


# 从字体文件中获得字形数据用来备用待对比
def getGlyphCoordinates(filename):
    """
    获取字体轮廓坐标,手动修改 key 值为对应数字
    """
    font = TTFont("woff/"+f'{filename}')  # 自动带上了 woff 文件夹
    # font.saveXML("bd10f635.xml")
    glyfList = list(font['glyf'].keys())
    data = dict()
    for key in glyfList:
        # 剔除非数字的字体
        if key[0:3] == 'uni':
            data[key] = list(font['glyf'][key].coordinates)
    return data


def getFontData(font_url):
    # 获取字体信息
    filename = os.path.basename(font_url)
    font_data = None
    if os.path.exists("woff/"+filename):
        # 直接读取
        font_data = getGlyphCoordinates(filename)  # 读取时候自带 woff 文件夹
    else:
        # 先下载再读取
        download_font(font_url, filename, path=None)
        font_data = getGlyphCoordinates(filename)
    if font_data == None:
        print("字题文件读取出错，请检查")
    else:
        #       print(font_data)
        return font_data

# 自动分割大写形式的价格
```

```
def splitABC(price_unicode):
    raw_price = price_unicode.split("&amp;")
    temp_price_unicode = []
    for x in raw_price:
        if x != "":
            temp_price_unicode.append(x.upper().replace("#X", "").replace(";", ""))
    return temp_price_unicode  # 提取出简化大写的价格，例如 400 是原价，折扣价是 280

def getBothSplit(UserJson):
    UserJson = UserJson.replace("\\", "").replace("'", '"')
    result_price = []
    result_discountprice = []
    try:
        price_unicode = re.findall('(?<=price\\"\\:\\").*?(?=\\"\\,)', UserJson)[0]
        # 原价数字 400
        result_price = splitABC(price_unicode)
    except Exception as e:
        print("没有找到价格")
        print(e)

    try:  # 可能没有找到，那就会有乱码符号
        discountprice_unicode = re.findall('(?<=discountPrice\\"\\:\\").
            *?(?=\\"\\,)', UserJson)[0]  # 原价数字 400
        result_discountprice = splitABC(discountprice_unicode)
    except Exception as e:
        print("没有找到折扣价")
        print(e)
    if result_discountprice == [] and result_price != []:
        result_discountprice = result_price  # 如果折扣价为 0 的话，那么就等于原价
    return result_price,result_discountprice  # 没有折扣返回的价格编码

    def pickdict(dict):  # 序列化这个字典
    with open(os.path.join(os.path.abspath('.'),"label_dict.pickle"), "wb") as f:
        pickle.dump(dict, f)
```

## 7.2.5 下载器中间件

下载器中间件是在引擎及下载器之间的特定钩子(specific hook)，处理 Downloader 传递给引擎的 response(也包括引擎传递给下载器的 Request)。其提供了一个简便的机制，通过插入自定义代码来扩展 Scrapy 功能。在本项目中的下载器中间件文件 middlewares.py 中，主要实现了在线代理 IP 功能。具体实现流程如下所示。

(1) 准备好基础工作，先创建类 EnvironmentIP 和 EnvironmentFlag，对应实现代码如下所示：

```
class EnvironmentIP:                    # 设置一个全局变量，单例模式
    _env = None

    def __init__(self):
        self.IP = 0                     # 存储 IP 属性

    @classmethod
```

```
    def get_instance(cls):
        """
        返回单例 Environment 对象
        """
        if EnvironmentIP._env is None:
            cls._env == cls()
        return cls._env

    def set_flag(self, IP):  # 里面放的是数字
        self.IP = IP

    def get_flag(self):
        return self.IP

envVarIP = EnvironmentIP()  #是否切换使用的代理

class EnvironmentFlag:        # 设置一个全局变量，单例模式
    _env = None
    def __init__(self):
        self.flag = False     # 默认不使用代理

    @classmethod
    def get_instance(cls):
        """
        返回单例 Environment 对象
        """
        if EnvironmentFlag._env is None:
            cls._env == cls()
        return cls._env

    def set_flag(self, flag):
        self.flag = flag

    def get_flag(self):
        return self.flag

envVarFlag = EnvironmentFlag()  # 是否切换使用的代理

class Environment:              # 设置一个全局变量，单例模式
    _env = None
    def __init__(self):
        self.countTime = datetime.datetime.now()

    @classmethod
    def get_instance(cls):
        """
        返回单例 Environment 对象
        """
        if Environment._env is None:
            cls._env == cls()
        return cls._env

    def set_countTime(self, time):
        self.countTime = time
```

```
    def get_countTime(self):
        return self.countTime

envVar = Environment()  # 初始化一个默认的
```

(2) 定义类 RandomUserAgent，实现随机生成 IP 功能，通过函数 process_request()和 process_response()及时获取响应信息，这样可以判断这个 IP 是否可用。对应实现代码如下所示：

```
class RandomUserAgent(object):  # ua中间件
    # def __init__(self):

    @classmethod
    def from_crawler(cls, crawler):
        s = cls()
        crawler.signals.connect(s.spider_opened, signal=signals.spider_opened)
        return s

    def process_request(self, request, spider):
        ua = UserAgent()
        print(ua.random)
        request.headers['User-Agent'] = ua.random
        return None

    def process_response(self, request, response, spider):
        # Called with the response returned from the downloader.
        print(f"请求的状态码是 {response.status}")
        print("调试ing")
        print(request.url)
        HTML = response.body.decode("utf-8")
        # print(HTML)
        print(HTML[:200])
        try:
            # print('进来中间件调试')
            if HTML.find("code")!=-1:
                if re.findall('(?<=code\\"\\:).*?(?=\\,)',HTML)[0]=='406':
                  # 转义保留双引号
                    print("正在重新请求(网络不好)")
                    return request
        except Exception as e:
            print(request.url)
            print(e)

        try:
            temp = json.loads(HTML)
            if temp['code'] == 406:  #
                print("正在重新请求(网络不好)状态码406")
                request.meta["code"] = 406
                return request    # 重新发给调度器，重新请求
        except Exception as e:
            print(e)
        return response
```

```
    def process_exception(self, request, exception, spider):
        pass

    def spider_opened(self, spider):
        spider.logger.info('Spider opened: %s' % spider.name)
```

(3) 定义类 proxyMiddleware，实现在线代理 IP 功能，同时创建了 redis 代理连接池，用列表 remote_iplist 中的 IP 轮询访问，并打印输出对应的响应信息。对应实现代码如下所示。

```
class proxyMiddleware(object):  # 代理中间件
    # MYTIME = 0   # 类变量 Mytime 用来设定切换 IP 代理的频率

    def __init__(self):
        # self.count = 0
        from redis import StrictRedis, ConnectionPool
        # 使用默认方式连接到数据库
        pool = ConnectionPool(host='localhost', port=6378, db=0,password=
               'Zz123zxc')
        self.redis = StrictRedis(connection_pool=pool)

    @classmethod
    def from_crawler(cls, crawler):
        # This method is used by Scrapy to create your spiders.
        s = cls()
        crawler.signals.connect(s.spider_opened, signal=signals.spider_opened)
        return s

    def get_proxy_address(self):
        proxyTempList = list(self.redis.hgetall("useful_proxy"))
        # proxyTempList = list(redis.hgetall("useful_proxy"))
        return str(random.choice(list(proxyTempList)), encoding="utf-8")

    def process_request(self, request, spider):
        # 设置使用几个代理
        remote_iplist = ['125.105.70.77:4376', '58.241.203.162:4386',
                         '17.5.181.109:4358', '14.134.186.95:4372',
                         '125.111.150.25:4305', '122.246.173.161:4375']

        print()
        print("proxyMiddleware")
        now = datetime.datetime.now()
        print("flag")
        print("time")
        print(f"现在时间{now}")
        print(f"变量内时间{envVar.get_countTime()}")
        print("变量状态{True}才使用代理")
        print(envVarFlag.get_flag())
        print("相减少后的结果")
        print((now-envVar.get_countTime()).seconds / 40)
        if envVarFlag.get_flag() ==True:  #envVarFlag.get_flag() == True:
            if (now-envVar.get_countTime()).seconds / 20 >= 1:
                envVarFlag.set_flag(not envVarFlag.get_flag())  # 切换为使用代理
```

```
            envVar.set_countTime(now)
        print("使用代理中池中的 ip")
        proxy_address = None
        try:
            proxy_address = self.get_proxy_address()
            if proxy_address is not None:
                print(f'代理 IP -- {proxy_address}')
                request.meta['proxy'] = f"http://{proxy_address}"
                # 如果出现了 302 错误，有可能是因为代理的类型不对
            else:
                print("代理池中没有代理 ip 存在")
        except Exception as e:
            print("检查到代理池里面已经没有 ip 了,使用本地")
    else:  # 不使用代理,这儿轮流使用本地 ip 和外面的 ip
        if (now-envVar.get_countTime()).seconds/40>=1: #传进来的是切换状态的
            envVarFlag.set_flag(not envVarFlag.get_flag())  # 切换为使用代理
            envVar.set_countTime(now)

        if envVarIP.get_flag() <= len(remote_iplist)-1:   ## 直接使用本地 IP
            remoteip = remote_iplist[envVarIP.get_flag()]
            print(f'使用远程 ip -- {remoteip}')
            request.meta['proxy'] = f"http://{remoteip}"
            envVarIP.set_flag(envVarIP.get_flag()+1)
        else:
            envVarIP.set_flag(0)  # 把这个 ip 设置成 0,表示使用本地的 ip
            print("使用到本地 ip")
        pass
```

## 7.2.6 保存爬虫信息

编写实例文件 pipelines.py，功能是将爬取的民宿房源信息保存到本地数据库中。具体实现流程如下所示：

(1) 编写类 urlItemPipeline，保存房源的 URL 信息，对应实现代码如下所示。

```
class urlItemPipeline(object):                  # master 专用管道
    def __init__(self):
        self.redis_url = "redis://Zz123zxc:@localhost:6379/"
        # 设置使用本地 redis
        self.r = redis.Redis.from_url(self.redis_url,decode_response=True)

    def process_item(self, item, spider):
        if isinstance(item, urlItem):
            print("urlItem item")
            try :
                # item.save()
                self.r.lpush("Meituan:start_urls",item['url'])
            except Exception as e:
                print(e)
        return item
```

(2) 编写类 cityItemPipeline，保存房源的城市信息，对应实现代码如下所示：

```
class cityItemPipeline(object):
```

```
    def process_item(self, item, spider):
        if isinstance(item, CityItem):
            print("CityItem item")
            try :
                item.save()
            except Exception as e:
                print(e)
        return item
```

(3) 编写类houseItemPipeline，保存房源的详细信息，主要包括Labels、Facility和Host等信息。对应实现代码如下所示：

```
class houseItemPipeline(object):
    def __init__(self):
        pass

    def process_item(self, item, spider):
        if isinstance(item, HouseItem):
            print("HouseItem item")
            house = None
            try:
                house = item.save()  # 保存信息
                print("hosuse 保存成功")

            except Exception as e:
                print("hosuse 保存失败，后面的跳过保存")
                print(e)
                print(item)
                return item

            jsonString = item.get("jsonString")
            labelsList = jsonString['Labels']
            facilityList = jsonString['Facility']
            hostInfos = jsonString['Host']
            # 查询房源的具体信息，多条件查询
            for onetype in labelsList:  # 1.添加所有标签的
                for one in labelsList[onetype]:  # one 都是一个标签
                    try:  # 先找找有没有，然后把已经有的添加进来
                        label = Labels.objects.filter(**{'label_name':one[0],
                         "label_desc":one[1]})  # 找到的话就直接加入另一个 Meiju 对象中
                        # print("长度")
                        if len(label) == 0:
                            # print("需要创建后添加")
                            l = Labels()
                            l.label_name = one[0]
                            l.label_desc = one[1]
                            # print("检查 label")
                            # print(f"onetype:{onetype}")
                            # print(one)
                            if onetype == "1":  # 优惠标签
                                l.label_type = 1  # 标签类型
                            else:
                                l.label_type = 0
                            l.save()
                            house.house_labels.add(l)  # 添加标签
                        else:
```

```
                # print("找到有直接添加")
                # print(label)
                house.house_labels.add(label.first()) #添加到第一个位置
                # print("添加成功")
            except Exception as e:
                print(e)
                print("label 已存在，跳过插入")
                # print(e)


    # 写入 Facility
    # print(facilityList)
    # 遍历 facilityList 列表
    for facility in facilityList:if 'metaValue' in facility:print()
    # 执行先检查后添加
            try:  # 先验证是否存在，然后把已经有的添加进来
                fac = Facility.objects.filter(**{'facility_name':
                  facility['value']})  # 找到的话就直接加入另一个 Meiju 对象中
                # print("长度")
                if len(fac) == 0:
                    # print("需要创建后添加")
                    l = Facility()
                    l.facility_name = facility['value']
                    l.save()
                    house.house_facility.add(l)
                else:
                    house.house_facility.add(fac.first()) # 添加到第一个位置
            except Exception as e:
                print("facility 已经存在，跳过插入")
                # print(e)

    print("下面开始 host 信息的添加")
    '''{'hostId': '36438164',
        'host_RoomNum': '51',
        'host_commentNum': '991',
        'host_name': 'hostname
        'host_replayRate': '100'}'''
    print(hostInfos)

    try:  # 先找找有没有，然后把已经有的添加进来,fixing
        hosts = Host.objects.filter(**{
            'host_id':hostInfos['hostId'],
        'host_updateDate':house.house_date})
        # 找到的话就直接加入另一个 Meiju 对象中
        print("长度")
        print("输出查找到的结果")
        print(hosts)
        try:
            # print("需要创建后添加")
            l = Host()
            l.host_name = hostInfos['host_name']
            l.host_id = hostInfos['hostId']
            l.host_RoomNum = hostInfos['host_RoomNum']
            l.host_commentNum = hostInfos['host_commentNum']
            l.host_replayRate = hostInfos['host_replayRate']
            l.save()  # 保存处理
```

```
                # fac
                # print("__label_")
                # print(l)
                house.house_host.add(l)
            except Exception as e:
                print(e)
            # else:
                # print("不为空找到了")
                # print("__label_")
                # print(fac)
                house.house_host.add(hosts.first())  # 添加到第一个位置
                print("已有的情况下添加成功")
        except Exception as e:
            print("以有这个 host 跳过插入")
            print(e)

        return item  # 返回 item
    return item
```

通过如下命令运行爬虫程序：

```
scrapy crawl hotel
```

爬虫数据被保存在 MySQL 数据库中，如图 7-1 所示。

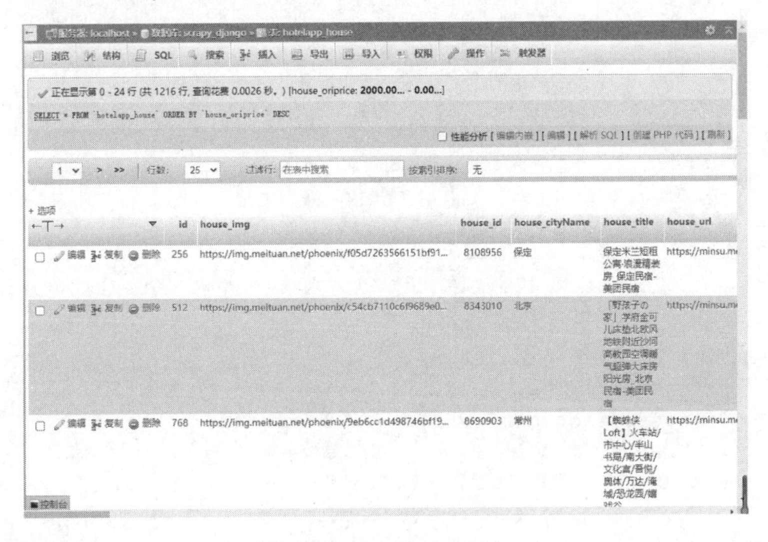


图 7-1　数据库中的爬虫数据

## 7.3　数据可视化

扫码观看视频讲解

本项目使用 Django 框架实现可视化功能，提取在 MySQL 数据库中保存的民宿数据信息，然后使用 Echarts 实现数据可视化功能。本节将详细讲解实现数据可视化功能的具体过程。

## 7.3.1 数据库设计

编写文件 models.py，实现数据库模型设计功能。在此文件中，每个类和 MySQL 数据库中的表一一对应，每个变量和数据库表中的字段一一对应。代码如下：

```
class City(models.Model):
    city_nm = models.CharField(max_length=50,unique=True)   # 城市名字
    city_pynm = models.CharField(max_length=50,unique=True) # 拼音名字
    city_statas = models.BooleanField(default=False)

    def __str__(self):
        return self.city_nm + "  " + self.city_pynm


class Labels(models.Model):
    TYPE_CHOICE = (
        (0, "普通标签"),
        (1, "优惠标签"),
        )
    label_type = models.IntegerField(choices=TYPE_CHOICE)  # 类型
    label_name = models.CharField(max_length=171,unique=True)
    label_desc = models.CharField(max_length=171,unique=True)  # 降序显示

    def __str__(self):
        return str(self.label_type)+"  "+self.label_name+"  "+self.label_desc

    class Meta:
        # 设置约束
        unique_together = ('label_name',"label_desc")


class Host(models.Model):
    '''自己创建一个 id'''
    host_name = models.CharField(max_length=171)  # 房东名字
    host_id = models.IntegerField()  # 房东 id
    host_replayRate = models.IntegerField(default=0)  # 回复率
    host_commentNum = models.IntegerField(default=0)  # 评价总数
    host_RoomNum = models.IntegerField(default=0)  # 不同时间段的房子的数量
    host_updateDate = models.DateField(default=timezone.now) # 自动创建时间，不可修改

    def __str__(self):
        return str(self.host_id)+ " "+self.host_name

    class Meta:
            # 各一个房子一天最多一天数据(一个价格)
        unique_together = ('host_id',"host_updateDate")

class Facility(models.Model):
    ''' 设施类型，本系统功能设置了 85 个设施
        django 建议不用外键约束，因为这样可以更高效
    '''
    facility_name = models.CharField(max_length=50,unique=True)   # 设施名字
```

```
    def __str__(self):
        return self.facility_name


# 表示酒店式公寓的model
class House(models.Model):
    '''
    house_id 为主键
    '''

house_img = models.CharField(max_length=171,default="static/media/default.jpg")
# 房子默认预览图

    house_id =  models.IntegerField(default=0)   # 用来标识唯一的房子
    house_cityName = models.CharField(max_length=50,default="未知城市")
    house_title = models.CharField(max_length=171)   # 标题
    house_url = models.CharField(max_length=171) # 地址信息
    house_date = models.DateField(default=timezone.now)   # 爬取时间
    house_firstOnSale = models.DateTimeField(default=datetime.datetime(1770, 1,
                            1, 1, 1, 1, 499454)) # 发布时间
    house_favcount = models.IntegerField(default=0)  # 房子页面的点赞数
    house_commentNum = models.IntegerField(default=0)  # 评分人数(也是评论人数)

    house_descScore = models.FloatField(default=0)  # 对这个房子评分最高的四个分数
    house_talkScore = models.FloatField(default=0)
    house_hygieneScore = models.FloatField(default=0)
    house_positionScore = models.FloatField(default=0)
    house_avarageScore = models.FloatField(default=0)
    # 平均分，满分5.0，0分表示未被评价

    # 房子的具体内容信息
    house_type = models.CharField(max_length=50,default="未分类")  # 整套/单间/合住
    house_area = models.IntegerField(default=0)   # 房子的面积，单位m²
    house_kitchen = models.IntegerField(default=0)  # 厨房数量,0就表示没有
    house_living_room =  models.IntegerField(default=0)  # 客厅数量、
    house_toilet = models.IntegerField(default=0)  # 卫生间数量
    house_bedroom = models.IntegerField(default=0)  # 卧室数量
    house_capacity = models.IntegerField()  # 可以容纳的人数
    house_bed = models.IntegerField(default=1)  # 床的数量

    # 房子的价格信息
    house_oriprice = models.DecimalField(max_digits=16,decimal_places=2)
    # 刚发布价格
    house_discountprice = models.DecimalField(max_digits=16,decimal_places=2,
                            default=0.00)
    # 折扣价格，如果没有discountPrice，那么表示原价就是折扣价

    # 房源位置
    house_location_text = models.CharField(max_length=171)
    # 因为使用utf8mb4格式，char最长为171，四个字节为一个字符
    house_location_lat = models.DecimalField(max_digits=16,decimal_places=6)
    # 纬度，取值小数点后6位
```

```
    house_location_lng = models.DecimalField(max_digits=16,decimal_places=6)
    # 经度

    #房源设施
    house_facility = models.ManyToManyField(Facility)  # 房屋设置信息

    #房东信息
    house_host = models.ManyToManyField(Host)  # 多对多，一个房东可能有多套房子

    #普通标签和优惠标签
    house_labels = models.ManyToManyField(Labels)

    earliestCheckinTime = models.TimeField(default="00:00")
    # 可以最早入住时间

    def __str__(self):
        return str(self.house_id)+":" +f"{str(self.house_cityName)}" + ":"+ \
f"{str(self.house_title[0:15])}..." + ":"+str(self.house_oriprice) + "¥/晚"

    class Meta:
            # 约束
        unique_together = ('house_id',"house_date")
        # 一个房子一天最多一个价格


class Favourite(models.Model):  # 收藏夹
    user =  models.OneToOneField(User,unique=True,on_delete=models.CASCADE)

    fav_city = models.ForeignKey(City, on_delete=models.CASCADE)
    # 偏好城市，默认是广州
    fav_houses = models.ManyToManyField(House)

    def __str__(self):
        return str(self.user.username) + ":" + str(self.fav_city)
```

## 7.3.2 视图显示

在本项目中，数据可视化功能是通过 View 视图文件和模板文件实现的。本项目的 View 视图文件是 drawviews.py，具体实现流程如下所示。

(1) 编写函数 bar_base()，获取系统内的爬虫数量，分别显示房源总数和城市房源数量。代码如下：

```
def bar_base() -> Bar:  # 返回给前端用来显示图的json设置，按城市分组来统计数量
    nowdate = time.strftime('%Y-%m-%d', time.localtime(time.time()))
     count_total_city = House.objects.filter(house_date=nowdate).values
    ("house_cityName").annotate(count=Count("house_cityName")).order_by
    ("-count")
    # for i in count_total_city:
    #    print(i['house_cityName']," ",str(i['count']))
    c = (
        Bar(init_opts=opts.InitOpts(theme=ThemeType.WONDERLAND))
            .add_xaxis([city['house_cityName'] for city in count_total_city])
```

```
        .add_yaxis("房源数量", [city['count'] for city in count_total_city])
        .set_global_opts(title_opts=opts.TitleOpts(title="今天城市房源数量",
                          subtitle="如图"),
                      xaxis_opts=opts.AxisOpts(axislabel_opts=opts.
                                  LabelOpts(rotate=-90)),
                      )
        .set_global_opts(
        datazoom_opts={'max_': 2, 'orient': "horizontal", 'range_start': 10,
                      'range_end': 20, 'type_': "inside"})
        .dump_options_with_quotes()
    )
    return c
```

(2) 编写类 PieView，统计数据库中的房型数据并绘制饼图，代码如下：

```
class PieView(APIView):
    def get(self, request, *args, **kwargs):
        result = fetchall_sql(
            "select house_type,count(house_type) from (select distinct house_id ,
              house_type from hotelapp_house  group by house_id,house_type ) hello
              group by house_type")
        c = (
            Pie()
                .add("", [z for z in zip([i[0] for i in result], [i[1] for i in result])])
                # .add("",[list(z) for z in zip([x['house_type'] for x in
                  house_type_count],[x['count'] for x in house_type_count])])
                .set_global_opts(title_opts=opts.TitleOpts(title="总房屋类型"))
                .set_series_opts(label_opts=opts.LabelOpts(
                formatter="{b}: {c} | {d}%",
            ))
                .dump_options_with_quotes()
        )
        return JsonResponse(json.loads(c))
```

(3) 编写类 getMonthPostTime、getMonthPostTime2 和 timeLineView，获取数据库中保存的发布时间信息，并绘制发布时间折线图和最近 7 天的折线图。

(4) 编写类 drawMap，绘制房源分布热力图，代码如下：

```
class drawMap(APIView):  # 参数 apiview，绘制美团房源数量热力图
    def get(self, request, *args, **kwargs):
        from pyecharts import options as opts
        from pyecharts.charts import Map
        from pyecharts.faker import Faker

        result = cache.get('house_city', None)  # 使用缓存实现共享
        if result is None:  # 如果在缓存中无相关数据，则在数据库查询数据
            print("使用缓存房源城市统计")
            result = fetchall_sql(
                """select house_cityName,count(house_cityName) as count from
(SELECT distinct(house_id),house_cityName FROM  hotelapp_house) hello group by
house_cityName""")
            cache.set('house_city', result, 3600 * 12)  # 设置缓存

        else:
            pass
```

```
        c = (
            Geo()
                .add_schema(maptype="china")
                .add(
                "房源",
                [z for z in zip([i[0] for i in result], [i[1] for i in result])],
                type_=ChartType.HEATMAP,
            )
                .set_series_opts(label_opts=opts.LabelOpts(is_show=False))
                .set_global_opts(
                visualmap_opts=opts.VisualMapOpts(),
                title_opts=opts.TitleOpts(title="美团民宿房源热力图"),
            )
                .dump_options_with_quotes()
        )

        return JsonResponse(json.loads(c))  # 返回 JSON 结果
```

为了节省本书篇幅，本项目介绍到此为止。有关本项目的具体实现流程，请参考本书的配套源码。执行本项目可视化模块后的效果如图 7-2～图 7-6 所示。

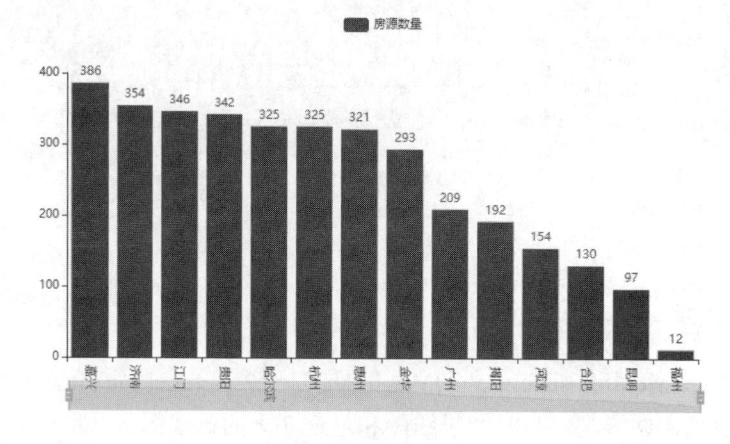


图 7-2　数据概览

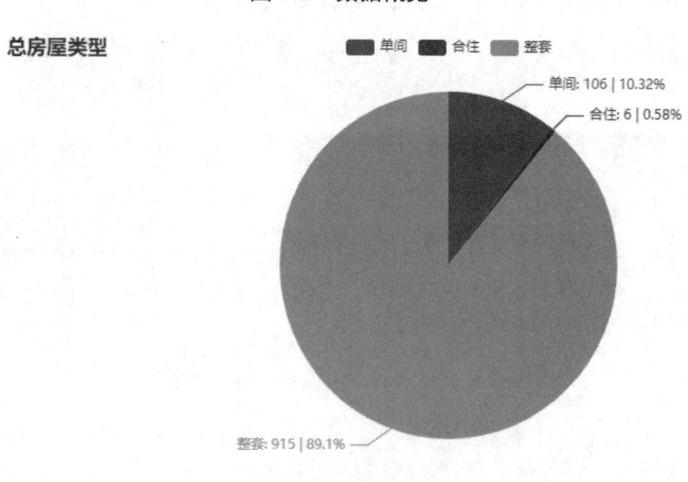


图 7-3　房屋类型

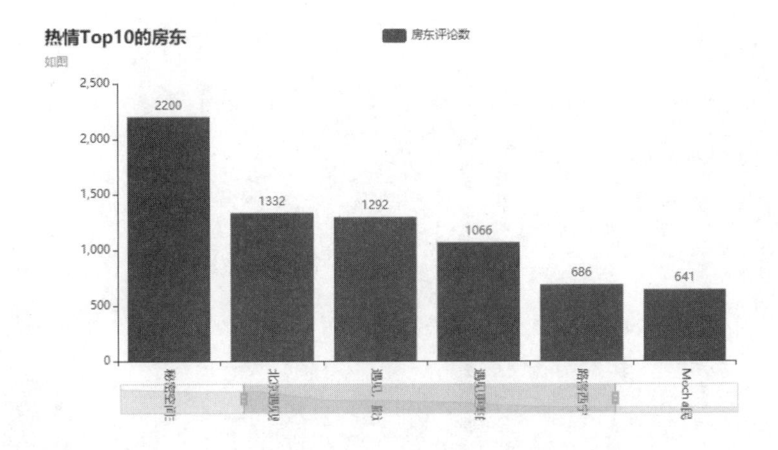


图 7-4　热情 Top10 的房东

图 7-5　房屋设施分析

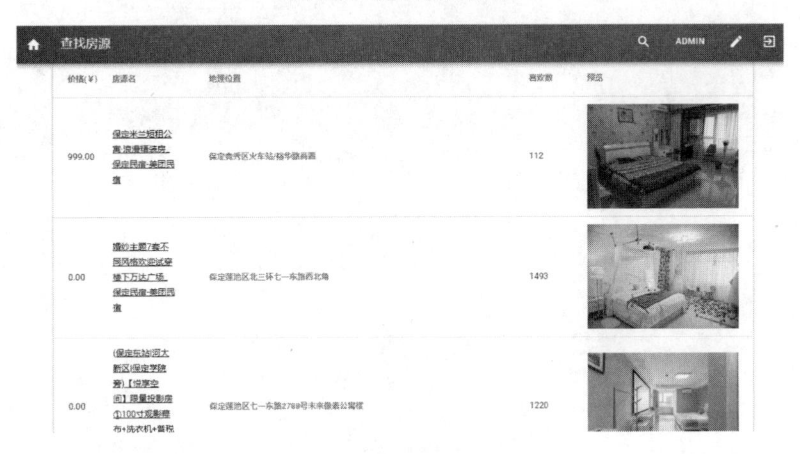


图 7-6　搜索房源

# 第 8 章

# 实时疫情监控系统

## （腾讯 API 接口+Seaborn+matplotlib 实现）

2020 年的新冠病毒对我们的日常生活和学习带来了极大冲击，新型冠状病毒肺炎疫情的暴发无疑是近些年最让人心痛的黑天鹅。面对疫情，如何利用全面、有效、及时的数据和可视化技术准确感知疫情态势，为决策者、管理人员提供宏观数据依据？本章将详细讲解用 Python 程序开发一个实时疫情监控系统的方法，为读者步入本书后面知识的学习打下基础。

## 8.1 背景介绍

鼠年伊始，在世界多个国家中暴发了新型冠状病毒引发的肺炎。新型冠状病毒肺炎简称“新冠肺炎”，世界卫生组织命名为“2019 冠状病毒病”，是指 2019 新型冠状病毒感染导致的肺炎。2020 年 2 月 11 日，世界卫生组织总干事谭德塞在瑞士日内瓦宣布，将新型冠状病毒感染的肺炎命名为 COVID-19。2020 年 2 月 21 日，国家卫生健康委发布了关于修订新型冠状病毒肺炎英文命名事宜的通知，决定将“新型冠状病毒肺炎”英文名称修订为 COVID-19，与世界卫生组织命名保持一致，中文名称保持不变。

## 8.2 系统分析

自从 COVID-19 疫情暴发以来，每天的疫情数据牵动人心，政府疾病控制中心每天都发布各地疫情数据，新闻媒体通过各种新媒体技术以新闻方式实时发布并呈现疫情数据，人们可以通过数据可视化感知疫情现状和流行病传染扩散趋势及救治情况。为了节省决策时间，让数据可视化成为管理者和时间赛跑的帮手，是快速打赢这场“战疫”的关键。

### 8.2.1 需求分析

此实时疫情监控系统主要面向各行各业的全世界人民，其中大部分是中国的高知人群，在学历、社会地位、经济收入、家庭背景都占有相当的优势。

在系统需求分析中，要做到以用户为中心，场景化思考，需清楚数据大屏不同于其他信息管理系统的特点，主要包括以下几个方面。

- 面积巨大——用户站远才能看全内容。
- 深色背景——紧张感强，让视觉更好地聚焦，避免过分的视觉刺激。
- 不可操作——大屏主要用来给来用户看的，一般不会直接操作大屏。
- 空间局限——大屏不像网页有滚动条，它的长和宽都是固定的。
- 单独主题——每块大屏都有具体想向用户表达的某个主题。

以上特点决定了大屏没有复杂的业务流程和任务流程，需求分析的重点在范围定义和信息架构设计及非功能性需求。目前已有的实时疫情监控系统，主要是对累计和当日的确诊、疑似及死亡人数的统计，统计范围达到市级的粒度。

### 8.2.2 数据分析

对需求分析结果中涉及的数据进行分析，可以为后面的设计和开发避开很多坑。其中需要思考以下 8 个方面。

- 可以公开哪些数据？很多信息很敏感。
- 数据来源有哪些？如果接第三方数据，接口是否稳定？
- 能获取的数据精度怎样？精度与数据分析指标息息相关。
- 预先评估数据量的级别有多大？
- 如何实时刷新大批量数据？
- 维度会是用户都想看的吗？
- 应该使用哪种可视化方式？
- 展现这些数据有意义吗？

目前国内能收集到的，并且能持续更新的新冠疫情数据如表 8-1 所示。

表 8-1 疫情数据来源

| 序 号 | 主要内容 | 提供方 |
| --- | --- | --- |
| 1 | 全国新型冠状病毒肺炎确诊病例分布图 | 中国疾病预防控制中心 |
| 2 | 实时更新疫情通报 | 中国疾病预防控制中心周报 |
| 3 | 世卫组织的最新数据 | 中国疾病预防控制中心 |
| 4 | 疫情防控进展，疫情新闻报道 | 国家卫生健康委员会统计信息中心 |
| 5 | 实时更新疫情通报 | 国家卫生健康委员会 |
| 6 | 疫情数据、人口迁移地图、实时新闻播报 | 国家卫健委、各省市区卫健委、各省市区政府 |
| 7 | 一线网络服务商 | 腾讯、阿里和百度都提供了 API 接口 |

本系统将使用腾讯提供的免费 API 接口作为数据来源，可视化展示实时疫情数据。

## 8.3 具体实现

扫码观看视频讲解

本节将使用腾讯提供的 API 接口获取实时疫情信息，并使用 matplotlib 和 Seaborn 绘制可视化统计图，帮助大家更直观地了解当前的疫情信息。

### 8.3.1 列出统计的省和地区的名字

编写实例文件 test01-spider.py，功能是获取腾讯 API 接口中提供的 JSON 数据，使用 for 循环打印输出腾讯疫情平台中国内省份和地区的信息。代码如下：

```
import time, json, requests
# 抓取腾讯疫情实时 json 数据
url = 'https://view.inews.qq.com/g2/getOnsInfo?
\name=disease_h5&callback=&_= %d'%int(time.time()*1000)
data = json.loads(requests.get(url=url).json()['data'])
print(data)
print(data.keys())

# 统计省份信息(34 个省份 湖北 广东 河南 浙江 湖南 安徽....)
num = data['areaTree'][0]['children']
```

```
print(len(num))
for item in num:
   print(item['name'],end=" ")   # 不换行
else:
   print("\n")                   # 换行
```

执行代码后会输出：

```
34
北京 香港 上海 四川 甘肃 河北 陕西 广东 辽宁 台湾 重庆 福建 浙江 澳门 天津 江苏 云南 湖南 海南 吉林 江西 黑龙江 山西 河南 湖北 西藏 贵州 安徽 内蒙古 宁夏 山东 广西 新疆 青海
```

## 8.3.2 查询并显示各地的实时确诊数据

编写实例文件 test03-spider.py，功能是获取腾讯 API 接口中提供的 JSON 数据，使用 for 循环打印输出国内各个省份和地区的实时确诊数据。代码如下：

```
import time, json, requests
# 抓取腾讯疫情实时 json 数据
url = 'https://view.inews.qq.com/g2/getOnsInfo?
\name=disease_h5&callback=&_= %d'%int(time.time()*1000)
data = json.loads(requests.get(url=url).json()['data'])
print(data)
print(data.keys())

# 统计省份信息(34 个省份 湖北 广东 河南 浙江 湖南 安徽....)
num = data['areaTree'][0]['children']
print(len(num))
for item in num:
   print(item['name'],end=" ")   # 不换行
else:
   print("\n")                   # 换行


# 解析数据(确诊 疑似 死亡 治愈)
total_data = {}
for item in num:
   if item['name'] not in total_data:
      total_data.update({item['name']:0})
   for city_data in item['children']:
      total_data[item['name']] +=int(city_data['total']['confirm'])
print(total_data)
```

执行后会输出：

```
34
北京 香港 上海 四川 甘肃 陕西 河北 广东 辽宁 台湾 重庆 福建 云南 江苏 浙江 澳门 天津 西藏 河南 江西 山东 新疆 宁夏 贵州 湖南 黑龙江 山西 安徽 广西 湖北 吉林 青海 内蒙古 海南

{'北京': 926, '香港': 1247, '上海': 715, '四川': 595, '甘肃': 164, '陕西': 320, '河北': 349, '广东': 1643, '辽宁': 156, '台湾': 449, '重庆': 582, '福建': 363, '云南': 186, '江苏': 654, '浙江': 1269, '澳门': 46, '天津': 198, '西藏': 1, '河南': 1276, '江西': 932, '山东': 792, '新疆': 76, '宁夏': 75, '贵州': 147, '湖南': 1019,
```

```
'黑龙江': 947, '山西': 198, '安徽': 991, '广西': 254, '湖北': 68135, '吉林': 155,
'青海': 18, '内蒙古': 238, '海南': 171}
```

### 8.3.3　绘制实时全国疫情确诊数对比图

编写实例文件 test04-matplotlib.py，功能是获取腾讯 API 接口中提供的 JSON 数据，然后使用 matplotlib 绘制实时全国疫情确诊数对比图。代码如下：

```
import time, json, requests
# 抓取腾讯疫情实时 json 数据
url = 'https://view.inews.qq.com/g2/getOnsInfo?
name=disease_h5&callback=&_=%d'%int(time.time()*1000)
data = json.loads(requests.get(url=url).json()['data'])
print(data)
print(data.keys())

# 统计省份信息(34 个省份 湖北 广东 河南 浙江 湖南 安徽....)
num = data['areaTree'][0]['children']
print(len(num))
for item in num:
    print(item['name'],end=" ")   # 不换行
else:
    print("\n")                   # 换行

# 解析数据(确诊 疑似 死亡 治愈)
total_data = {}
for item in num:
    if item['name'] not in total_data:
        total_data.update({item['name']:0})
    for city_data in item['children']:
        total_data[item['name']] +=int(city_data['total']['confirm'])
print(total_data)
# {'湖北': 48206, '广东': 1241, '河南': 1169, '浙江': 1145, '湖南': 968, ..., '
澳门': 10, '西藏': 1}

#-----------------------------------------------------------------------------
# 第二步：绘制柱状图
#-----------------------------------------------------------------------------
import matplotlib.pyplot as plt
import numpy as np

plt.rcParams['font.sans-serif'] = ['SimHei'] #设置字体，确保在图标中能正常显示中文标签
plt.rcParams['axes.unicode_minus'] = False #用来正常显示负号

#获取数据
names = total_data.keys()
nums = total_data.values()
print(names)
print(nums)

# 绘图
plt.figure(figsize=[10,6])
plt.bar(names, nums, width=0.3, color='green')
```

```
# 设置标题
plt.xlabel("地区", fontproperties='SimHei', size=12)
plt.ylabel("人数", fontproperties='SimHei', rotation=90, size=12)
plt.title("全国疫情确诊数对比图", fontproperties='SimHei', size=16)
plt.xticks(list(names), fontproperties='SimHei', rotation=-45, size=10)
# 显示数字
for a, b in zip(list(names), list(nums)):
    plt.text(a, b, b, ha='center', va='bottom', size=6)
plt.show()
```

执行效果如图 8-1 所示。

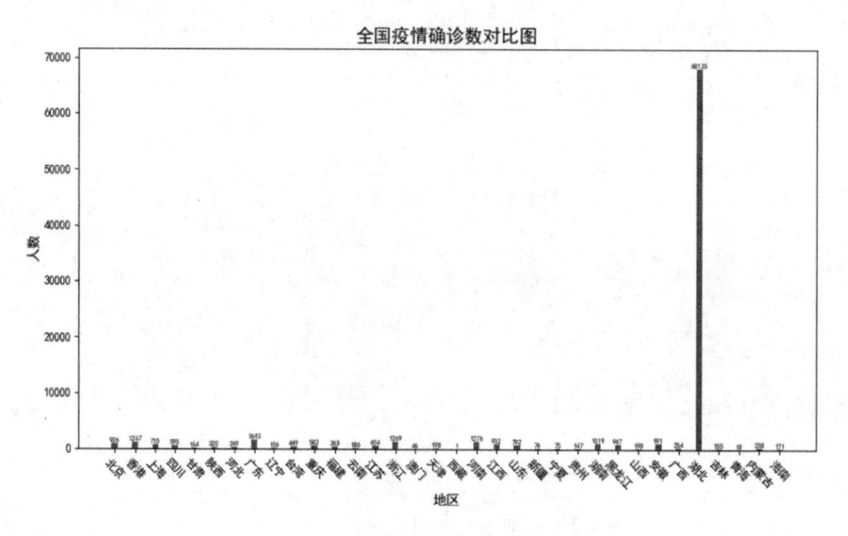


图 8-1　实时全国疫情确诊数对比图

## 8.3.4　绘制实时确诊人数、新增确诊人数、死亡人数、治愈人数对比图

编写实例文件 test05-matplotlib.py，功能是获取腾讯 API 接口中提供的 JSON 数据，然后使用 matplotlib 绘制国内各地实时确诊人数、新增确诊人数、死亡人数、治愈人数对比图。

(1)　首先抓取数据腾讯 API 的实时数据信息，加载显示获取的 JSON 数据，代码如下：

```
import time, json, requests
# 抓取腾讯疫情实时 json 数据
url = 'https://view.inews.qq.com/g2/getOnsInfo? \
name=disease_h5&callback=&_=%d'%int(time.time()*1000)
data = json.loads(requests.get(url=url).json()['data'])
print(data)
print(data.keys())
```

(2)　统计国内的省份和地区信息，代码如下：

```
num = data['areaTree'][0]['children']
print(len(num))
for item in num:
    print(item['name'],end=" ")   # 不换行
```

```
else:
    print("\n")                         # 换行
```

(3) 提取各地确诊人数数据，代码如下：

```
# 解析确诊数据
total_data = {}
for item in num:
    if item['name'] not in total_data:
        total_data.update({item['name']:0})
    for city_data in item['children']:
        total_data[item['name']] +=int(city_data['total']['confirm'])
print(total_data)
# {'湖北': 48206, '广东': 1241, '河南': 1169, '浙江': 1145, '湖南': 968, ...,
# '澳门': 10, '西藏': 1}
```

(4) 提取各地疑似人数数据，代码如下：

```
# 解析疑似数据
total_suspect_data = {}
for item in num:
    if item['name'] not in total_suspect_data:
        total_suspect_data.update({item['name']:0})
    for city_data in item['children']:
        total_suspect_data[item['name']] +=int(city_data['total']['suspect'])
print(total_suspect_data)
```

(5) 提取各地死亡人数数据，代码如下：

```
# 解析死亡数据
total_dead_data = {}
for item in num:
    if item['name'] not in total_dead_data:
        total_dead_data.update({item['name']:0})
    for city_data in item['children']:
        total_dead_data[item['name']] +=int(city_data['total']['dead'])
print(total_dead_data)
```

(6) 提取各地治愈人数数据，代码如下：

```
# 解析治愈数据
total_heal_data = {}
for item in num:
    if item['name'] not in total_heal_data:
        total_heal_data.update({item['name']:0})
    for city_data in item['children']:
        total_heal_data[item['name']] +=int(city_data['total']['heal'])
print(total_heal_data)
```

(7) 提取各地新增确诊人数数据，代码如下：

```
# 解析新增确诊数据
total_new_data = {}
for item in num:
    if item['name'] not in total_new_data:
        total_new_data.update({item['name']:0})
    for city_data in item['children']:
        total_new_data[item['name']] +=int(city_data['today']['confirm'])
        # today
print(total_new_data)
```

(8) 使用 matplotlib 分别绘制实时确诊人数、新增确诊人数、死亡人数、治愈人数对比图，代码如下：

```
#---------------------------------------------------------------------------
# 第二步：绘制柱状图
#---------------------------------------------------------------------------
import matplotlib.pyplot as plt
import numpy as np

plt.figure(figsize=[10,6])
plt.rcParams['font.sans-serif'] = ['SimHei']  #用来正常显示中文标签
plt.rcParams['axes.unicode_minus'] = False    #用来正常显示负号

#--------------------------1.绘制确诊数据-----------------------------------
p1 = plt.subplot(221)

# 获取数据
names = total_data.keys()
nums = total_data.values()
print(names)
print(nums)
print(total_data)
plt.bar(names, nums, width=0.3, color='green')

# 设置标题
plt.ylabel("确诊人数", rotation=90)
plt.xticks(list(names), rotation=-60, size=8)
# 显示数字
for a, b in zip(list(names), list(nums)):
   plt.text(a, b, b, ha='center', va='bottom', size=6)
plt.sca(p1)

#--------------------------2.绘制新增确诊数据------------------------------------
p2 = plt.subplot(222)
names = total_new_data.keys()
nums = total_new_data.values()
print(names)
print(nums)
plt.bar(names, nums, width=0.3, color='yellow')
plt.ylabel("新增确诊人数", rotation=90)
plt.xticks(list(names), rotation=-60, size=8)
# 显示数字
for a, b in zip(list(names), list(nums)):
   plt.text(a, b, b, ha='center', va='bottom', size=6)
plt.sca(p2)

#--------------------------3.绘制死亡数据-----------------------------------------
p3 = plt.subplot(223)
names = total_dead_data.keys()
nums = total_dead_data.values()
print(names)
print(nums)
plt.bar(names, nums, width=0.3, color='blue')
plt.xlabel("地区")
plt.ylabel("死亡人数", rotation=90)
plt.xticks(list(names), rotation=-60, size=8)
```

```
for a, b in zip(list(names), list(nums)):
   plt.text(a, b, b, ha='center', va='bottom', size=6)
plt.sca(p3)

#---------------------------4.绘制治愈数据---------------------------------
p4 = plt.subplot(224)
names = total_heal_data.keys()
nums = total_heal_data.values()
print(names)
print(nums)
plt.bar(names, nums, width=0.3, color='red')
plt.xlabel("地区")
plt.ylabel("治愈人数", rotation=90)
plt.xticks(list(names), rotation=-60, size=8)
for a, b in zip(list(names), list(nums)):
   plt.text(a, b, b, ha='center', va='bottom', size=6)
plt.sca(p4)
plt.show()
```

执行效果如图 8-2 所示。

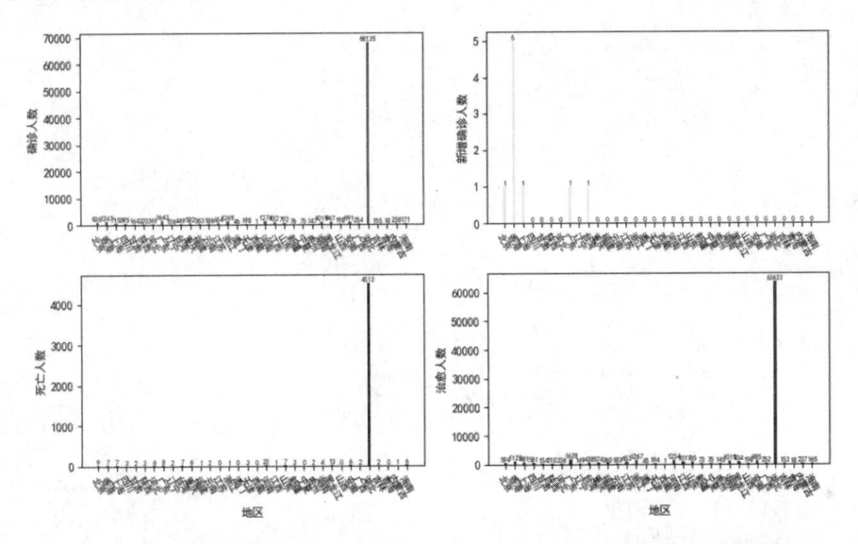


图 8-2　实时确诊人数、新增确诊人数、死亡人数、治愈人数对比图

## 8.3.5　将实时疫情数据保存到 CSV 文件

编写实例文件 test06-seaborn-write.py，功能是获取腾讯 API 接口中提供的 JSON 数据，然后将抓取的国内各地的实时疫情数据保存到 CSV 文件 2020-07-04-all.csv 中，其中文件名中的“2020-07-04”不是固定的，和当前日期相对应。

```
import time, json, requests
# 抓取腾讯疫情实时 json 数据
url = 'https://view.inews.qq.com/g2/getOnsInfo? \
name=disease_h5&callback= &_=%d'%int(time.time()*1000)
data = json.loads(requests.get(url=url).json()['data'])
print(data)
print(data.keys())
```

```
# 统计省份信息(34 个省份 湖北 广东 河南 浙江 湖南 安徽....)
num = data['areaTree'][0]['children']
print(len(num))
for item in num:
    print(item['name'],end=" ")   # 不换行
else:
    print("\n")                   # 换行


# 解析确诊数据
total_data = {}
for item in num:
    if item['name'] not in total_data:
        total_data.update({item['name']:0})
    for city_data in item['children']:
        total_data[item['name']] +=int(city_data['total']['confirm'])
print(total_data)
# {'湖北': 48206, '广东': 1241, '河南': 1169, '浙江': 1145, '湖南': 968, ...,
# '澳门': 10, '西藏': 1}

# 解析疑似数据
total_suspect_data = {}
for item in num:
    if item['name'] not in total_suspect_data:
        total_suspect_data.update({item['name']:0})
    for city_data in item['children']:
        total_suspect_data[item['name']] +=int(city_data['total']['suspect'])
print(total_suspect_data)

# 解析死亡数据
total_dead_data = {}
for item in num:
    if item['name'] not in total_dead_data:
        total_dead_data.update({item['name']:0})
    for city_data in item['children']:
        total_dead_data[item['name']] +=int(city_data['total']['dead'])
print(total_dead_data)

# 解析治愈数据
total_heal_data = {}
for item in num:
    if item['name'] not in total_heal_data:
        total_heal_data.update({item['name']:0})
    for city_data in item['children']:
        total_heal_data[item['name']] +=int(city_data['total']['heal'])
print(total_heal_data)

# 解析新增确诊数据
total_new_data = {}
for item in num:
    if item['name'] not in total_new_data:
        total_new_data.update({item['name']:0})
    for city_data in item['children']:
        total_new_data[item['name']] +=int(city_data['today']['confirm'])
        # today
print(total_new_data)

#------------------------------------------------------------------------------
```

```
# 第二步：存储数据至 CSV 文件
#----------------------------------------------------------------------------
names = list(total_data.keys())         # 省份名称
num1 = list(total_data.values())        # 确诊数据
num2 = list(total_suspect_data.values()) # 疑似数据(全为 0)
num3 = list(total_dead_data.values())    # 死亡数据
num4 = list(total_heal_data.values())    # 治愈数据
num5 = list(total_new_data.values())     # 新增确诊病例
print(names)
print(num1)
print(num2)
print(num3)
print(num4)
print(num5)

# 获取当前日期，并使用这个日期 CSV 文件命名
n = time.strftime("%Y-%m-%d") + "-all.csv"
fw = open(n, 'w', encoding='utf-8')
fw.write('province,confirm,dead,heal,new_confirm\n')
i = 0
while i<len(names):
   fw.write(names[i]+','+str(num1[i])+','+str(num3[i])+','+str(num4[i])+',
           '+str(num5[i])+'\n')
   i = i + 1
else:
   print("Over write file!")
   fw.close()
```

执行代码后，会创建一个以当前日期命名的 CSV 文件，例如作者执行文件的时间是“2020-07-04”，所以会创建文件 2020-07-04-all.csv，在文件里面保存了当前国内疫情的实时数据，如图 8-3 所示。

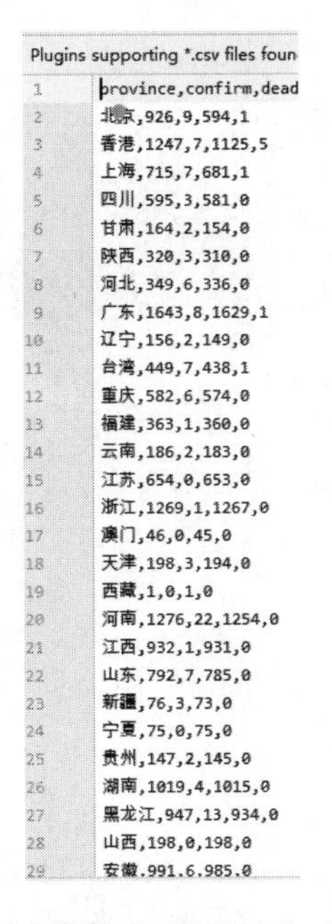
Plugins supporting *.csv files foun
1 province,confirm,dead
2 北京,926,9,594,1
3 香港,1247,7,1125,5
4 上海,715,7,681,1
5 四川,595,3,581,0
6 甘肃,164,2,154,0
7 陕西,320,3,310,0
8 河北,349,6,336,0
9 广东,1643,8,1629,1
10 辽宁,156,2,149,0
11 台湾,449,7,438,1
12 重庆,582,6,574,0
13 福建,363,1,360,0
14 云南,186,2,183,0
15 江苏,654,0,653,0
16 浙江,1269,1,1267,0
17 澳门,46,0,45,0
18 天津,198,3,194,0
19 西藏,1,0,1,0
20 河南,1276,22,1254,0
21 江西,932,1,931,0
22 山东,792,7,785,0
23 新疆,76,3,73,0
24 宁夏,75,0,75,0
25 贵州,147,2,145,0
26 湖南,1019,4,1015,0
27 黑龙江,947,13,934,0
28 山西,198,0,198,0
29 安徽,991,6,985,0

图 8-3　文件 2020-07-04-all.csv

### 8.3.6 绘制国内实时疫情统计图

编写实例文件 test07-seaborn-write.py，功能是提取刚才创建的 CSV 文件 2020-07-04-all.csv 中的数据，使用 Seaborn 绘制国内实时疫情统计图。

```
import time
import matplotlib
import numpy as np
import seaborn as sns
import pandas as pd
import matplotlib.pyplot as plt

# 读取数据
n = time.strftime("%Y-%m-%d") + "-all.csv"
data = pd.read_csv(n)

# 设置窗口
fig, ax = plt.subplots(1,1)
print(data['province'])

# 设置绘图风格及字体
sns.set_style("whitegrid",{'font.sans-serif':['simhei','Arial']})

# 绘制柱状图
g = sns.barplot(x="province", y="confirm", data=data, ax=ax,
         palette=sns.color_palette("hls", 8))

# 在柱状图上显示数字
i = 0
for index, b in zip(list(data['province']), list(data['confirm'])):
   g.text(i+0.05, b+0.05, b, color="black", ha="center", va='bottom', size=6)
   i = i + 1

# 设置 Axes 的标题
ax.set_title('全国疫情最新情况')

# 设置坐标轴文字方向
ax.set_xticklabels(ax.get_xticklabels(), rotation=-60)

# 设置坐标轴刻度的字体大小
ax.tick_params(axis='x',labelsize=8)
ax.tick_params(axis='y',labelsize=8)

plt.show()
```

执行效果如图 8-4 所示。

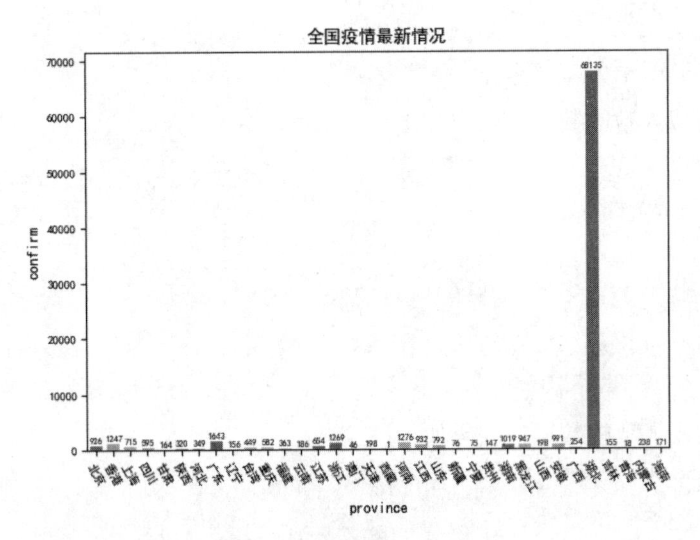


图 8-4　国内实时疫情统计图

### 8.3.7　可视化实时疫情的详细数据

编写实例文件 test08-seaborn-write-4db.py，功能是获取腾讯 API 接口中提供的 JSON 数据，然后将抓取的国内各地的实时疫情详细数据保存到 CSV 文件 2020-07-04-all-4db.csv 中。其中文件名中不是固定的，和当前日期相对应。详细数据包括确诊人数、治愈人数、死亡人数和新增确诊人数，最后使用 Seaborn 绘制实时疫情详细数据的统计图。

(1) 抓取腾讯网 API 中的 JSON 数据，代码如下：

```
import time, json, requests
# 抓取腾讯疫情实时 json 数据
url = 'https://view.inews.qq.com/g2/getOnsInfo? \
name=disease_h5&callback=&_=%d'%int(time.time()*1000)
data = json.loads(requests.get(url=url).json()['data'])
print(data)
print(data.keys())
```

(2) 统计国内各个省份和地区的信息，代码如下：

```
num = data['areaTree'][0]['children']
print(len(num))
for item in num:
    print(item['name'],end=" ")    # 不换行
else:
    print("\n")                    # 换行
```

(3) 提取确诊人数的数据，代码如下：

```
# 解析确诊数据
total_data = {}
for item in num:
    if item['name'] not in total_data:
        total_data.update({item['name']:0})
```

```
    for city_data in item['children']:
        total_data[item['name']] +=int(city_data['total']['confirm'])
print(total_data)
```

(4) 提取疑似人数的数据，代码如下：

```
# 解析疑似数据
total_suspect_data = {}
for item in num:
    if item['name'] not in total_suspect_data:
        total_suspect_data.update({item['name']:0})
    for city_data in item['children']:
        total_suspect_data[item['name']] +=int(city_data['total']['suspect'])
print(total_suspect_data)
```

(5) 提取死亡人数的数据，代码如下：

```
# 解析死亡数据
total_dead_data = {}
for item in num:
    if item['name'] not in total_dead_data:
        total_dead_data.update({item['name']:0})
    for city_data in item['children']:
        total_dead_data[item['name']] +=int(city_data['total']['dead'])
print(total_dead_data)
```

(6) 提取治愈人数的数据，代码如下：

```
# 解析治愈数据
total_heal_data = {}
for item in num:
    if item['name'] not in total_heal_data:
        total_heal_data.update({item['name']:0})
    for city_data in item['children']:
        total_heal_data[item['name']] +=int(city_data['total']['heal'])
print(total_heal_data)
```

(7) 提取新增确诊人数的数据，代码如下：

```
# 解析新增确诊数据
total_new_data = {}
for item in num:
    if item['name'] not in total_new_data:
        total_new_data.update({item['name']:0})
    for city_data in item['children']:
        total_new_data[item['name']] +=int(city_data['today']['confirm'])
        # today
print(total_new_data)
```

(8) 分别将提取的各地确诊人数、治愈人数、死亡人数和新增确诊人数信息保存到CSV文件中，代码如下：

```
#-----------------------------------------------------------------------------
# 第二步：存储数据至CSV文件
#-----------------------------------------------------------------------------
names = list(total_data.keys())          # 省份名称
```

```
num1 = list(total_data.values())          # 确诊数据
num2 = list(total_suspect_data.values())  # 疑似数据(全为0)
num3 = list(total_dead_data.values())     # 死亡数据
num4 = list(total_heal_data.values())     # 治愈数据
num5 = list(total_new_data.values())      # 新增确诊病例
print(names)
print(num1)
print(num2)
print(num3)
print(num4)
print(num5)

# 获取当前日期，并使用这个日期为CSV 文件命名
n = time.strftime("%Y-%m-%d") + "-all-4db.csv"
fw = open(n, 'w', encoding='utf-8')
fw.write('province,tpye,data\n')
i = 0
while i<len(names):
    fw.write(names[i]+',confirm,'+str(num1[i])+'\n')
    fw.write(names[i]+',dead,'+str(num3[i])+'\n')
    fw.write(names[i]+',heal,'+str(num4[i])+'\n')
    fw.write(names[i]+',new_confirm,'+str(num5[i])+'\n')
    i = i + 1
else:
    print("Over write file!")
    fw.close()
```

执行代码后，会创建一个以当前日期命名的 CSV 文件，例如作者执行文件的时间是“2020-07-04”，所以会创建文件 2020-07-04-all-4db.csv，在文件里面保存了当前国内疫情的实时详细数据，如图 8-5 所示。

```
1   province,tpye,data
2   北京,confirm,926
3   北京,dead,9
4   北京,heal,594
5   北京,new_confirm,1
6   香港,confirm,1247
7   香港,dead,7
8   香港,heal,1125
9   香港,new_confirm,5
10  上海,confirm,715
11  上海,dead,7
12  上海,heal,681
13  上海,new_confirm,1
14  四川,confirm,595
15  四川,dead,3
16  四川,heal,581
17  四川,new_confirm,0
18  甘肃,confirm,164
19  甘肃,dead,2
20  甘肃,heal,154
21  甘肃,new_confirm,0
```

图 8-5　文件 2020-07-04-all-4db.csv

## 8.3.8 绘制实时疫情信息统计图

编写实例文件 test09-seaborn-write.py，功能是根据 CSV 文件 2020-07-04-all-4db.csv 中的数据，使用 Seaborn 绘制国内各地实时疫情信息统计图。代码如下：

```
# 读取数据
data = pd.read_csv("2020-07-04-all-4db.csv")

# 设置窗口
fig, ax = plt.subplots(1,1)
print(data['province'])

# 设置绘图风格及字体
sns.set_style("whitegrid",{'font.sans-serif':['simhei','Arial']})

# 绘制柱状图
g = sns.barplot(x="province", y="data", hue="tpye", data=data, ax=ax,
         palette=sns.color_palette("hls", 8))

# 设置 Axes 的标题
ax.set_title('全国疫情最新情况')

# 设置坐标轴文字方向
ax.set_xticklabels(ax.get_xticklabels())

# 设置坐标轴刻度的字体大小
ax.tick_params(axis='x',labelsize=8)
ax.tick_params(axis='y',labelsize=8)

plt.show()
```

执行效果如图 8-6 所示。

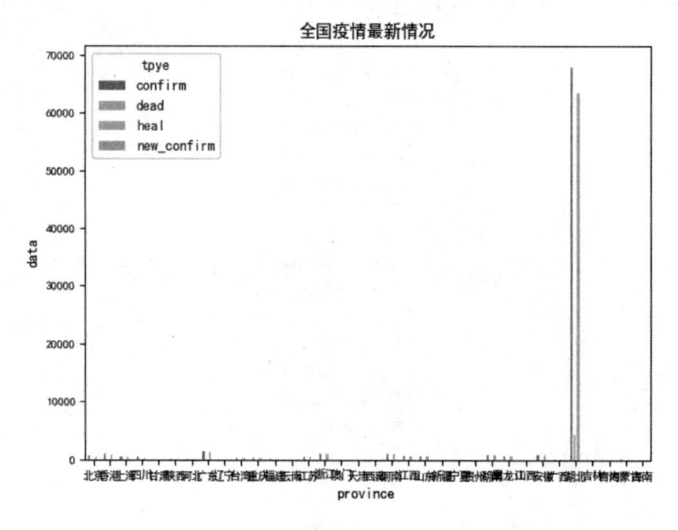


图 8-6　国内各地实时疫情信息统计图

## 8.3.9　绘制本年度国内疫情曲线图

编写实例文件 test8-qushi.py，功能是获取腾讯 API 接口中提供的 JSON 数据，然后使用 matplotlib 绘制本年度国内疫情曲线图，时间在 1 月开始，到当前时间结束。最后将绘制的图片保存为 "nCoV 疫情曲线.png"。代码如下：

```
# 抓取腾讯疫情实时 json 数据
def catch_daily():
    url = 'https://view.inews.qq.com/g2/getOnsInfo? \
    name=wuwei_ww_cn_day_counts&callback=&_=%d'%int(time.time()*1000)
    data = json.loads(requests.get(url=url).json()['data'])
    data.sort(key=lambda x:x['date'])

    date_list = list() # 日期
    confirm_list = list() # 确诊
    suspect_list = list() # 疑似
    dead_list = list() # 死亡
    heal_list = list() # 治愈
    for item in data:
        month, day = item['date'].split('/')
        date_list.append(datetime.strptime('2020-%s-%s'%(month, day), '%Y-%m-%d'))
        confirm_list.append(int(item['confirm']))
        suspect_list.append(int(item['suspect']))
        dead_list.append(int(item['dead']))
        heal_list.append(int(item['heal']))
    return date_list, confirm_list, suspect_list, dead_list, heal_list

# 绘制每日确诊和死亡数据
def plot_daily():


    date_list, confirm_list, suspect_list, dead_list, heal_list = catch_daily()
# 获取数据

    plt.figure('疫情统计图表', facecolor='#f4f4f4', figsize=(10, 8))
    plt.title('nCoV 疫情曲线', fontsize=20)

    plt.rcParams['font.sans-serif'] = ['SimHei']  #用来正常显示中文标签
    plt.rcParams['axes.unicode_minus'] = False     #用来正常显示负号

    plt.plot(date_list, confirm_list, 'r-', label='确诊')
    plt.plot(date_list, confirm_list, 'rs')
    plt.plot(date_list, suspect_list, 'b-',label='疑似')
    plt.plot(date_list, suspect_list, 'b*')
    plt.plot(date_list, dead_list, 'y-', label='死亡')
    plt.plot(date_list, dead_list, 'y+')
    plt.plot(date_list, heal_list, 'g-', label='治愈')
    plt.plot(date_list, heal_list, 'gd')

    plt.gca().xaxis.set_major_formatter(mdates.DateFormatter('%m-%d'))
    # 格式化时间轴标注
    plt.gcf().autofmt_xdate() # 优化标注(自动倾斜)
```

```
    plt.grid(linestyle=':') # 显示网格
    plt.legend(loc='best') # 显示图例
    plt.savefig('nCoV疫情曲线.png') # 保存为图片
    plt.show()

if __name__ == '__main__':
    plot_daily()
```

执行效果如图 8-7 所示。

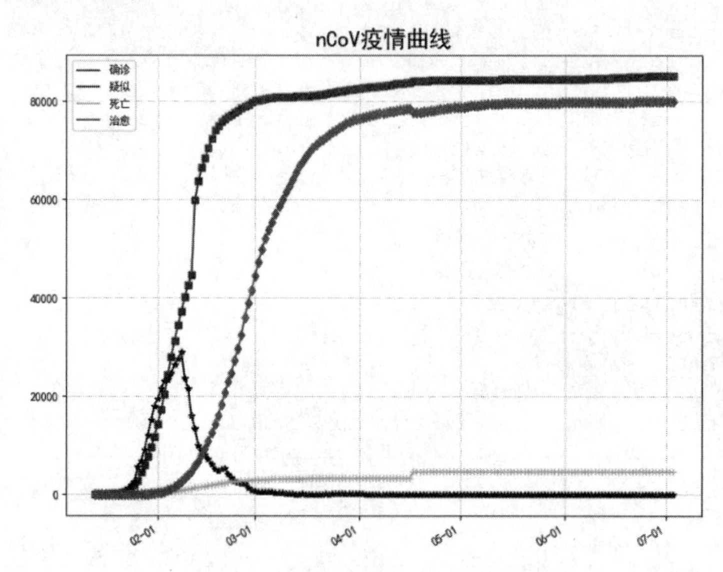


图 8-7　本年度国内疫情曲线图

## 8.3.10　统计山东省的实时疫情数据

编写实例文件 test13-spider-shandong.py，功能是获取腾讯 API 接口中提供的 JSON 数据，然后将抓取的山东省各地的实时疫情详细数据保存到 CSV 文件 2020-07-04-sd.csv 中。其中文件名中不是固定的，和当前日期相对应。

(1)　抓取腾讯网 API 中的 JSON 数据，代码如下：

```
import time, json, requests
# 抓取腾讯疫情实时 json 数据
url = 'https://view.inews.qq.com/g2/getOnsInfo? \
name=disease_h5&callback=&_=%d'%int(time.time()*1000)
data = json.loads(requests.get(url=url).json()['data'])
print(data)
print(data.keys())

# 统计省份信息(34 个省份 湖北 广东 河南 浙江 湖南 安徽....)
num = data['areaTree'][0]['children']
print(len(num))
```

(2)　提取 JSON 数据中省份为“山东”的数据，代码如下：

```
k = 0
for item in num:
```

```
    print(item['name'],end=" ")   # 不换行
    if item['name'] in "山东":
        print("")
        print(item['name'], k)
        break
    k = k + 1
print("") # 换行
```

(3) 分别获取确诊数据、疑似数据、死亡数据、治愈数据和新增确诊数据，代码如下：

```
#-----------------------------------------------------------------------
# 解析确诊数据
total_data = {}
for item in gz:
    if item['name'] not in total_data:
        total_data.update({item['name']:0})
    total_data[item['name']] = item['total']['confirm']
print('确诊人数')
print(total_data)


# 解析疑似数据
total_suspect_data = {}
for item in gz:
    if item['name'] not in total_suspect_data:
        total_suspect_data.update({item['name']:0})
    total_suspect_data[item['name']] = item['total']['suspect']
print('疑似人数')
print(total_suspect_data)

# 解析死亡数据
total_dead_data = {}
for item in gz:
    if item['name'] not in total_dead_data:
        total_dead_data.update({item['name']:0})
    total_dead_data[item['name']] = item['total']['dead']
print('死亡人数')
print(total_dead_data)

# 解析治愈数据
total_heal_data = {}
for item in gz:
    if item['name'] not in total_heal_data:
        total_heal_data.update({item['name']:0})
    total_heal_data[item['name']] = item['total']['heal']
print('治愈人数')
print(total_heal_data)

# 解析新增确诊数据
total_new_data = {}
for item in gz:
    if item['name'] not in total_new_data:
        total_new_data.update({item['name']:0})
```

```
    total_new_data[item['name']] = item['today']['confirm'] # today
print('新增确诊人数')
print(total_new_data)
```

(4) 将提取的数据信息保存到CSV文件中，代码如下：

```
names = list(total_data.keys())          # 省份名称
num1 = list(total_data.values())         # 确诊数据
num2 = list(total_suspect_data.values()) # 疑似数据(全为0)
num3 = list(total_dead_data.values())    # 死亡数据
num4 = list(total_heal_data.values())    # 治愈数据
num5 = list(total_new_data.values())     # 新增确诊病例
print(names)
print(num1)
print(num2)
print(num3)
print(num4)
print(num5)

n = time.strftime("%Y-%m-%d") + "-sd.csv"
fw = open(n, 'w', encoding='utf-8')
fw.write('province,confirm,dead,heal,new_confirm\n')
i = 0
while i<len(names):
    fw.write(names[i]+','+str(num1[i])+','+str(num3[i])+','+str(num4[i])+',
    '+str(num5[i])+'\n')
    i = i + 1
else:
    print("Over write file!")
    fw.close()
```

执行后会创建一个以当前日期命名的CSV文件，例如作者执行上述实例文件的时间是“2020-07-04”，所以会创建文件2020-07-04-sd.csv，在文件里面保存了当前山东省各地疫情的实时数据，如图8-8所示。

```
Plugins supporting *.csv files found.
1  province,confirm,dead,heal,new_confirm
2  境外输入,29,0,29,0
3  青岛,65,1,64,0
4  济宁,260,0,260,0
5  济南,47,0,47,0
6  潍坊,44,0,44,0
7  枣庄,24,0,24,0
8  淄博,30,1,29,0
9  烟台,47,0,47,0
10 泰安,35,2,33,0
11 威海,38,1,37,0
12 日照,16,0,16,0
13 临沂,49,0,49,0
14 德州,37,2,35,0
15 聊城,38,0,38,0
16 滨州,15,0,15,0
17 菏泽,18,0,18,0
```

图8-8　文件2020-07-04-all-4db.csv

## 8.3.11　绘制山东省实时疫情数据统计图

编写实例文件 test14-spider-shandong.py，功能是获取腾讯 API 接口中提供的 JSON 数据，然后将抓取的山东省各地的实时疫情详细数据保存到 CSV 文件 2020-07-04-all-4db.csv 中。其中文件名中不是固定的，和当前日期相对应。详细数据包括确诊人数、治愈人数、死亡人数和新增确诊人数，最后使用 Seaborn 绘制山东省实时疫情详细数据的统计图。

(1)　抓取腾讯网 API 中的 JSON 数据，代码如下：

```
import time, json, requests
# 抓取腾讯疫情实时 json 数据
url = 'https://view.inews.qq.com/g2/getOnsInfo? \
name=disease_h5&callback=&_=%d'%int(time.time()*1000)
data = json.loads(requests.get(url=url).json()['data'])
print(data)
print(data.keys())
```

(2)　提取 JSON 数据中省份为“山东”的数据，代码如下：

```
# 统计省份信息(34 个省份 湖北 广东 河南 浙江 湖南 安徽....)
num = data['areaTree'][0]['children']
print(len(num))

# 获取山东数据
k = 0
for item in num:
    print(item['name'],end=" ")   # 不换行
    if item['name'] in "山东":
        print("")
        print(item['name'], k)
        break
    k = k + 1
print("") # 换行

# 显示山东省数据
gz = num[k]['children']
for item in gz:
    print(item)
else:
    print("\n")
```

(3)　分别获取山东省各地的确诊数据、疑似数据、死亡数据、治愈数据和新增确诊数据，代码如下：

```
total_data = {}
for item in gz:
    if item['name'] not in total_data:
        total_data.update({item['name']:0})
    total_data[item['name']] = item['total']['confirm']
print('确诊人数')
print(total_data)
```

```
# 解析疑似数据
total_suspect_data = {}
for item in gz:
    if item['name'] not in total_suspect_data:
        total_suspect_data.update({item['name']:0})
    total_suspect_data[item['name']] = item['total']['suspect']
print('疑似人数')
print(total_suspect_data)

# 解析死亡数据
total_dead_data = {}
for item in gz:
    if item['name'] not in total_dead_data:
        total_dead_data.update({item['name']:0})
    total_dead_data[item['name']] = item['total']['dead']
print('死亡人数')
print(total_dead_data)

# 解析治愈数据
total_heal_data = {}
for item in gz:
    if item['name'] not in total_heal_data:
        total_heal_data.update({item['name']:0})
    total_heal_data[item['name']] = item['total']['heal']
print('治愈人数')
print(total_heal_data)

# 解析新增确诊数据
total_new_data = {}
for item in gz:
    if item['name'] not in total_new_data:
        total_new_data.update({item['name']:0})
    total_new_data[item['name']] = item['today']['confirm'] # today
print('新增确诊人数')
print(total_new_data)

#----------------------------------------------------------------------------
```

(4) 将山东省各地的实时疫情数据信息保存到CSV文件中，代码如下：

```
names = list(total_data.keys())          # 省份名称
num1 = list(total_data.values())          # 确诊数据
num2 = list(total_suspect_data.values()) # 疑似数据(全为0)
num3 = list(total_dead_data.values())     # 死亡数据
num4 = list(total_heal_data.values())     # 治愈数据
num5 = list(total_new_data.values())      # 新增确诊病例
print(names)
print(num1)
print(num2)
print(num3)
print(num4)
print(num5)

# 获取当前日期命名(2020-02-13-gz.csv)
```

```
n = time.strftime("%Y-%m-%d") + "-sd-4db.csv"
fw = open(n, 'w', encoding='utf-8')
fw.write('province,type,data\n')
i = 0
while i<len(names):
    fw.write(names[i]+',confirm,'+str(num1[i])+'\n')
    fw.write(names[i]+',dead,'+str(num3[i])+'\n')
    fw.write(names[i]+',heal,'+str(num4[i])+'\n')
    fw.write(names[i]+',new_confirm,'+str(num5[i])+'\n')
    i = i + 1
else:
    print("Over write file!")
    fw.close()
```

(5) 调用 Seaborn 绘制山东省实时疫情数据统计图，代码如下：

```
import time
import matplotlib
import numpy as np
import seaborn as sns
import pandas as pd
import matplotlib.pyplot as plt

# 读取数据
n = time.strftime("%Y-%m-%d") + "-sd-4db.csv"
data = pd.read_csv(n)

# 设置窗口
fig, ax = plt.subplots(1,1)
print(data['province'])

# 设置绘图风格及字体
sns.set_style("whitegrid",{'font.sans-serif':['simhei','Arial']})

# 绘制柱状图
g = sns.barplot(x="province", y="data", hue="type", data=data, ax=ax,
          palette=sns.color_palette("hls", 8))

# 设置 Axes 的标题
ax.set_title('山东疫情最新情况')

# 设置坐标轴文字方向
ax.set_xticklabels(ax.get_xticklabels(), rotation=-60)

# 设置坐标轴刻度的字体大小
ax.tick_params(axis='x',labelsize=8)
ax.tick_params(axis='y',labelsize=8)

plt.show()
```

执行代码后，会创建一个以当前日期命名的 CSV 文件，例如作者执行文件的时间是“2020-07-04”，所以会创建文件 2020-07-04-all-4db.csv，在文件里面保存了当前山东省各地疫情的实时数据，如图 8-9 所示。并且还绘制了山东省各地确诊人数、治愈人数、死亡人数和新增确诊人数的统计图，如图 8-10 所示。

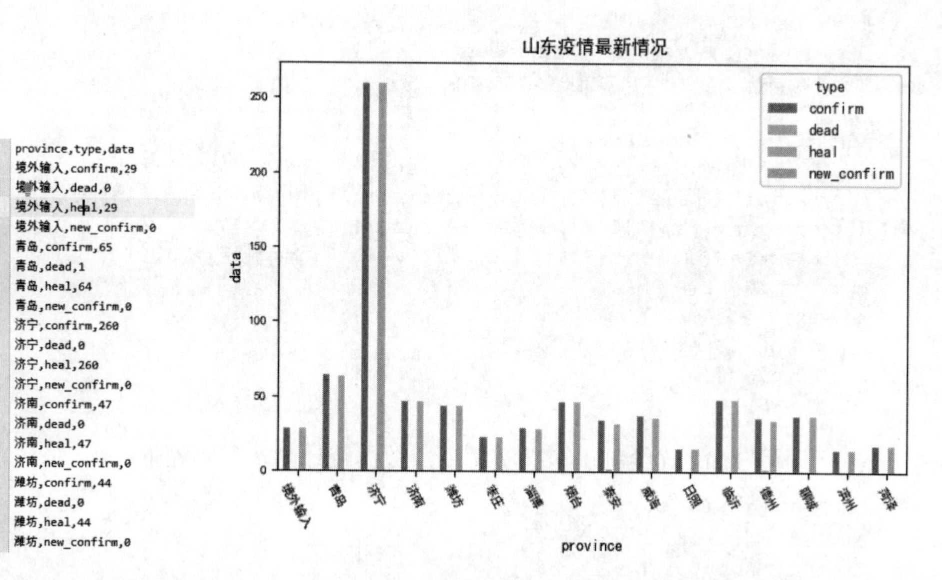

```
1  province,type,data
2  境外输入,confirm,29
3  境外输入,dead,0
4  境外输入,heal,29
5  境外输入,new_confirm,0
6  青岛,confirm,65
7  青岛,dead,1
8  青岛,heal,64
9  青岛,new_confirm,0
10 济宁,confirm,260
11 济宁,dead,0
12 济宁,heal,260
13 济宁,new_confirm,0
14 济南,confirm,47
15 济南,dead,0
16 济南,heal,47
17 济南,new_confirm,0
18 潍坊,confirm,44
19 潍坊,dead,0
20 潍坊,heal,44
21 潍坊,new_confirm,0
```



图 8-9 文件 2020-07-04-all-4db.csv

图 8-10 山东省实时疫情数据统计图

# 第 9 章

# 个人博客系统（Flask+TinyDB 实现）

在当今的信息时代，网络已经成为人们工作和学习的一部分，不断充实和改变着人们的生活。在网络中，构建一个个性化的日志系统，可以通过发布文章展示个人才能，抒发个人情感；网友则可以根据主题发表个人的意见，表达自己的想法，与博主进行思想交流。本章将详细讲解使用 Flask 开发个人博客系统的知识，对本书前面所学的 Flask 知识进行回顾。

## 9.1 博客系统介绍

扫码观看视频讲解

网络博客，又被称为网络日志、部落格或部落阁等，是一种通常由个人管理、不定期张贴新文章的网站。博客系统上的文章通常根据张贴时间，以倒序方式由新到旧排列。许多博客专注对特定的课题提供评论或新闻，其他则被作为比较个人化的日记。一个典型的博客结合了文字、图像、其他博客或网站的链接及其他与主题相关的媒体。让读者以互动的方式留下意见，是许多博客的重要目的。博客是社会媒体网络的一部分。

博客系统，是指使用计算机语言编写，并便于用户安装和使用，在互联网上建立个人博客的一整套系统。博客最初的名称是 Weblog，由 Web 和 Log 两个单词组成，按字面意思就是网络日记，后来喜欢新名词的人把这个词的发音故意改了一下，读成 we blog，由此，blog 这个词被创造出来。在中国大陆，有人往往也将 Blog 本身和 blogger(即博客作者)均音译为“博客”。“博客”有较深的涵义：“博”为“广博”；“客”不单是 blogger 更有“好客”之意，看 Blog 的人都是“客”。而在中国台湾，则分别音译成“部落格”(或“部落阁”)及“部落客”，认为 Blog 本身有社群群组的意含，借由 Blog 可以将网络上的网友集结成一个大博客，成为另一个具有影响力的自由媒体。

## 9.2 可行性分析

扫码观看视频讲解

作为软件开发必不可少的关键环节，可行性分析是第一个关键步骤，主要工作是根据项目需求分析使用的开发技术。本节将详细讲解本项目可行性分析的知识。

### 9.2.1 技术可行性分析：使用 TinyDB

TinyDB 是使用纯 Python 编写的 NoSQL 数据库，和 SQLite 数据库对应。SQLite 是小型、嵌入式的关系型数据库；而 TinyDB 是小型、嵌入式的 NoSQL 数据库，它不需要外部服务器也没有任何依赖，使用 JSON 文件存储数据。

TinyDB 源代码位于 https://github.com/msiemens/tinydb，TinyDB 官方文档位于 http://tinydb.readthedocs.io/en/latest/。

在现实应用中，只需使用如下所示的命令即可安装 TinyDB：

```
pip install tinydb
```

例如在下面的代码中，演示了创建、插入、查询、删除和更新 TinyDB 数据库中数据的过程：

```
from tinydb import TinyDB, Query, where
db = TinyDB('db.json')
```

```
# 插入两条记录
db.insert({'name': 'John', 'age': 22})
db.insert({'name': 'apple', 'age': 7})
# 输出所有记录
print(db.all())
# [{u'age': 22, u'name': u'John'}, {u'age': 7, u'name': u'apple'}]
# 查询
User = Query()
print(db.search(User.name == 'apple'))
# [{u'age': 7, u'name': u'apple'}]
# 查询
print(db.search(where('name') == 'apple'))
# 更新记录
db.update({'age': 10}, where('name') == 'apple')
# [{u'age': 10, u'name': u'apple'}]
# 删除 age 大于 20 的记录
db.remove(where('age') > 20)
# 清空数据库
db.purge()
```

执行代码后，会输出：

```
[{'name': 'John', 'age': 22}, {'name': 'apple', 'age': 7}]
[{'name': 'apple', 'age': 7}]
[{'name': 'apple', 'age': 7}]
```

## 9.2.2　系统基本要求

### 1. 采用架构

本项目采用浏览器/服务器(B/S)架构，服务器端使用 Falsk，数据库连接方式采用 TinyDB。

### 2. 主要功能要求

主要分为四大模块功能，即系统设置、后台管理、登录认证管理和日志展示。

- 系统设置：为了便于系统维护，将一些常用的功能放在系统设置文件中实现，这样做可以提高系统的健壮性。
- 后台管理：在管理目录中保存和后台管理相关的程序文件，实现了前台和后台的分离，提高了系统的安全性。
- 登录认证管理：确保只有系统合法用户才可以登录系统发布信息，提高系统的安全性。
- 日志展示：在前台显示系统中的博客信息。

## 9.2.3　可行性分析总结

上述可行性分析，参考现有开发资料、文档等资源，个人博客系统的体系结构比较完善，开发要求相比于复杂的系统较低，具备进一步进行需求分析与后续开发的条件。

## 9.3 具体实现

在接下来的内容中，将详细介绍使用 Flask+TinyDB 实现个人博客系统的过程。

### 9.3.1 系统设置

为了便于系统维护，将一些常用的功能放在系统设置文件中实现。具体来说，本项目主要涉及如下所示的系统设置文件。

(1) 初始化文件__init__.py，功能是设置用户登录模块，导入指定模块的视图文件，通过 app.config 命令设置系统加密信息和邮件服务器信息。具体实现代码如下所示：

```
app = Flask(__name__)

# Register blueprints
from app.admin import admin
app.register_blueprint(admin, url_prefix='/admin')
from app.auth import auth
app.register_blueprint(auth, url_prefix='/admin')
from app.main import main
app.register_blueprint(main, url_prefix='/')

from app.main.views import *
from app.auth.views import *
from app.main.errors import *

# Configure app
app.config['SECRET_KEY'] = os.environ.get('you-will-never-guess')
app.config['MAIL_SERVER'] = os.environ.get('MAIL_SERVER')
app.config['MAIL_PORT'] = 587
app.config['MAIL_USE_TLS'] = True
app.config['MAIL_USERNAME'] = os.environ.get('MAIL_USERNAME')
app.config['MAIL_PASSWORD'] = os.environ.get('MAIL_PASSWORD')
app.config['MAIL_SUBJECT_PREFIX'] = '[Chronoflask]'
app.config['MAIL_SENDER'] = 'Chronoflask <admin@chronoflask.com>'
app.config['DEFAULT_NAME'] = 'Chronoflask'
app.config['DEFAULT_AUTHOR'] = 'Chronologist'


# Set up Mail
mail = Mail(app)

# Set up Bootstrap
bootstrap = Bootstrap(app)
```

(2) 编写文件 db.py，实现 TinyDB 数据库操作功能，包括数据获取、数据检索、添加数据和数据更新功能。具体实现代码如下所示：

```
from tinydb import TinyDB, Query
```

```
import ujson

def get_db():
    db = TinyDB('db.json')
    return db

def get_table(table_name):
    table = get_db().table(table_name)
    return table

def get_record(table_name, query):
    result = get_table(table_name).get(query)
    return result

def search_records(table_name, query):
    results = get_table(table_name).search(query)
    return results

def insert_record(table_name, record):
    # Inserting a record returns the element id of the new record.
    element_id = get_table(table_name).insert(record)
    return element_id

def update_record(table_name, field, query):
    get_table(table_name).update(field, query)
    return True

def get_element_id(table_name, query):
    element = get_record(table_name, query)
    element_id = element.eid
    return element_id
```

(3)　编写文件 pagination.py，实现日志信息分页显示功能，具体实现代码如下所示：

```
def update_pagination():
    ''' Creates a table of entries organized by page
    for use when browsing entries. '''
    results = search_records('entries', \
        Query().creator_id == session.get('user_id'))
    all_entries = results[::-1]
    limit = len(all_entries)
    total = 0 # counter for total number of entries processed
    count = 0 # counter for number of entries—loop at 10
    page = 1 # will be key for dictionary of pages
    p_entries = list()
    get_table('pagination').purge() # clear the old pagination table
    for entry in all_entries:
        total += 1
        count += 1
        p_entries.append(entry)
        if count == 10 or total == limit:
            insert_record('pagination', {'page': page, 'entries': p_entries})
            del p_entries[:] # start the list fresh
            page += 1
            count = 0
    return True
```

```
# Get entries for the given page
def get_entries_for_page(page):
    results = get_record('pagination', Query().page == page)
    if not results:
        return abort(404)
    else:
        return results['entries']

def check_next_page(page):
    next_page = page + 1
    if not get_record('pagination', Query().page == next_page):
        return None
    else:
        return next_page
```

(4) 编写解析文件parse.py，功能是根据不同的节点显示日志信息，例如可以根据tags参数来发布新信息，根据days参数来浏览某一天的日志信息，根据raw_entry参数显示不同的日志信息界面或管理界面。具体实现代码如下所示：

```
def parse_input(raw_entry, current_time):
    '''Parse input and either create a new entry using the input
    or call a function (that may take part of the input as an argument).'''
    if raw_entry == 'browse all':
        return redirect(url_for('main.browse_all_entries'))
    elif raw_entry[:3] == 't: ':
        return redirect(url_for('main.view_single_entry', \
                            timestamp=raw_entry[3:]))
    elif raw_entry[:5] == 'tag: ':
        return redirect(url_for('main.view_entries_for_tag', \
                            tag=raw_entry[5:]))
    elif raw_entry[:5] == 'day: ':
        return redirect(url_for('view_entries_for_day', day=raw_entry[5:]))
    elif raw_entry == 'login':
        return redirect(url_for('auth.login'))
    elif raw_entry == 'logout':
        return redirect(url_for('auth.logout'))
    elif raw_entry == 'change email':
        return redirect(url_for('auth.change_email'))
    elif raw_entry == 'change password':
        return redirect(url_for('auth.change_password'))
    elif raw_entry == 'about':
        return redirect(url_for('admin.get_details'))
    elif raw_entry == 'rename chrono':
        return redirect(url_for('admin.rename_chronofile'))
    elif raw_entry == 'rename author':
        return redirect(url_for('admin.rename_author'))
    else:
        return process_entry(raw_entry, current_time)


def process_entry(raw_entry, current_time):
    '''Take the user's input and UTC datetime and return a clean, formatted
    entry, a timestamp as a string, and a list of tags'''
```

```
    raw_tags = find_and_process_tags(raw_entry)
    clean_entry = clean_up_entry(raw_entry, raw_tags)
    clean_tags = clean_up_tags(raw_tags)
    timestamp = create_timestamp(current_time)
    return create_new_entry(clean_entry, timestamp, clean_tags)


def clean_up_entry(raw_entry, raw_tags):
    '''Strip tags from end of entry.'''
    bag_of_words = raw_entry.split()
    if bag_of_words == raw_tags:
        clean_entry = bag_of_words[0]
    else:
        while bag_of_words[-1] in raw_tags:
            bag_of_words.pop()
        stripped_entry = ' '.join(bag_of_words)
        # Uppercase first letter
        clean_entry = stripped_entry[0].upper() + stripped_entry[1:]
    return clean_entry


def create_timestamp(current_time):
    '''TinyDB can't handle datetime objects; convert datetime to string.'''
    timestamp = datetime.strftime(current_time, '%Y-%m-%d %H:%M:%S')
    return timestamp
```

(5) 因为本项目用到了信息加密认证功能，所以需要编写独立文件 config.py 来保存 SECRET_KEY 信息。在编写 Flask Web 项目的时候，如果没有独立设置 SECRET_KEY，则会显示“Must provide secret_key to use csrf”错误提示。不能将 SECRET_KEY 写在程序代码中，需要单独放在独立文件 config.py 中，具体实现代码如下所示：

```
CSRF_ENABLED = True
SECRET_KEY = 'you-will-never-guess'
```

(6) 编写文件 run.py 作为 Flask 项目的启动文件，在里面调用了文件 config.py 中的 SECRET_KEY 信息，具体实现代码如下所示：

```
app.config.from_object('config')

# Run app
if __name__ == '__main__':
    app.run(debug=app.config['DEBUG'])
```

(7) 编写文件 mail.py，实现邮件发送功能，邮件服务器的设置信息在文件 app__init__.py 中实现，具体实现代码如下所示：

```
mail = Mail(app)

def send_async_email(app, msg):1
    with app.app_context():
        mail.send(msg)

def send_email(to, subject, template, **kwargs):
    msg = Message(app.config['MAIL_SUBJECT_PREFIX'] + ' ' + subject,\
```

```
            sender=app.config['MAIL_SENDER'], recipients=[to])
    msg.body = render_template(template + '.txt', **kwargs)
    msg.html = render_template(template + '.html', **kwargs)
    thr = Thread(target=send_async_email, args=[app, msg])
    thr.start()
    return thr
```

## 9.3.2 后台管理

在 admin 目录中保存了和后台管理相关的程序文件，接下来将详细介绍这部分程序文件的具体实现。

(1) 编写程序文件 forms.py，功能是分别修改系统名字和系统作者名字，具体实现代码如下所示：

```
class RenameChronofileForm(Form):
    new_name = StringField('Enter new name for chronofile:', \
                    validators=[Required()])
    submit = SubmitField('Rename chronofile')


class RenameAuthorForm(Form):
    new_name = StringField('Enter new author name:', \
                    validators=[Required()])
    submit = SubmitField('Rename author')
```

(2) 编写程序文件 views.py，功能是根据用户操作来到指定的后台管理页面，分别实现修改系统名称和系统作者名字的功能。具体实现代码如下所示：

```
@admin.route('/')
@login_required
def view_admin():
    '''Display name of site, author, etc. as well as links to edit
    those details and change email, password, etc.'''
    details = get_details()
    return render_template('admin.html', details=details)


@admin.route('/rename_chronofile', methods=['GET', 'POST'])
@login_required
def rename_chronofile():
    details = get_details()
    form = RenameChronofileForm()
    if form.validate_on_submit():
        update_record('admin', {'chronofile_name': form.new_name.data}, \
                Query().creator_id == session.get('user_id'))
        flash('Chronfile name updated.')
        return redirect(url_for('admin.view_admin'))
    return render_template('rename_chronofile.html', \
                    form=form, details=details)


@admin.route('/rename_author', methods=['GET', 'POST'])
@login_required
```

```
def rename_author():
    details = get_details()
    form = RenameAuthorForm()
    if form.validate_on_submit():
        test=update_record('admin', {'author_name': form.new_name.data}, \
                        Query().creator_id == session.get('user_id'))
        flash('Author name updated.')
        return redirect(url_for('admin.view_admin'))
    return render_template('rename_author.html', form=form, details=details)
```

## 9.3.3　登录认证管理

在 auth 目录中保存了和用户登录认证相关的程序文件，接下来将详细介绍这部分程序文件的具体实现。

(1) 编写程序文件 forms.py，功能是分别实现账号检测、邮箱检测、登录验证、注册信息验证、登录表单信息处理、邮箱设置、重设邮箱、修改密码和重置密码等功能。具体实现代码如下所示：

```
# 基本验证
def account_exists(form, field):
    user = get_record('auth', Query().email == field.data)
    if not user:
        raise ValidationError('Create an account first.')


def email_exists(form, field):
    user = get_record('auth', Query().email == field.data)
    if not user:
        raise ValidationError('Please verify that you typed your email \
            correctly.')


def authorized(form, field):
    '''Verify user through password.'''
    user = get_table('auth').get(eid=session.get('user_id'))
    if not pwd_context.verify(field.data, user['password_hash']):
        raise ValidationError('Invalid login credentials. Please try again.')


def has_digits(form, field):
    if not bool(re.search(r'\d', field.data)):
        raise ValidationError('Your password must contain at least one \
            number.')


def has_special_char(form, field):
    if not bool(re.search(r'[^\w\*]', field.data)):
        raise ValidationError('Your password must contain at least one \
            special character.')


class PasswordCorrect(object):
```

```
    '''Verify email/password combo before validating form.'''
    def __init__(self, fieldname):
        self.fieldname = fieldname

    def __call__(self, form, field):
        try:
            email = form[self.fieldname]
        except KeyError:
            raise ValidationError(field.gettext("Invalid field name '%s'.") \
                % self.fieldname)
        user = get_record('auth', Query().email == email.data)
        if not pwd_context.verify(field.data, user['password_hash']):
            raise ValidationError('Invalid password. Please try again.')


class LoginForm(Form):
    email = StringField('Email:', validators=[Required(), Email(), \
        account_exists])
    password = PasswordField('Password:', validators=[Required(), \
        PasswordCorrect('email')])
    submit = SubmitField('Log in')


class RegistrationForm(Form):
    email = StringField('Enter email address:', \
        validators=[Required(), Email()])
    password = PasswordField('Enter password: ' +\
        '(min 12 char., must incl. number and special character)', \
        validators=[Required(), Length(min=12), has_digits, has_special_char])
    submit = SubmitField('Create account')


class ChangeEmailForm(Form):
    password = PasswordField('Enter your password:', validators=[Required(), \
        authorized])
    new_email = StringField('New email address:', \
        validators=[Required(), Email(), EqualTo('verify_email', \
        message='Emails must match')])
    verify_email = StringField('Re-enter new email address:', \
        validators=[Required(), Email()])
    submit = SubmitField('Change email')


class ChangePasswordForm(Form):
    current_password = PasswordField('Your current password:', \
        validators=[Required(), authorized])
    new_password = PasswordField('New password: ' +\
        '(min 12 char., must incl. number and special character)', \
        validators=[Required(), Length(min=12), EqualTo('verify_password', \
        message='New passwords must match.'), has_digits, has_special_char])
    verify_password = PasswordField('Re-enter new password:', \
        validators=[Required(), Length(min=12)])
    submit = SubmitField('Change password')
```

```
class ResetPasswordForm(Form):
    email = StringField('Your registered email address:',
        validators=[Required(), Email(), email_exists])
    submit = SubmitField('Request password reset link')


class SetNewPasswordForm(Form):
    new_password = PasswordField('New password: ' +\
        '(min 12 char., must incl. number and special character)', \
        validators=[Required(), Length(min=12), EqualTo('verify_password', \
        message='New passwords must match.'), has_digits, has_special_char])
    verify_password = PasswordField('Re-enter new password:', \
        validators=[Required(), Length(min=12)])
    submit = SubmitField('Set new password')
```

(2) 编写程序文件 views.py，功能是根据用户操作来到指定的认证页面，分别实现登录验证、登录注销、密码重置和邮箱修改等视图功能。具体实现代码如下所示：

```
@auth.route('/login', methods=['GET', 'POST'])
def login():
    details = get_details()
    if not get_record('auth', Query().email.exists()):
        flash('You need to register first.')
        return redirect(url_for('auth.register'))
    if session.get('logged_in'):
        return redirect(url_for('main.browse_all_entries'))
    form = LoginForm()
    if form.validate_on_submit():
        session['logged_in'] = True
        user_id = get_element_id('auth', Query().email == form.email.data)
        session['user_id'] = user_id
        if request.args.get('next'):
            return redirect(request.args.get('next'))
        else:
            return redirect(url_for('main.browse_all_entries'))
    return render_template('login.html', form=form, details=details)


@auth.route('/logout')
@login_required
def logout():
    session['logged_in'] = None
    flash('You have been logged out.')
    return redirect(url_for('main.browse_all_entries'))


@auth.route('/register', methods=['GET', 'POST'])
def register():
    ''' Register user and create pagination table with one page
    and no entries.'''
    details = get_details()
    if details:
        flash('A user is already registered. Log in.')
        return redirect(url_for('auth.login'))
    details = {'chronofile_name': current_app.config['DEFAULT_NAME'], \
```

```
            'author_name': current_app.config['DEFAULT_AUTHOR']}
    register = True
    form = RegistrationForm()
    if form.validate_on_submit():
        password_hash = pwd_context.hash(form.password.data)
        # Create account and get creator id
        creator_id = insert_record('auth', {'email': form.email.data, \
                                  'password_hash': password_hash})
        insert_record('admin', {'chronofile_name': \
                               current_app.config['DEFAULT_NAME'], \
                               'author_name': \
                               current_app.config['DEFAULT_AUTHOR'], \
                               'creator_id': creator_id})
        insert_record('pagination', {'page': 1, 'entries': None})
        flash('Registration successful. You can login now.')
        return redirect(url_for('auth.login'))
    return render_template('register.html', form=form, details=details, \
                          register=register)


@auth.route('/reset_password', methods=['GET', 'POST'])
def request_reset():
    details = get_details()
    if not details:
        return abort(404)
    form = ResetPasswordForm()
    if form.validate_on_submit():
        email = form.email.data
        user_id = get_element_id('auth', Query().email == email)
        token = generate_confirmation_token(user_id)
        send_email(email, 'Link to reset your password',
                   'email/reset_password', token=token)
        flash('Your password reset token has been sent.')
        return redirect(url_for('auth.login'))
    return render_template('reset_password.html', form=form, details=details)


@auth.route('/reset_password/<token>', methods=['GET', 'POST'])
def confirm_password_reset(token):
    details = get_details()
    if not details:
        return abort(404)
    s = Serializer(current_app.config['SECRET_KEY'])
    try:
        data = s.loads(token)
    except:
        flash('The password reset link is invalid or has expired.')
        return redirect(url_for('auth.request_reset'))
    if not data.get('confirm'):
        flash('The password reset link is invalid or has expired.')
        return redirect(url_for('auth.request_reset'))
    user_id = data.get('confirm')
    form = SetNewPasswordForm()
    if form.validate_on_submit():
        new_password_hash = pwd_context.hash(form.new_password.data)
```

```
        get_table('auth').update({'password_hash': new_password_hash}, \
                         eids=[user_id])
        flash('Password updated—you can now log in.')
        return redirect(url_for('auth.login'))
    return render_template('set_new_password.html', form=form, token=token, \
                     details=details)


@auth.route('/change_email', methods=['GET', 'POST'])
@login_required
def change_email():
    details = get_details()
    form = ChangeEmailForm()
    if form.validate_on_submit():
        new_email = form.new_email.data
        user_id = session.get('user_id')
        get_table('auth').update({'email': new_email}, eids=user_id)
        flash('Your email address has been updated.')
        return redirect(url_for('admin.view_admin'))
    return render_template('change_email.html', form=form, details=details)


@auth.route('/change_password', methods=['GET', 'POST'])
@login_required
def change_password():
    details = get_details()
    form = ChangePasswordForm()
    if form.validate_on_submit():
        new_password_hash = pwd_context.hash(form.new_password.data)
        get_table('auth').update({'password_hash': new_password_hash}, \
                         eids=[session.get('user_id')])
        flash('Your password has been updated.')
        return redirect(url_for('admin.view_admin'))
    return render_template('change_password.html', form=form, details=details)


def generate_confirmation_token(user_id, expiration=3600):
    serial = Serializer(current_app.config['SECRET_KEY'], expiration)
    return serial.dumps({'confirm': user_id})
```

## 9.3.4　前台日志展示

在 main 目录中保存了和用户登录认证相关的程序文件，接下来将详细介绍这部分程序文件的具体实现。

(1)　编写程序文件 forms.py，功能是分别实现日志信息发布和修改功能。具体实现代码如下所示：

```
class RawEntryForm(Form):
    raw_entry = StringField('New entry in chronofile:',
validators=[Required()])
    submit = SubmitField('Post entry')

def no_hashtags(form, field):
```

```
    if bool(re.search(r'#', field.data)):
        raise ValidationError(Markup("Don't start tags with \
            <code>#</code> here."))

def use_commas(form, field):
    tags = field.data.split(' ')
    items = len(tags)
    print(items)
    if items != 1:
        count = 1
        for i in tags:
            print(i)
            print(i[-1])
            if i[-1] == ',':
                print(i[-1])
                count += 1
        print(count)
        if count < items:
            raise ValidationError(Markup('Separate tags \
                with commas like this: <code>tag1, tag2</code>.'))


class EditEntryForm(Form):
    new_entry = StringField('Edit entry text:', validators=[Required()])
    new_tags = StringField("Edit tags: (separate with commas, don't use #)", \
        validators=[no_hashtags, use_commas])
    submit = SubmitField('Save edited entry')
```

(2) 编写程序文件 views.py，功能是根据用户操作来到指定的前台展示页面，分别实现发布新日志信息、浏览日志信息、浏览全部日志信息和浏览某日信息功能。具体实现代码如下所示：

```
@main.route('/', methods=['GET', 'POST'])
def browse_all_entries():
    '''Returns all entries (most recent entry at the top of the page).'''
    details = get_details()
    if not session.get('logged_in'):
        if details:
            register = False
        else:
            register = True
            details = {'chronofile_name': current_app.config['DEFAULT_NAME'], \
                    'author_name': current_app.config['DEFAULT_AUTHOR']}
        return render_template('welcome.html', details=details, \
                        register=register)
    form = RawEntryForm()
    # Try to validate form and create a new entry
    if form.validate_on_submit():
        return parse_input(form.raw_entry.data, datetime.utcnow())
    # Otherwise, show the latest entries
    page = 1
    # Get entries for the given page
    entries_for_page = get_entries_for_page(page)
    # Check if there's another page, returns None if not
    next_page = check_next_page(page)
    return render_template('home.html', entries_for_page=entries_for_page, \
```

```
        form=form, details=details, next_page=next_page)


@main.route('page/<page>', methods=['GET', 'POST'])
@login_required
def view_entries_for_page(page):
    '''Returns entries for given page in reverse chronological order.'''
    try:
        int(page)
    except:
        TypeError
        return abort(404)
    page = int(page)
    if page == 1:
        return redirect(url_for('main.browse_all_entries'))
    details = get_details()
    form = RawEntryForm()
    if form.validate_on_submit():
        return parse_input(form.raw_entry.data, datetime.utcnow())
    # Get entries for the given page
    entries_for_page = get_entries_for_page(page)
    # Check if there's another page, returns None if not
    next_page = check_next_page(page)
    prev_page = page - 1
    return render_template('page.html', form=form, \
        entries_for_page=entries_for_page, details=details, \
        page=page, next_page=next_page, prev_page=prev_page)


@main.route('day/<day>', methods=['GET', 'POST'])
@login_required
def view_entries_for_day(day):
    '''Returns entries for given day in chronological order.'''
    details = get_details()
    form = RawEntryForm()
    if form.validate_on_submit():
        return parse_input(form.raw_entry.data, datetime.utcnow())
    entries_for_day = search_records('entries', Query().timestamp.all([day]))
    if not entries_for_day:
        return abort(404)
    return render_template('day.html', form=form, day=day, \
                      entries_for_day=entries_for_day, details=details)


@main.route('timestamp/<timestamp>', methods=['GET', 'POST'])
@login_required
def view_single_entry(timestamp):
    '''Return a single entry based on given timestamp.'''
    entry = get_record('entries', Query().timestamp == timestamp)
    if not entry:
        return abort(404)
    form = RawEntryForm()
    if form.validate_on_submit():
        return parse_input(form.raw_entry.data, datetime.utcnow())
    details = get_details()
    return render_template('entry.html', form=form, timestamp=timestamp, \
                      entry=entry, details=details)
```

```
@main.route('timestamp/<timestamp>/edit', methods=['GET', 'POST'])
@login_required
def edit_entry(timestamp):
    '''Edit an entry and return a view of the edited entry'''
    entry = get_record('entries', Query().timestamp == timestamp)
    if not entry:
        return abort(404)
    form = EditEntryForm()
    if form.validate_on_submit():
        # Split comma-delimted string of tags into a list
        # Delete spaces at the start of tags if necessary
        tags = form.new_tags.data.split(", ")
        update_record('entries', {'entry': form.new_entry.data, \
            'tags': tags}, (Query().creator_id == 1) & \
            (Query().timestamp == timestamp))
        flash('Entry updated.')
        update_pagination()
        return redirect(url_for('main.view_single_entry',
timestamp=timestamp))
    form.new_entry.default = entry['entry']
    form.new_tags.default = ', '.join(entry['tags'])
    form.process()
    details = get_details()
    return render_template('edit_entry.html', form=form,
    \timestamp=timestamp, details=details)


@main.route('tags', methods=['GET', 'POST'])
@login_required
def view_all_tags():
    details = get_details()
    form = RawEntryForm()
    if form.validate_on_submit():
        return parse_input(form.raw_entry.data, datetime.utcnow())
    all_entries = search_records('entries', \
                            Query().creator_id == session.get('user_id'))
    all_tags = list()
    for entry in all_entries:
        for tag in entry['tags']:
            if tag not in all_tags:
                all_tags.append(tag)
    all_tags.sort()
    return render_template('tags.html', all_tags=all_tags, form=form, \
                        details=details)


@main.route('days', methods=['GET', 'POST'])
@login_required
def view_all_days():
    details = get_details()
    form = RawEntryForm()
    if form.validate_on_submit():
        return parse_input(form.raw_entry.data, datetime.utcnow())
    all_entries = search_records('entries', \
                            Query().creator_id == session.get('user_id'))
    all_days = list()
    for entry in all_entries:
```

```
        if entry['timestamp'][:10] not in all_days:
            all_days.append(entry['timestamp'][:10])
    return render_template('days.html', all_days=all_days, form=form, \
                    details=details)


@main.route('tags/<tag>', methods=['GET', 'POST'])
@login_required
def view_entries_for_tag(tag):
    '''Return entries for given tag in chronological order.'''
    details = get_details()
    form = RawEntryForm()
    if form.validate_on_submit():
        return parse_input(form.raw_entry.data, datetime.utcnow())
    entries_for_tag = search_records('entries', Query().tags.all([tag]))
    if not entries_for_tag:
        return abort(404)
    return render_template('tag.html', form=form, tag=tag, \
                    entries_for_tag=entries_for_tag, details=details)
```

(3) 编写程序文件 errors.py，功能是如果获取数据库信息出错，则跳转到指定的 HTML 页面。具体实现代码如下所示：

```
@main.app_errorhandler(404)
def page_not_found(e):
    details = get_details()
    return render_template('404.html', details=details), 404


@main.app_errorhandler(500)
def internal_server_error(e):
    details = get_details()
    return render_template('500.html', details=details), 500
```

## 9.3.5　系统模板

在 templates 目录中保存了 Flask 系统模板文件，接下来将详细介绍这部分程序文件的具体实现。

(1) 编写程序文件 welcome.html，功能是显示欢迎信息，提供登录链接供用户登录系统。具体实现代码如下所示：

```
{% extends "base.html" %}
{% block page_content %}

    <div class="page-header">
        <h1>Hello</h1>
    </div>

    <div>
        {% if register == True %}
        <p>Chronoflask is a minimalist diary/journal application using Python 3,
Flask, and TinyDB inspired by Warren Ellis's Chronofile Minimal and Buckminster
Fuller's Dymaxion Chronofile.</p></br>
```

```
        <p>Add new entries (with or witout hashtags) in a single input field. Each
entry is stored with a UTC timestamp.</p></br>
        <p>View recent entries, view all entries for a single day or date-range
(chronologically), view a single entry, view all entries associated with a tag,
and view a list of tags.</p></br>
        </p>Private by default. Please <a
href="{{ url_for('auth.register') }}">register</a> to begin using
Chronoflask.</p>
        {% else %}
        <h2>Please <a href="{{ url_for('auth.login') }}">log in</a>.</h2>
        {% endif %}
    </div>
```

执行效果如图 9-1 所示。

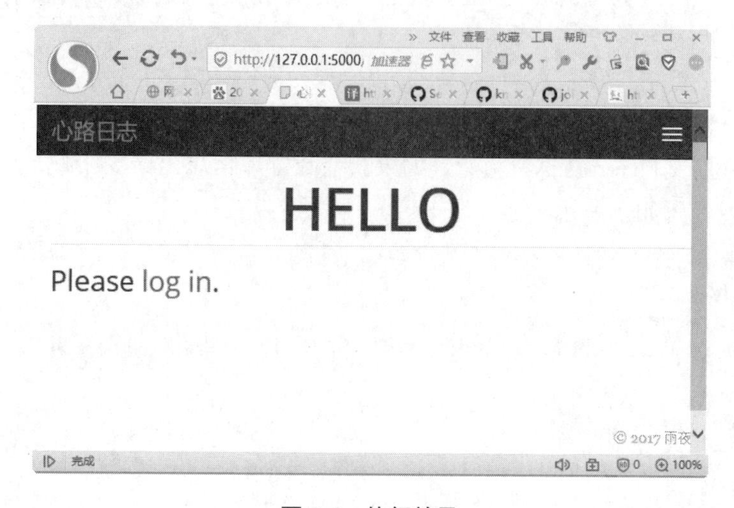


图 9-1　执行效果

(2) 编写程序文件 login.html，功能是提供登录表单让用户登录系统，并提供找回密码链接。具体实现代码如下所示：

```
{% extends "base.html" %}
{% block page_content %}

    <div class="page-header">
        <h1>Login</h1>
    </div>

    <div>
        {{ wtf.quick_form(form) }}
        <p>Forgot your password? Click <a
href="{{ url_for('auth.request_reset') }}">here</a> to reset it.</p>
    </div>

{% endblock %}
```

执行效果如图 9-2 所示。

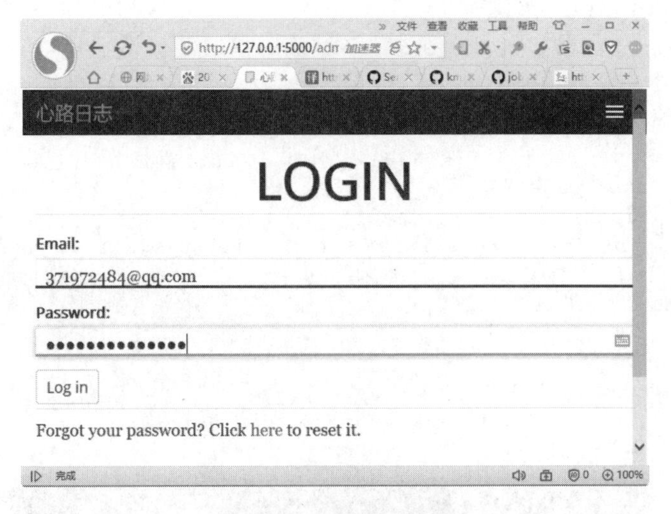

图 9-2　用户登录界面

(3) 编写程序文件 home.html，功能是实现系统前台页面信息展示功能，将以分页的样式展示系统数据库内所有的日志信息。具体实现代码如下所示：

```
{% extends "base.html" %}
{% block page_content %}

   <div class="page-header">
      <h1>Stream</h1>
   </div>

   <div>
      {{ wtf.quick_form(form) }}
   </div>

   <div>
   {% if not entries_for_page %}
      <p>Nothing in your chronofile yet.</p>
   {% else %}
      <ul>
      {% for entry in entries_for_page %}
         <li><p><a href="{{ url_for('main.view_entries_for_day',
day=entry['timestamp'][:10],
_external=True) }}">{{ entry['timestamp'][:10] }}</a> at <a
href="{{ url_for('main.view_single_entry', timestamp=entry['timestamp'],
_external=True) }}">{{ entry['timestamp'][11:] }}</a>:</br>{{entry['entry']
}}</br>
         Tags:
            {% if entry['tags']|length != 0 %}
               {% for tag in entry['tags'] %}
                  <a href="{{ url_for('main.view_entries_for_tag', tag=tag,
_external=True) }}">#{{ tag }}</a>
               {% endfor %}
            {% else %}
               none
            {% endif %}
            </p></li>
```

```
        {% endfor %}
        </ul>
    {% endif %}
    </div>
    </br>
    <div>
        {% if next_page %}
        <p><a class="btn btn-default"
href="{{ url_for('main.view_entries_for_page', page=next_page,
_external=True) }}">Older entries</a></p>
        {% endif %}
    </div>

{% endblock %}
```

执行效果如图 9-3 所示。

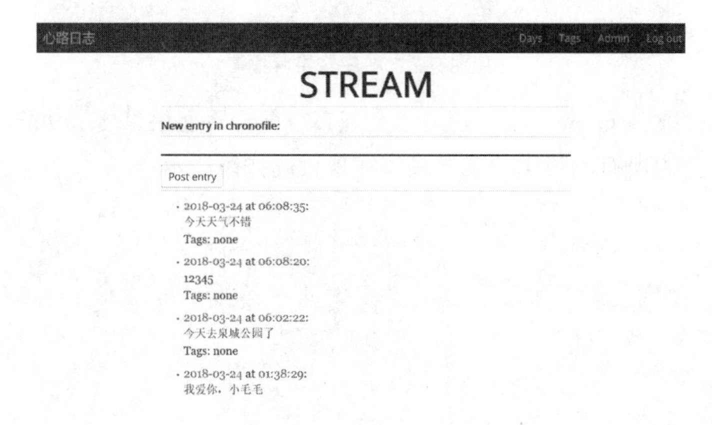


图 9-3　系统前后页面

(4) 编写程序文件 days.html，功能是以“日”为单位来查看系统数据库内的日志信息。具体实现代码如下所示：

```
{% extends "base.html" %}
{% block page_content %}

    <div class="page-header">
        <h1>Days</h1>
    </div>

    <div>
        {{ wtf.quick_form(form) }}
    </div>

    <div>
    {% if all_days|length == 0 %}
        <p>No entries yet.</p>
    {% else %}
        <ul>
        {% for day in all_days %}
            <li><p>
            <a href="{{ url_for('main.view_entries_for_day', day=day,
_external=True) }}">{{ day }}</a>
```

```
            </p></li>
        {% endfor %}
        </ul>
    {% endif %}
    </div>

{% endblock %}
```

执行效果如图 9-4 所示。

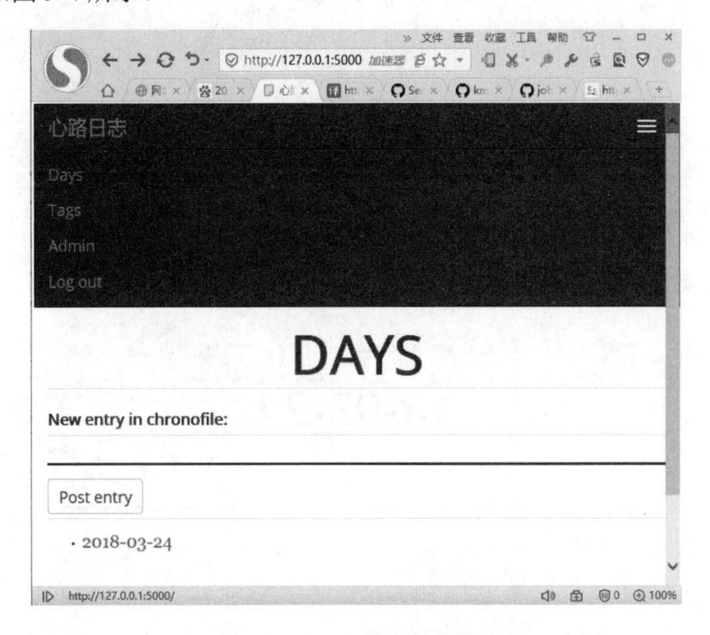


图 9-4　以“日”为单位查看日志

(5)　编写程序文件 day.html，功能是浏览系统数据库内某日的日志信息。具体实现代码如下所示：

```
{% extends "base.html" %}
{% block page_content %}

    <div class="page-header">
        {% if entries_for_day|length == 0 %}
            <h1>No Entries</h1>
        {% else %}
            <h1>{{ day[2:] }}</h1>
        {% endif %}
    </div>

    <div>
        {{ wtf.quick_form(form) }}
    </div>

    <div>
        <ul>
        {% for entry in entries_for_day %}
            <li><p><a href="{{ url_for('main.view_single_entry',
timestamp=entry['timestamp'],
```

```
_external=True) }}">{{ entry['timestamp'][11:] }}</a>:
{{entry['entry']}}</br>
          Tags:
              {% if entry['tags']|length != 0 %}
                 {% for tag in entry['tags'] %}
                    <a href="{{ url_for('main.view_entries_for_tag', tag=tag,
_external=True) }}">#{{ tag }}</a>
                 {% endfor %}
              {% else %}
                 none
              {% endif %}
              </p></li>
       {% endfor %}
       </ul>
   </div>

{% endblock %}
```

执行效果如图 9-5 所示。

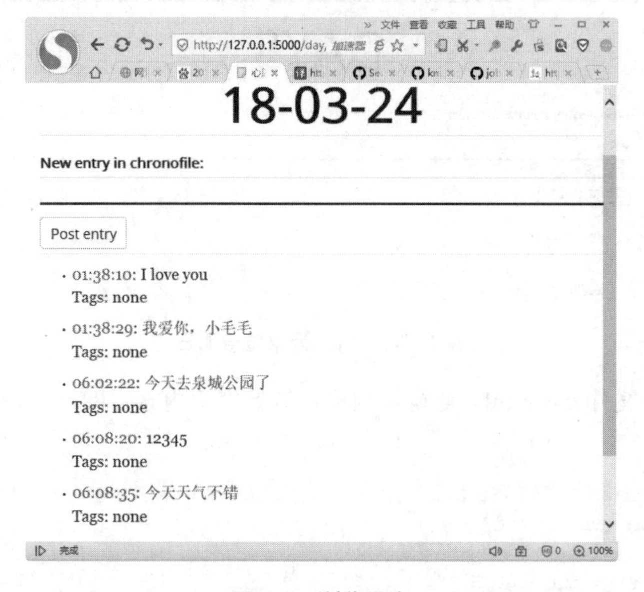


图 9-5 浏览日志

(6) 编写程序文件 admin.html，功能是显示系统管理主页，提供了如下所示的功能按钮和链接。

- Chronofile name：修改系统名称。
- Author name：修改系统作者名字。
- Change email address：修改邮箱地址。
- Change password：修改账户密码。

文件 admin.html 的具体实现代码如下所示：

```
{% extends "base.html" %}
{% block page_content %}

   <div class="page-header">
```

```
        <h1>Admin</h1>
    </div>

    <div>
        <p>Chronofile name:</p>
        <p>{{ details['chronofile_name'] }} <a class="btn btn-default"
href="{{ url_for('admin.rename_chronofile') }}">Edit</a></p>
    </div>
    </br>
    <div>
        <p>Author name:</p>
        <p>{{ details['author_name'] }} <a class="btn btn-default"
href="{{ url_for('admin.rename_author') }}">Edit</a></p>
    </div>
    </br>
    <div>
        <p>Change email address:</p>
        <p><a class="btn btn-default"
href="{{ url_for('auth.change_email') }}">Click here</a></p>
    </div>
    </br>
    <div>
        <p>Change password:</p>
        <p><a class="btn btn-default"
href="{{ url_for('auth.change_password') }}">Click here</a></p>
    </div>

{% endblock %}
```

执行效果如图 9-6 所示。

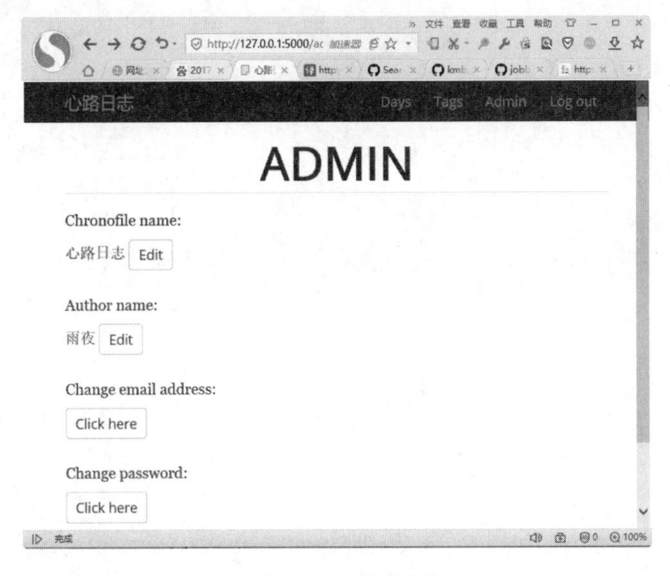


图 9-6　系统管理主页

**注　意**

为节省本书篇幅，对本项目中的其他模板文件不再详细讲解。读者只需查看本书配套源码即可，相信大家一看便懂。

# 第 10 章

# 电影票房数据可视化系统

## （网络爬虫+MySQL+Pandas 实现）

在当前的市场环境下，去影院看电影仍是消费者休闲娱乐的主要方式之一，这一点可以从近些年电影市场的高速发展和私人影院的迅速崛起得到佐证。大数据能分析电影票房并提取出有关资料，这对于电影行业从业者尤为重要。本章将详细讲解提取某专业电影网站电影数据，并根据提取的数据分析电影票房和其他相关资料的过程。

## 10.1 需求分析

电影这种娱乐载体，几乎可以出现在所有的消费场景之中：情侣约会、朋友聚会、闺蜜小聚、公司团建、家庭周末娱乐，甚至打发时间。现如今，电影产业的发展更多依靠的是票房和电影院市场的带动。

本项目将抓取 XX 网的电影信息，并爬虫提取 2018 年的全年数据和 2019 年的部分(2 月 14 日 22 点之前)数据，然后进行分析。通过使用本系统、可以分析如下所示的信息。

- 电影票房 TOP10：显示本年度总票房前 10 名的电影信息。
- 电影评分 TOP10：显示本年度评分前 10 名的电影信息。
- 电影人气 TOP10：显示本年度点评数量前 10 名的电影信息。
- 每月电影上映数量：显示本年度每月电影的上映数量。
- 每月电影票房：显示本年度每月电影的总票房。
- 中外票房对比：显示中外票房对比。
- 名利双收 TOP10：显示本年度名利双收前 10 名的电影信息。
- 叫座不叫好 TOP10：显示本年度叫座不叫好前 10 名的电影信息。
- 电影类型分布：显示本年度所有电影类型的统计信息。

## 10.2 模块架构

在开发一个大型应用程序时，模块架构是一个非常重要的前期准备工作，是关系到整个项目的实现流程是否顺利完成的关键。本节将根据严格的市场需求分析，得出我们这个项目的模块架构。本电影票房可视化系统的基本模块架构如图 10-1 所示。

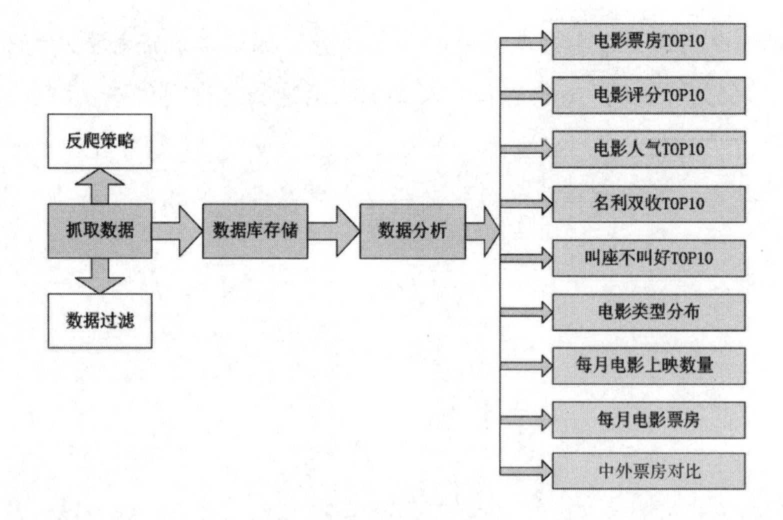


图 10-1　模块架构

扫码观看视频讲解

## 10.3　爬虫抓取数据

本节将详细讲解爬取 XX 网电影信息的过程，包括分别爬取 2018 年全年和 2019 年部分(2 月 14 日 22 点之前)数据的方法。

### 10.3.1　分析网页

XX 网的 2018 年电影信息 URL 网页地址是：
https://.域名主页 com/films?showType=3&yearId=13&sortId=3&offset=0
XX 网的 2019 年电影信息 URL 网页地址是：
https://.域名主页 com/films?showType=3&yearId=14&sortId=3&offset=0
通过对上述两个分页的分析可以得出如下结论。

- 2018 年电影信息有 184 个分页，每个分页有 30 部电影，但是有评分的只有 10 个分页。
- 2019 年电影信息有 184 个分页，每个分页有 30 部电影，但是有评分的只有 10 个分页。
- 分页参数是 offset，其中第一个分页的值是 30，第 2 个分页的值是 30，第 3 个值是 60，以此类推。
- 2018 年电影信息 URL 地址和 2019 年电影信息 URL 地址的区别是 yearId 编号值，其实 13 表示 2018 年，14 表示 2019 年。

XX 网某部电影详情页面的 URL 地址是：

```
https://.域名主页 com/films/1200486
```

在上述地址中，数字“1200486”是这部电影的编号，XX 网中的每一部电影都有自己的编号，上面的数字“1200486”是电影《我不是药神》的编号。按下 F12，进入浏览器的开发模式，会发现对评分、评分人数和累计票房等数据进行了文字反爬处理，这些数据都显示为“口口口”的形式，不能直接抓取，如图 10-2 所示。

```
▼<div class="movie-stats-container">
  ▼<div class="movie-index">
     <p class="movie-index-title">用户评分</p>
   ▼<div class="movie-index-content score normal-score">
     ▼<span class="index-left info-num ">
        <span class="stonefont">□.□</span>
      </span>
     ▼<div class="index-right">
       ▼<div class="star-wrapper">
          <div class="star-on" style="width:96%;"></div>
        </div>
       ▼<span class="score-num">
          <span class="stonefont">□□□万</span>
          "人评分"
        </span>
      </div>
    </div>
  </div>
  ▼<div class="movie-index">
     <p class="movie-index-title">累计票房</p>
   ▼<div class="movie-index-content box">
      <span class="stonefont">□□.□□</span>
      <span class="unit">亿</span>
    </div>
```

图 10-2　关键数据反爬

### 10.3.2 破解反爬

打开电影详情页面 https://.域名主页 com/films/1200486，右击，查看此网页的源码，然后查找关键字 font-face，找到如下所示代码：

```
@font-face {
  font-family: stonefont;
  src:
url('//vfile.meituan.net/colorstone/793c4d16ee74ce2c792b9d2fe1d0f4fb3168.eot');
  src:
url('//vfile.meituan.net/colorstone/793c4d16ee74ce2c792b9d2fe1d0f4fb3168.eot?#iefix') format('embedded-opentype'),

url('//vfile.meituan.net/colorstone/4a604c119c4aa8f9585e794730a83fbf2088.woff') format('woff');
}
```

在上述代码中，因为在每次刷新网页后，3 个 url 网址都会发生变化，所以现在还无法直接匹配信息。接下来需要下载“.woff”格式的文字文件，对其进行破解匹配。破解匹配的基本思路如下所示。

(1) 首先下载一个字体文件并保存到本地(比如上面代码中的“.woff”格式文件)，比如命名为 base.woff，然后人工找出每一个数字对应的编码。

(2) 当我们重新访问网页时，同样也可以把新的字体文件下载下来并保存到本地，比如叫 base1.woff。网页中的一个数字的编码比如为 AAAA，如何确定 AAAA 对应的数字。我们先通过编码 AAAA 找到这个字符在 base1.woff 中的对象，并且把它和 base.woff 中的对象逐个对比，直到找到相同的对象，然后获取这个对象在 base.woff 中的编码，再通过编码确认是哪个数字。

例如将上述代码中的//vfile.meituan.net/colorstone/4a604c119c4aa8f9585e794730a83fbf2088.woff 输入到浏览器地址，浏览器会自动下载文件 4a604c119c4aa8f9585e794730a83fbf2088.woff。将下载到的文件重命名为 base.woff，然后打开网址 http://fontstore.baidu.com/static/editor/index.html，通过此网页打开刚刚下载的文件 base.woff，此时会显示这个文件中的字体对应关系，如图 10-3 所示。

图 10-3 字体对应关系

这说明 uniE05B 代表数字 9，uniF09B 代表数字 0，以此类推。

接下来开始进行编码，实现破解反爬功能，具体实现流程如下所示。

(1) 编写文件 font_change.py，功能是将下载的“.woff”格式的文字文件转换为 XML 文件，这样可以获取爬虫时需要用到的 HTML 标签。文件 font_change.py 的具体实现代码如下所示：

```
from fontTools.ttLib import TTFont
font = TTFont('base.woff')
font.saveXML('.域名主页 xml')
```

执行代码后，会解析文件 base.woff 的内容，并生成 XML 文件.域名主页 xml。文件.域名主页 xml 的主要内容如下所示：

```
<GlyphOrder>
  <!-- The 'id' attribute is only for humans; it is ignored when parsed. -->
  <GlyphID id="0" name="glyph00000"/>
  <GlyphID id="1" name="x"/>
  <GlyphID id="2" name="uniE3DE"/>
  <GlyphID id="3" name="uniE88E"/>
  <GlyphID id="4" name="uniE63E"/>
  <GlyphID id="5" name="uniE82E"/>
  <GlyphID id="6" name="uniE94D"/>
  <GlyphID id="7" name="uniF786"/>
  <GlyphID id="8" name="uniE5E6"/>
  <GlyphID id="9" name="uniEEC6"/>
  <GlyphID id="10" name="uniF243"/>
  <GlyphID id="11" name="uniE5C7"/>
</GlyphOrder>

<glyf>
  <TTGlyph name="glyph00000"/><!-- contains no outline data -->

  <TTGlyph name="uniE3DE" xMin="0" yMin="-12" xMax="512" yMax="719">
    <contour>
      <pt x="139" y="173" on="1"/>
      <pt x="150" y="113" on="0"/>
      <pt x="210" y="60" on="0"/>
      <pt x="258" y="60" on="1"/>
      <pt x="300" y="60" on="0"/>
      <pt x="359" y="97" on="0"/>
      <pt x="398" y="159" on="0"/>
      <pt x="412" y="212" on="1"/>
      <pt x="418" y="238" on="0"/>
      <pt x="425" y="292" on="0"/>
      <pt x="425" y="319" on="1"/>
      <pt x="425" y="327" on="1"/>
      <pt x="425" y="331" on="0"/>
      <pt x="424" y="337" on="1"/>
      <pt x="399" y="295" on="0"/>
      <pt x="352" y="269" on="1"/>
      <pt x="308" y="243" on="0"/>
      <pt x="253" y="243" on="1"/>
      <pt x="164" y="243" on="0"/>
```

```
        <pt x="42" y="371" on="0"/>
        <pt x="42" y="477" on="1"/>
        <pt x="42" y="586" on="0"/>
        <pt x="169" y="719" on="0"/>
        <pt x="267" y="719" on="1"/>
        <pt x="335" y="719" on="0"/>
        <pt x="452" y="644" on="0"/>
        <pt x="512" y="503" on="0"/>
        <pt x="512" y="373" on="1"/>
        <pt x="512" y="235" on="0"/>
        <pt x="453" y="73" on="0"/>
        <pt x="335" y="-12" on="0"/>
        <pt x="256" y="-12" on="1"/>
        <pt x="171" y="-12" on="0"/>
        <pt x="119" y="34" on="1"/>
        <pt x="65" y="81" on="0"/>
        <pt x="55" y="166" on="1"/>
      </contour>
      <contour>
        <pt x="415" y="481" on="1"/>
        <pt x="415" y="557" on="0"/>
        <pt x="333" y="646" on="0"/>
        <pt x="277" y="646" on="1"/>
        <pt x="218" y="646" on="0"/>
        <pt x="132" y="552" on="0"/>
        <pt x="132" y="474" on="1"/>
        <pt x="132" y="404" on="0"/>
        <pt x="173" y="363" on="1"/>
        <pt x="215" y="320" on="0"/>
        <pt x="336" y="320" on="0"/>
        <pt x="415" y="407" on="0"/>
      </contour>
      <instructions/>
    </TTGlyph>
//省略后面的代码片段
  </glyf>
```

对上述代码的具体说明如下所示。

- 在标记<GlyphOrder>中的内容和图 10-3 中的字体对应关系是一一对应的。
- 在标记<glyf>中包含着每一个字符对象<TTGlyph>，同样第一个和最后一个不是0～9 的字符，需要删除。
- 在标记<TTGlyph>中包含了坐标点的信息，这些点的功能是描绘字体形状。

(2) 编写文件.域名主页 py，通过函数 get_numbers()对 XX 的文字进行破解，对应的实现代码如下所示。

```
def get_numbers(u):
    cmp = re.compile(",\n         url\('(//.*.woff)'\) format\('woff'\)")
    rst = cmp.findall(u)
    ttf = requests.get("http:" + rst[0], stream=True)
    with open(".域名主页 woff", "wb") as pdf:
        for chunk in ttf.iter_content(chunk_size=1024):
            if chunk:
                pdf.write(chunk)
```

```
    base_font = TTFont('base.woff')
    maoyanFont = TTFont('.域名主页 woff')
    maoyan_unicode_list = maoyanFont['cmap'].tables[0].ttFont.getGlyphOrder()
    maoyan_num_list = []
    base_num_list = ['.', '9', '0', '8', '2', '4', '5', '7', '3', '6', '1']
    base_unicode_list = ['x', 'uniE05B', 'uniF09B', 'uniF668', 'uniED4A',
'uniF140', 'uniE1B2', 'uniF48F', 'uniEB2A', 'uniED40', 'uniF50C']
    for i in range(1, 12):
        maoyan_glyph = maoyanFont['glyf'][maoyan_unicode_list[i]]
        for j in range(11):
            base_glyph = base_font['glyf'][base_unicode_list[j]]
            if maoyan_glyph == base_glyph:
                maoyan_num_list.append(base_num_list[j])
                break
    maoyan_unicode_list[1] = 'uni0078'
    utf8List = [eval(r"'\u" + uni[3:] + "'").encode("utf-8") for uni in
                maoyan_unicode_list[1:]]
    utf8last = []
    for i in range(len(utf8List)):
        utf8List[i] = str(utf8List[i], encoding='utf-8')
        utf8last.append(utf8List[i])
    return (maoyan_num_list, utf8last)
```

## 10.3.3 构造请求头

编写文件.域名主页 py，通过函数 str_to_dict()构造爬虫所需要的请求头，目的是获取浏览器的访问权限，具体实现代码如下所示：

```
head = """
Accept:text/html,application/xhtml+xml,application/xml;q=0.9,image/webp,ima
ge/apng,*/*;q=0.8
Accept-Encoding:gzip, deflate, br
Accept-Language:zh-CN,zh;q=0.8
Cache-Control:max-age=0
Connection:keep-alive
Host:.域名主页 com
Upgrade-Insecure-Requests:1
Content-Type:application/x-www-form-urlencoded; charset=UTF-8
User-Agent:Mozilla/5.0 (Windows NT 10.0; WOW64) AppleWebKit/5310.36 (KHTML,
like Gecko) Chrome/59.0.3071.86 Safari/5310.36
"""


def str_to_dict(header):
    """
    构造请求头,可以在不同函数里构造不同的请求头
    """
    header_dict = {}
    header = header.split('\n')
    for h in header:
        h = h.strip()
        if h:
            k, v = h.split(':', 1)
```

```
        header_dict[k] = v.strip()
    return header_dict
```

## 10.3.4 实现具体爬虫功能

编写文件.域名主页 py，首选通过函数 get_url()爬虫获取电影详情页链接，通过函数 get_message(url) 爬虫获取电影详情页里的信息。具体实现代码如下所示：

```
def get_url():
    for i in range(0, 300, 30):
        time.sleep(10)
    url = 'http://.域名主页 com/films?showType=3&yearId=13&sortId=3&offset='\
                + str(i)
        host = """http://.域名主页\com/films?showType=3&yearId=
                        13&sortId=3&offset=' + str(i)
        """
        header = head + host
        headers = str_to_dict(header)
        response = requests.get(url=url, headers=headers)
        html = response.text
        soup = BeautifulSoup(html, 'html.parser')
        data_1 = soup.find_all('div', {'class': 'channel-detail movie-item-title'})
        data_2 = soup.find_all('div', {'class': 'channel-detail channel-detail-
            orange'})
        num = 0
        for item in data_1:
            num += 1
            time.sleep(10)
            url_1 = item.select('a')[0]['href']
            if data_2[num-1].get_text() != '暂无评分':
                url = 'http://.域名主页 com' + url_1
                for message in get_message(url):
                    print(message)
                    to_mysql(message)
                print(url)
                print('---------------^^^Film_Message^^^-----------------')
            else:
                print('The Work Is Done')
                break


def get_message(url):
    """
    """
    time.sleep(10)
    data = {}
    host = """refer: http://.域名主页 com/news
    """
    header = head + host
    headers = str_to_dict(header)
    response = requests.get(url=url, headers=headers)
    u = response.text
```

```
# 破解 XX 文字反爬
(maoyan_num_list, utf8last) = get_numbers(u)
# 获取电影信息
soup = BeautifulSoup(u, "html.parser")
mw = soup.find_all('span', {'class': 'stonefont'})
score = soup.find_all('span', {'class': 'score-num'})
unit = soup.find_all('span', {'class': 'unit'})
ell = soup.find_all('li', {'class': 'ellipsis'})
name = soup.find_all('h3', {'class': 'name'})
# 返回电影信息
data["name"] = name[0].get_text()
data["type"] = ell[0].get_text()
data["country"] = ell[1].get_text().split('/')[0].strip().replace('\n', '')
data["length"] = ell[1].get_text().split('/')[1].strip().replace('\n', '')
data["released"] = ell[2].get_text()[:10]
# 因为会出现没有票房的电影,所以这里需要判断
if unit:
    bom = ['分', score[0].get_text().replace('.', '').replace('万', ''),
        unit[0].get_text()]
    for i in range(len(mw)):
        moviewish = mw[i].get_text().encode('utf-8')
        moviewish = str(moviewish, encoding='utf-8')
        # 通过比对获取反爬文字信息
        for j in range(len(utf8last)):
            moviewish = moviewish.replace(utf8last[j], maoyan_num_list[j])
        if i == 0:
            data["score"] = moviewish + bom[i]
        elif i == 1:
            if '万' in moviewish:
                data["people"] = int(float(moviewish.replace('万', '')) * 10000)
            else:
                data["people"] = int(float(moviewish))
        else:
            if '万' == bom[i]:
                data["box_office"] = int(float(moviewish) * 10000)
            else:
                data["box_office"] = int(float(moviewish) * 100000000)
else:
    bom = ['分', score[0].get_text().replace('.', '').replace('万', ''), 0]
    for i in range(len(mw)):
        moviewish = mw[i].get_text().encode('utf-8')
        moviewish = str(moviewish, encoding='utf-8')
        for j in range(len(utf8last)):
            moviewish = moviewish.replace(utf8last[j], maoyan_num_list[j])
        if i == 0:
            data["score"] = moviewish + bom[i]
        else:
            if '万' in moviewish:
                data["people"] = int(float(moviewish.replace('万', '')) * 10000)
            else:
                data["people"] = int(float(moviewish))
    data["box_office"] = bom[2]
yield data
```

在上述代码中，抓取的 URL 地址参数 yearId 的值是 13，这表示抓取的是 2018 年底电影信息。如果设置 yearId=14，则会抓取 2019 年的电影信息。

## 10.3.5 将爬取的信息保存到数据库

(1) 编写文件 maoyan_mysql_1.py，功能是在本地 MySQL 数据库中创建一个名为 maoyan 的数据库，具体实现代码如下所示：

```
import pymysql

db = pymysql.connect(host='1210.0.0.1', user='root', password='66688888',
port=3306)
cursor = db.cursor()
cursor.execute("CREATE DATABASE maoyan DEFAULT CHARACTER SET utf8")
db.close()
```

(2) 编写文件 maoyan_mysql_2.py，功能是在 MySQL 数据库 maoyan 中创建表 films，用于保存抓取到的 2018 年电影信息，具体实现代码如下所示：

```
import pymysql

db = pymysql.connect(host='1210.0.0.1', user='root', password='66688888',
port=3306, db='maoyan')
cursor = db.cursor()
sql = 'CREATE TABLE IF NOT EXISTS films (name VARCHAR(255) NOT NULL, type\
VARCHAR(255) NOT NULL, country VARCHAR(255) NOT NULL, length VARCHAR(255) NOT\
NULL, released VARCHAR(255) NOT NULL, score VARCHAR(255) NOT NULL, people INT\
NOT NULL, box_office BIGINT NOT NULL, PRIMARY KEY (name))'
cursor.execute(sql)
db.close()
```

(3) 编写文件 maoyan_mysql_2-1.py，功能是在 MySQL 数据库 maoyan 中创建表 films1，用于保存抓取到的 2019 年电影信息，具体实现代码如下所示：

```
import pymysql

db = pymysql.connect(host='1210.0.0.1', user='root', password='66688888',
port=3306, db='maoyan')
cursor = db.cursor()
sql = 'CREATE TABLE IF NOT EXISTS films1 (name VARCHAR(255) NOT NULL, type\
VARCHAR(255) NOT NULL, country VARCHAR(255) NOT NULL, length VARCHAR(255) NOT\
NULL, released VARCHAR(255) NOT NULL, score VARCHAR(255) NOT NULL, people INT\
NOT NULL, box_office BIGINT NOT NULL, PRIMARY KEY (name))'
cursor.execute(sql)
db.close()
```

(4) 在文件.域名主页 py 中编写函数 to_mysql(data)，功能是将抓取到电影信息保存到数据库中，具体实现代码如下所示：

```
def to_mysql(data):
    """
```

```
    信息写入 mysql
    """
    table = 'films'
    keys = ', '.join(data.keys())
    values = ', '.join(['%s'] * len(data))
    db = pymysql.connect(host='localhost', user='root', password='66688888',
        port=3306, db='maoyan',charset="utf8")
    cursor = db.cursor()
    sql = 'INSERT INTO {table}({keys}) VALUES ({values})'.format(table=table,
         keys=keys, values=values)
    try:
       if cursor.execute(sql, tuple(data.values())):
          print("Successful")
          db.commit()
    except:
       print('Failed')
       db.rollback()
    db.close()
```

上述代码将抓取到的电影信息保存到数据库表 films 中，当然也可以设置将抓取的数据保存到数据库表 films1 中，这个需要读者根据自己的需要来设置。例如 2018 年的数据被保存到数据库表 films 中，如图 10-4 所示。

正在显示第 0 - 99 行 (共 297 行, 查询花费 0.0043 秒。)

SELECT * FROM `films`

性能分析 [ 编辑内嵌 ] [ 编辑 ] [ 解析 SQL ] [ 创建 PHP 代码 ] [ 刷新 ]

1 > >> 显示全部 | 行数: 100 过滤行: 在表中搜索 按索引排序: 无

+ 选项

| ←T→ | name | type | country | length | released | score | people | box_office |
|---|---|---|---|---|---|---|---|---|
| 编辑 复制 删除 | 21克拉 | 喜剧,爱情 | 中国大陆 | 96分钟 | 2018-04-20 | 8.3分 | 124000 | 110000000 |
| 编辑 复制 删除 | 24小时：末路重生 | 犯罪,动作 | 美国,中国大陆,南非 | 93分钟 | 2018-01-26 | 7.9分 | 11542 | 9770000 |
| 编辑 复制 删除 | S4侠降魔记 | 喜剧 | 中国大陆 | 98分钟 | 2018-01-26 | 7.9分 | 2702 | 3640000 |
| 编辑 复制 删除 | 一个人的课堂 | 剧情,家庭 | 中国大陆 | 105分钟 | 2018-01-16 | 9.1分 | 1353 | 1370000 |
| 编辑 复制 删除 | 一出好戏 | 剧情,喜剧 | 中国大陆 | 134分钟 | 2018-08-10 | 8.2分 | 937000 | 1355000000 |
| 编辑 复制 删除 | 一百年很长吗 | 纪录片 | 中国大陆 | 106分钟 | 2018-12-01 | 8.1分 | 1605 | 1240000 |
| 编辑 复制 删除 | 一纸婚约 | 剧情,爱情 | 中国大陆 | 91分钟 | 2018-06-08 | 6.4分 | 2233 | 750000 |
| 编辑 复制 删除 | 七号公馆 | 喜剧,惊悚 | 中国大陆 | 94分钟 | 2018-05-11 | 5.1分 | 1428 | 620000 |
| 编辑 复制 删除 | 三国杀·幻 | 爱情,动作,悬疑 | 中国大陆 | 94分钟 | 2018-11-09 | 7.1分 | 2894 | 1050000 |
| 编辑 复制 删除 | 三块广告牌 | 剧情,犯罪 | 美国,英国 | 115分钟 | 2018-03-02 | 8.6分 | 34863 | 64990000 |
| 编辑 复制 删除 | 两个俏公主 | 动画,奇幻,家庭 | 中国大陆 | 89分钟 | 2018-11-10 | 7.6分 | 5267 | 2230000 |

图 10-4　在数据库中保存的电影信息

## 10.4　数据可视化分析

扫码观看视频讲解

本节将根据数据库中的电影数据实现大数据分析，逐一分析 2018 年的电影数据和 2019 年的部分(2 月 14 日 22 点之前)电影数据。

### 10.4.1　电影票房 TOP10

编写文件 movie_box_office_top10.py，功能是统计显示 2019 年部分(2 月 14 日 22 点之

前)总票房前 10 名的电影信息，具体实现代码如下所示。

```
from pyecharts import Bar
import pandas as pd
import numpy as np
import pymysql

conn = pymysql.connect(host='localhost', user='root', password='66688888',
port=3306, db='maoyan', charset='utf8')
cursor = conn.cursor()
sql = "select * from films1"
db = pd.read_sql(sql, conn)
df = db.sort_values(by="box_office", ascending=False)
dom = df[['name', 'box_office']]

attr = np.array(dom['name'][0:10])
v1 = np.array(dom['box_office'][0:10])
attr = ["{}".format(i.replace(': 无限战争', '')) for i in attr]
v1 = ["{}".format(float('%.2f' % (float(i) / 10000))) for i in v1]

bar = Bar("2019 年电影票房 TOP10(万元)(截止到 2 月 14 日)", title_pos='center',
          title_top='18', width=800, height=400)
bar.add("", attr, v1, is_convert=True, xaxis_min=10, yaxis_label_textsize=12,
        is_yaxis_boundarygap=True, yaxis_interval=0, is_label_show=True,
        is_legend_show=False, label_pos='right', is_yaxis_inverse=True,
        is_splitline_show=False)
bar.render("2019 年电影票房 TOP10.html")
```

执行代码后，会创建一个名为“2019 年电影票房 TOP10.html”的统计文件，打开此文件会显示对应的统计柱形图，如图 10-5 所示。

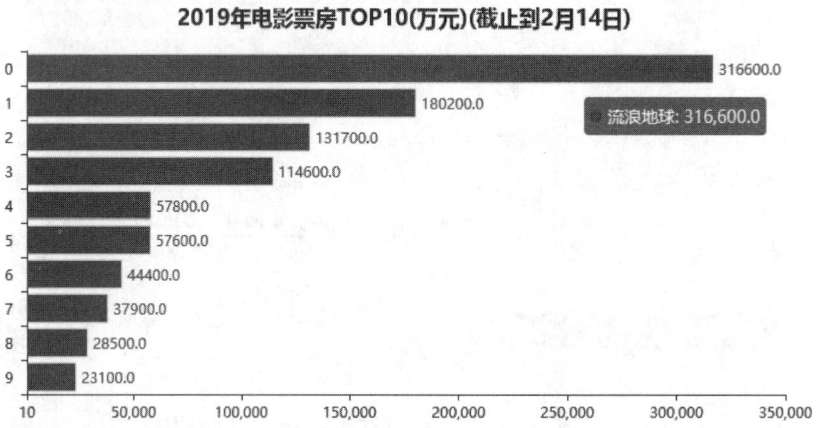

图 10-5　电影票房 TOP10 统计图(2019 年部分)

如果在上述代码中设置提取的数据库表是 films，则会统计显示 2018 年的电影票房 TOP10 数据，如图 10-6 所示。

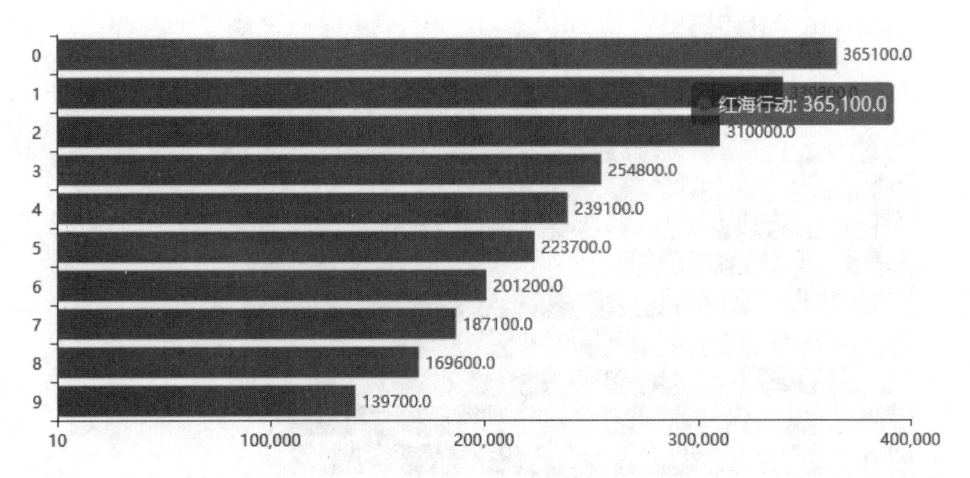


图 10-6　电影票房 TOP10 统计图(2018 年)

## 10.4.2　电影评分 TOP10

编写文件 movie_score_top10.py，功能是分析并显示年度评分前 10 名的电影信息，具体实现代码如下所示：

```
from pyecharts import Bar
import pandas as pd
import numpy as np
import pymysql

conn = pymysql.connect(host='localhost', user='root', password='66688888',
                       port=3306, db='maoyan', charset='utf8')
cursor = conn.cursor()
sql = "select * from films"
db = pd.read_sql(sql, conn)
df = db.sort_values(by="score", ascending=False)
dom = df[['name', 'score']]

v1 = []
for i in dom['score'][0:10]:
   number = float(i.replace('分', ''))
   v1.append(number)
attr = np.array(dom['name'][0:10])
attr = ["{}".format(i.replace(': 致命守护者', '')) for i in attr]

bar = Bar("2019 年电影评分 TOP10(截止到 2 月 14 日)", title_pos='center',
         title_top='18', width=800, height=400)
bar.add("", attr, v1, is_convert=True, xaxis_min=8, xaxis_max=9.8,
        yaxis_label_textsize=10, is_yaxis_boundarygap=True, yaxis_interval=0,
        is_label_show=True, is_legend_show=False, label_pos='right',
        is_yaxis_inverse=True, is_splitline_show=False)
bar.render("2019 年电影评分 TOP10.html")
```

执行代码后，会创建一个名为“2019 年电影评分 TOP10.html”的统计文件，打开此文件会显示对应的统计柱形图。其中 2019 年部分电影评分 TOP10 统计如图 10-7 所示，2018

年电影评分 TOP10 统计如图 10-8 所示。

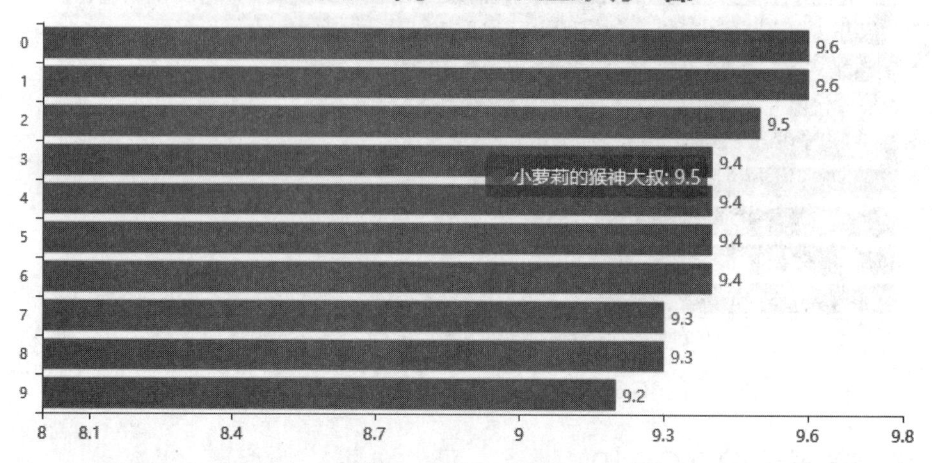


图 10-7　电影评分 TOP10 统计图(2019 年部分)

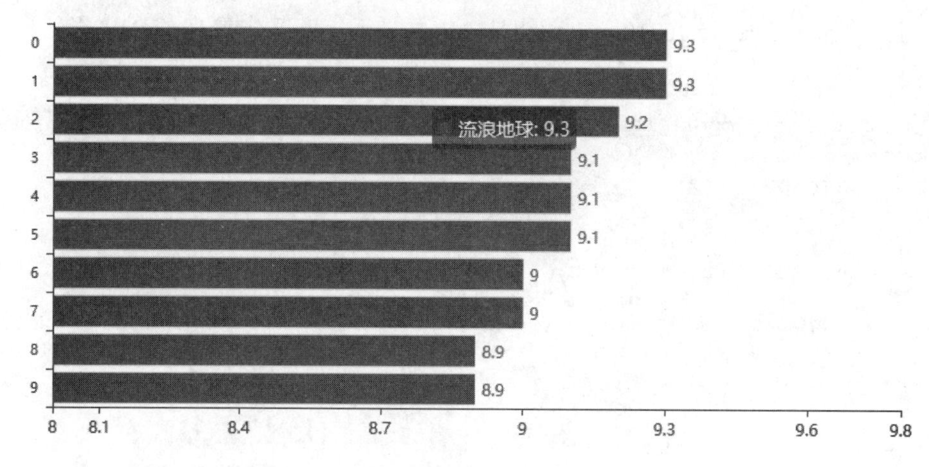


图 10-8　电影评分 TOP10 统计图(2018 年)

### 10.4.3　电影人气 TOP10

编写文件 movie_get_people_top10.py，功能是分析并显示年度点评数量前 10 名的电影信息，具体实现代码如下所示：

```
from pyecharts import Bar
import pandas as pd
import numpy as np
import pymysql

conn = pymysql.connect(host='localhost', user='root', password='66688888',
                       port=3306, db='maoyan', charset='utf8')
cursor = conn.cursor()
sql = "select * from films1"
```

```
db = pd.read_sql(sql, conn)
df = db.sort_values(by="people", ascending=False)
dom = df[['name', 'people']]

attr = np.array(dom['name'][0:10])
v1 = np.array(dom['people'][0:10])
attr = ["{}".format(i.replace(': 无限战争', '')) for i in attr]
v1 = ["{}".format(float('%.2f' % (float(i) / 1))) for i in v1]

bar = Bar("2019 年电影人气 TOP10)(截止到 2 月 14 日)", title_pos='center',
title_top='18', width=800, height=400)
bar.add("", attr, v1, is_convert=True, xaxis_min=10, yaxis_label_textsize=12,
        is_yaxis_boundarygap=True, yaxis_interval=0, is_label_show=True,
        is_legend_show=False, label_pos='right', is_yaxis_inverse=True,
        is_splitline_show=False)
bar.render("2019 年电影人气 TOP10.html")
```

执行代码后，会创建一个名为“2019 年电影人气 TOP10.html”的统计文件，打开此文件会显示对应的统计柱形图。其中 2019 年电影人气 TOP10 统计如图 10-9 所示，2018 年部分电影票房 TOP10 统计如图 10-10 所示。

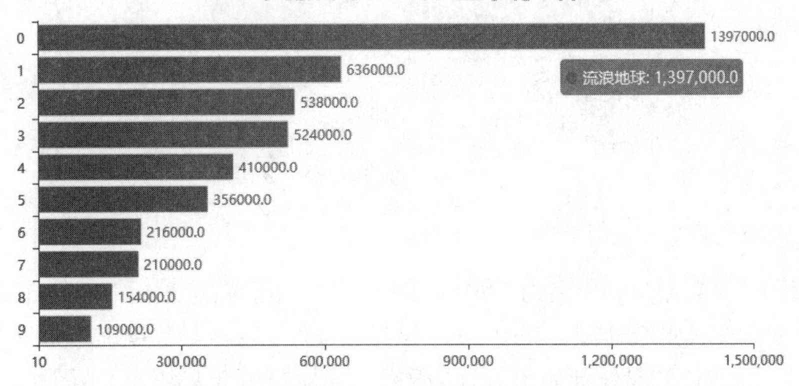


图 10-9　电影人气 TOP10 统计图(2019 年部分)

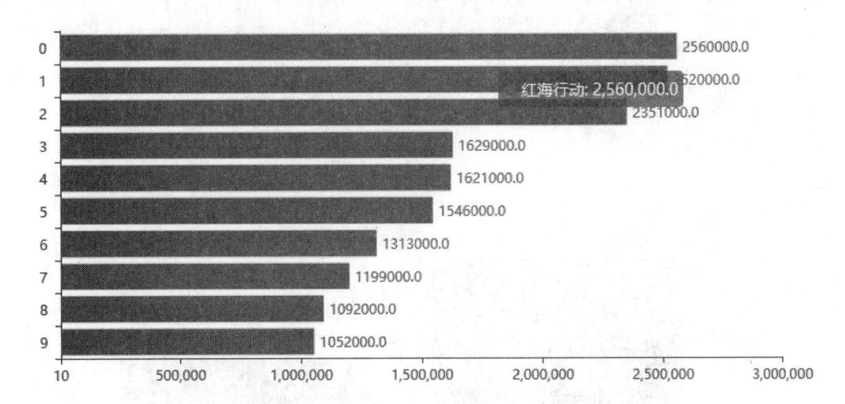


图 10-10　部分电影票房 TOP10 统计图(2018 年)

### 10.4.4 每月电影上映数量

编写文件 movie_month_update.py，功能是分析并显示本年度每月电影的上映数量，具体实现代码如下所示：

```
conn = pymysql.connect(host='localhost', user='root', password='66688888',
port=3306, db='maoyan', charset='utf8')
cursor = conn.cursor()
sql = "select * from films"
db = pd.read_sql(sql, conn)
df = db.sort_values(by="released", ascending=False)
dom = df[['name', 'released']]
list1 = []
for i in dom['released']:
   place = i.split('-')[1]
   list1.append(place)
db['month'] = list1

month_message = db.groupby(['month'])
month_com = month_message['month'].agg(['count'])
month_com.reset_index(inplace=True)
month_com_last = month_com.sort_index()

attr = ["{}".format(str(i) + '月') for i in range(1, 13)]
v1 = np.array(month_com_last['count'])
v1 = ["{}".format(i) for i in v1]

bar = Bar("2019 年每月上映电影数量(截止到 2 月 14 日)", title_pos='center',
          title_top='18', width=800, height=400)
bar.add("", attr, v1, is_stack=True, yaxis_max=40, is_label_show=True)
bar.render("2019 年每月上映电影数量.html")
```

执行代码后，会创建一个名为“2019 年每月上映电影数量.html”的统计文件，打开此文件会显示对应的柱形统计图。其中 2019 年部分每月上映电影数量统计如图 10-11 所示，2018 年每月上映电影数量统计如图 10-12 所示。因为 2019 年只统计了两个月的数据，所以上面的代码只能遍历 2 个月的数据，而不是 12 个月的，所以需要将遍历行代码改为：

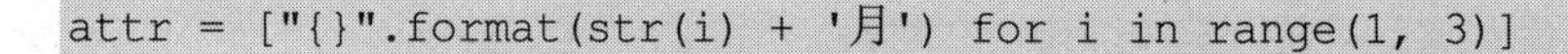

```
attr = ["{}".format(str(i) + '月') for i in range(1, 3)]
```

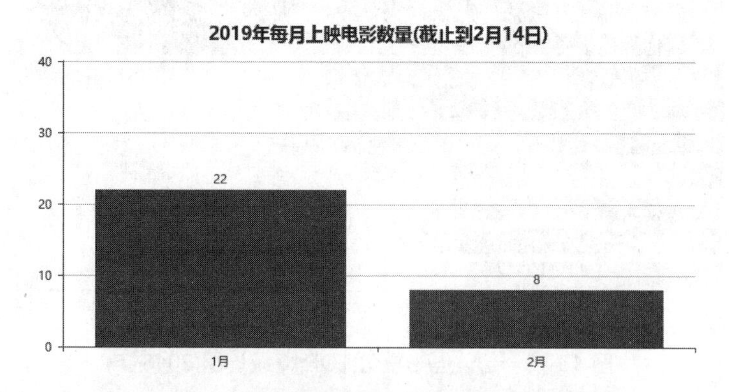


图 10-11 每月上映电影数量统计(2019 年部分)

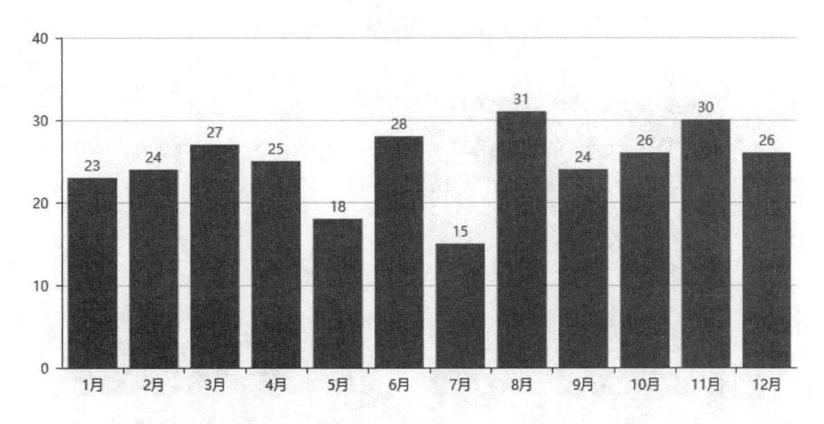


图 10-12　每月上映电影数量统计(2018 年)

## 10.4.5　每月电影票房

编写文件 movie_month_box_office.py，功能是分析并显示年度每月电影票房的电影信息，具体实现代码如下所示：

```
conn = pymysql.connect(host='localhost', user='root', password='66688888',
                       port=3306, db='maoyan', charset='utf8')
cursor = conn.cursor()
sql = "select * from films1"
db = pd.read_sql(sql, conn)
df = db.sort_values(by="released", ascending=False)
dom = df[['name', 'released']]
list1 = []
for i in dom['released']:
   time = i.split('-')[1]
   list1.append(time)
db['month'] = list1

month_message = db.groupby(['month'])
month_com = month_message['box_office'].agg(['sum'])
month_com.reset_index(inplace=True)
month_com_last = month_com.sort_index()

attr = ["{}".format(str(i) + '月') for i in range(1, 3)]
v1 = np.array(month_com_last['sum'])

v1 = ["{}".format(float('%.2f' % (float(i) / 100000000))) for i in v1]
bar = Bar("2019 年每月电影票房(亿元)(截止到 2 月 14 日)", title_pos='center',
          title_top='18', width=800, height=400)
bar.add("", attr, v1, is_stack=True, is_label_show=True)
bar.render("2019 年每月电影票房(亿元).html")
```

执行代码后，会创建一个名为“2019 年每月电影票房(亿元).html”的统计文件，打开此文件会显示对应的柱形统计图。其中 2019 年部分每月电影票房统计如图 10-13 所示，2018 年每月电影票房统计如图 10-14 所示。因为 2019 年只统计了两个月的数据，所以上面的代码只能遍历 2 个月的数据，而不是 12 个月的，所以需要将遍历行代码改为：

```
attr = ["{}".format(str(i) + '月') for i in range(1, 3)]
```

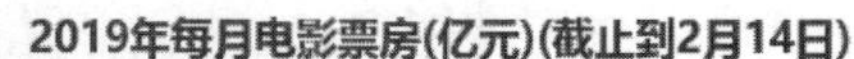


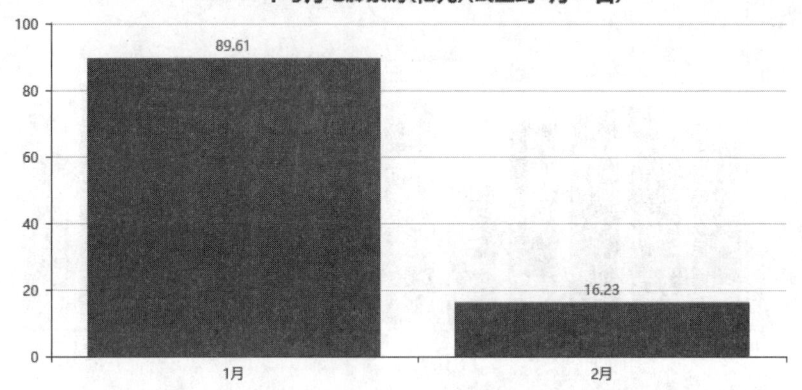


图 10-13　每月电影票房统计(2019 年部分)

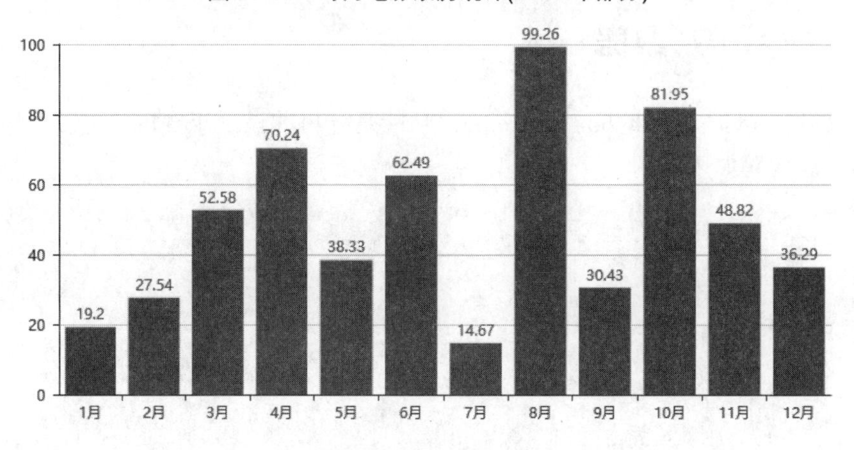


图 10-14　每月电影票房统计(2018 年)

## 10.4.6　中外票房对比

编写文件 movie_country_box_office.py，功能是分析并显示年度中外票房对比的信息，具体实现代码如下所示：

```
conn = pymysql.connect(host='localhost', user='root', password='66688888',
                       port=3306, db='maoyan', charset='utf8')
cursor = conn.cursor()
sql = "select * from films"
db = pd.read_sql(sql, conn)
list1 = []
for i in db['country']:
   type1 = i.split(',')[0]
   if type1 in ['中国大陆', '中国香港']:
      type1 = '中国'
   else:
      type1 = '外国'
   list1.append(type1)
db['country_type'] = list1

country_type_message = db.groupby(['country_type'])
```

```
country_type_com = country_type_message['box_office'].agg(['sum'])
country_type_com.reset_index(inplace=True)
country_type_com_last = country_type_com.sort_index()

attr = country_type_com_last['country_type']
v1 = np.array(country_type_com_last['sum'])
v1 = ["{}".format(float('%.2f' % (float(i) / 100000000))) for i in v1]

pie = Pie("2019 年中外电影票房对比(亿元)(截止到 2 月 14 日)", title_pos='center')
pie.add("", attr, v1, radius=[40, 75], label_pos='right', label_text_color=None,
        is_label_show=True, legend_orient="vertical",
        legend_pos="left",label_formatter='{c}')
pie.render("2019 年中外电影票房对比(亿元).html")
```

执行代码后，会创建一个名为“2019 年中外电影票房对比(亿元).html”的统计文件，打开此文件会显示对应的饼形统计图。其中 2019 年部分中外票房对比统计如图 10-15 所示，2018 年中外票房对比如图 10-16 所示。

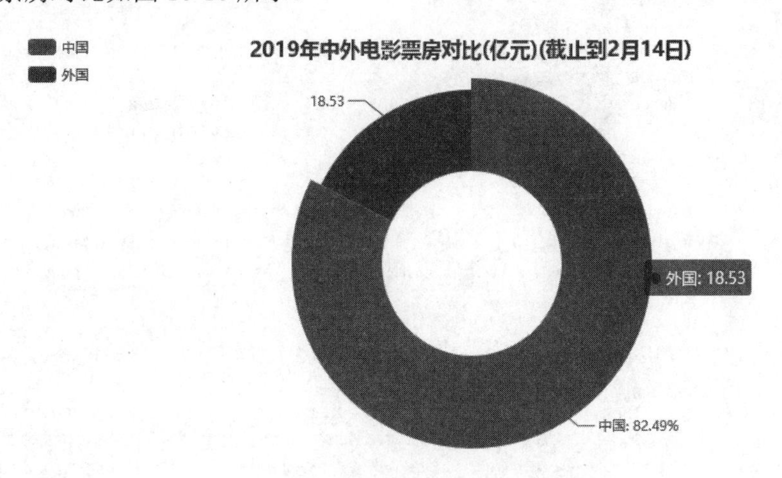


图 10-15　中外票房对比统计(2019 年部分)

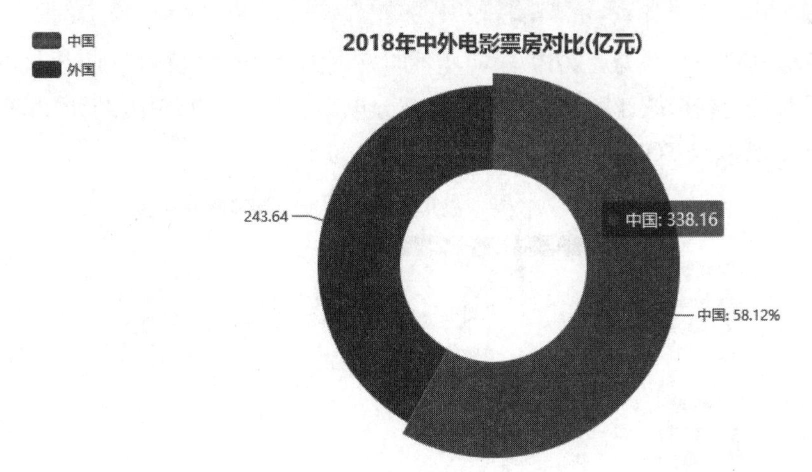


图 10-16　中外票房对比统计(2018 年)

## 10.4.7 名利双收 TOP10

编写文件 movie_get_double_top10.py，功能是分析并显示年度名利双收 TOP10 的电影信息。名利双收的计算公式是：

(某部电影的评分在所有电影评分中的排名+某部电影的票房在所有票房中的排)/电影总数

文件 movie_get_double_top10.py 的具体实现代码如下所示：

```
def my_sum(a, b, c):
   rate = (a + b) / c
   result = float('%.4f' % rate)
   return result

conn = pymysql.connect(host='localhost', user='root', password='66688888',
                       port=3306, db='maoyan', charset='utf8')
cursor = conn.cursor()
sql = "select * from films1"
db = pd.read_sql(sql, conn)
db['sort_num_money'] = db['box_office'].rank(ascending=0, method='dense')
db['sort_num_score'] = db['score'].rank(ascending=0, method='dense')
db['value'] = db.apply(lambda row: my_sum(row['sort_num_money'],
                       row['sort_num_score'], len(db.index)), axis=1)
df = db.sort_values(by="value", ascending=True)[0:10]

v1 = ["{}".format('%.2f' % ((1-i) * 100)) for i in df['value']]
attr = np.array(df['name'])
attr = ["{}".format(i.replace(': 无限战争', '').replace(': 全面瓦解', ''))
       for i in attr]

bar = Bar("2019 年电影名利双收 TOP10(%)(截止到 2 月 14 日)", title_pos='center',
          title_top='18', width=800, height=400)
bar.add("", attr, v1, is_convert=True, xaxis_min=85, xaxis_max=100,
        yaxis_label_textsize=12, is_yaxis_boundarygap=True, yaxis_interval=0,
        is_label_show=True, is_legend_show=False, label_pos='right',
        is_yaxis_inverse=True, is_splitline_show=False)
bar.render("2019 年电影名利双收 TOP10.html")
```

执行后会创建一个名为“2019 年叫座不叫好电影 TOP100.html”的统计文件，打开此文件会显示对应的柱形统计图。其中 2019 年部分电影名利双收 TOP10 统计如图 10-17 所示，2018 年电影名利双收 TOP10 如图 10-18 所示。

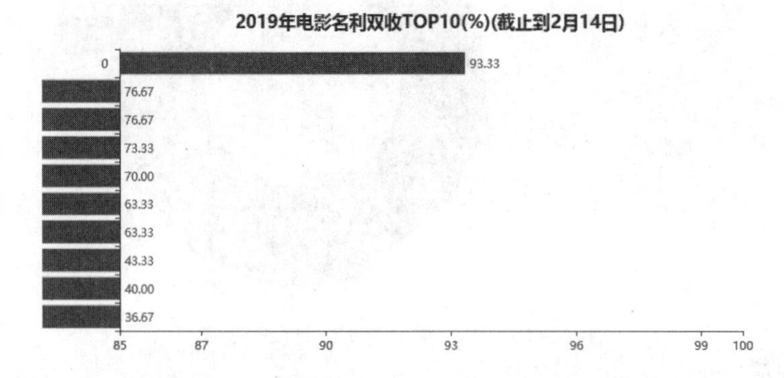


图 10-17 名利双收 TOP10 统计(2019 年部分)

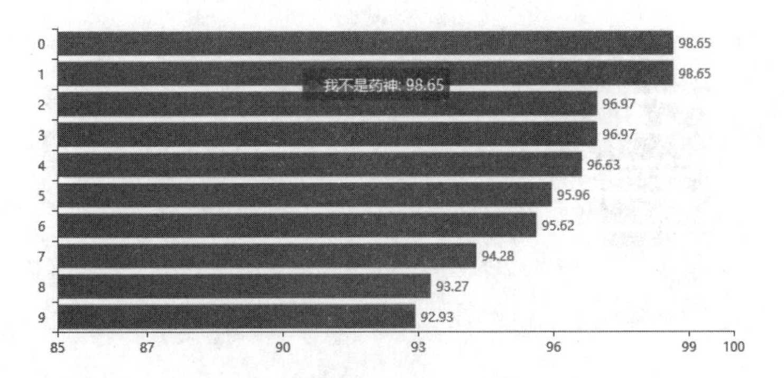


图 10-18　名利双收 TOP10 统计(2018 年)

## 10.4.8　叫座不叫好 TOP10

编写文件 movie_get_difference_top10.py，功能是分析并显示年度叫座不叫好 TOP10 的电影信息。叫座不叫好的计算公式是：

(某部电影的票房排名-某部电影的评分排名)/电影总数

文件 movie_get_difference_top10.py 的具体实现代码如下所示：

```
def my_difference(a, b, c):
   rate = (a - b) / c
   return rate

conn = pymysql.connect(host='localhost', user='root', password='66688888',
                       port=3306, db='maoyan', charset='utf8')
cursor = conn.cursor()
sql = "select * from films"
a = pd.read_sql(sql, conn)
a['sort_num_money'] = a['box_office'].rank(ascending=0, method='dense')
a['sort_num_score'] = a['score'].rank(ascending=0, method='dense')
a['value'] = a.apply(lambda row: my_difference(row['sort_num_money'],
                     row['sort_num_score'], len(a.index)), axis=1)
df = a.sort_values(by="value", ascending=True)[0:9]
```

执行代码后，会创建一个名为“2019 年电影叫座不叫好 TOP10.html”的统计文件，打开此文件会显示对应的柱形统计图。其中 2019 年部分电影叫座不叫好 TOP10 统计如图 10-19 所示，2018 年部分电影叫座不叫好 TOP10 如图 10-20 所示。

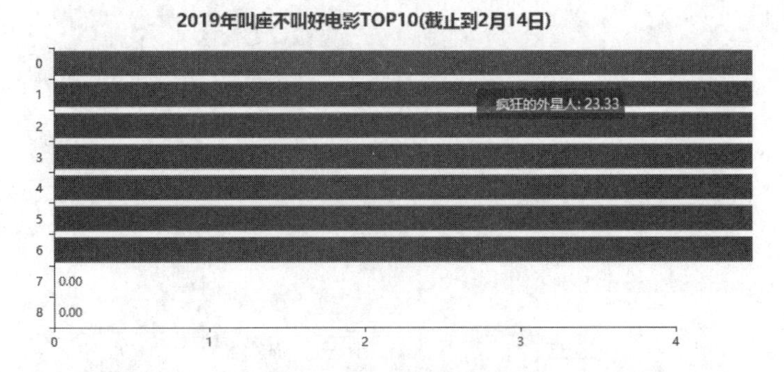


图 10-21　叫座不叫好 TOP10 统计(2019 年部分)

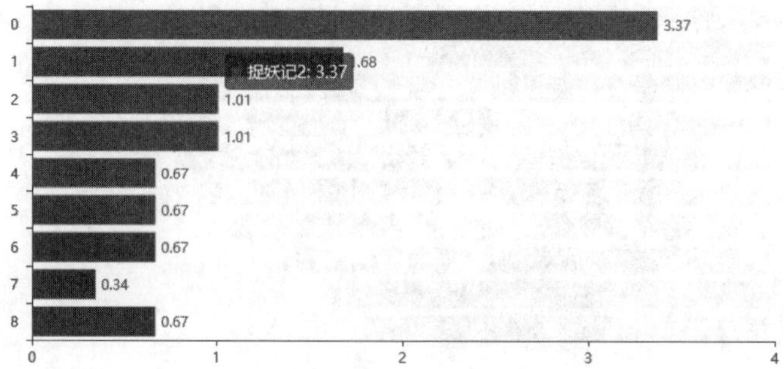

图 10-20　叫座不叫好 TOP10 统计(2018 年)

## 10.4.9　电影类型分布

编写文件 movie_type.py，功能是分析并显示年度电影类型分布的信息，具体实现代码如下所示：

```
conn = pymysql.connect(host='localhost', user='root', password='66688888',
                       port=3306, db='maoyan', charset='utf8')
cursor = conn.cursor()
sql = "select * from films"
db = pd.read_sql(sql, conn)

dom1 = []
for i in db['type']:
    type1 = i.split(',')
    for j in range(len(type1)):
        if type1[j] in dom1:
            continue
        else:
            dom1.append(type1[j])

dom2 = []
for item in dom1:
    num = 0
    for i in db['type']:
        type2 = i.split(',')
        for j in range(len(type2)):
            if type2[j] == item:
                num += 1
            else:
                continue
    dom2.append(num)


def message():
    for k in range(len(dom2)):
        data = {}
        data['name'] = dom1[k] + ' ' + str(dom2[k])
        data['value'] = dom2[k]
```

```
        yield data


data1 = message()
dom3 = []
for item in data1:
    dom3.append(item)

treemap = TreeMap("2019 年电影类型分布图(截止到 2 月 14 日)", title_pos='center',
                  title_top='5', width=800, height=400)
treemap.add('2019 年电影类型分布', dom3, is_label_show=True, label_pos='inside',
            is_legend_show=False)
treemap.render('2019 年电影类型分布图.html')
```

执行代码后，会创建一个名为“2019 年电影类型分布图.html”的统计文件，打开此文件会显示对应的分布统计图。其中 2019 年部分电影类型分布统计如图 10-22 所示，2018 年部分电影类型分布统计如图 10-22 所示。

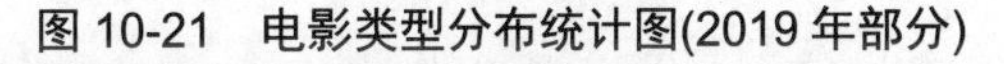


图 10-21　电影类型分布统计图(2019 年部分)

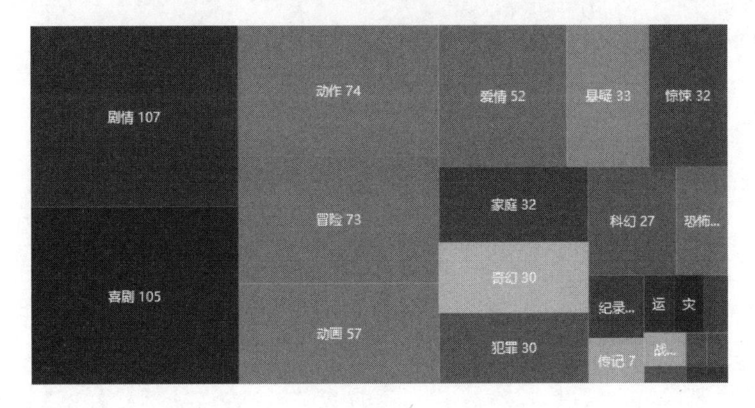


图 10-22　电影类型分布统计图(2018 年)

# 第 11 章

# 大型 3D 枪战类冒险游戏（Panda3D 实现）

阿凡达游戏是作者以前使用 Panda3D 开发的一个大型 3D 游戏，整个游戏的功能比较全面，包含选择穿戴搭配、场景切换、多方对战等功能。本章将截取游戏的前两个场景，详细讲解使用 Panda3D 开发大型游戏的过程。

扫码观看视频讲解

## 11.1 行业背景介绍

权威统计机构预测，在2020年，全球游戏市场的收入将超过1600亿美元，同比增长7.3%。2020年将会成为游戏行业总收入增长最低的年份之一，也是自2015年以来，主机游戏收入增长最低的一年。但是，索尼和微软将在2020年底推出下一代游戏主机，这无疑将会给市场带来新的增长。而看似有着大好前景的云游戏则不会对2020年的市场形势产生重大影响，因为该技术目前尚处于起步阶段。

移动游戏将会成为2020年发展最快的领域，同比增长可达11.6%。在过去的几年中，移动游戏的增长大多是由成熟市场推动的，例如中国、美国、日本、韩国和西欧。而在2020年，这一局面将被中东和北非、印度、东南亚等新兴市场改变，这些市场的游戏收入都将实现两位数的增长。不过这绝不意味着更成熟的市场将会后继乏力。西方和中国市场的订阅服务数量持续稳步攀升，势必使2020年及之后的市场产生振荡。

再看国内市场，据GPC数据显示，2020年上半年，我国游戏市场规模接近 1400 亿元，同比增速大幅提升。自年初新冠疫情发生以来，海内外地区为防止疫情扩散，大多采取了停工停学的措施，限制非必要的人员流动。在线下场景交流受阻碍的情况下，在线游戏/视频、远程办公/教学等线上需求出现集中式爆发。受此推动，我国游戏产业上半年实际销售收入达到 1394.93 亿元，同比增长 22.34%，自 2017 年以来重回双位数增长，增速较 2019 年提升了13.75 个百分点。

扫码观看视频讲解

## 11.2 功能模块介绍

本游戏是用Panda3D开发的大型游戏项目，功能模块如图11-1所示。

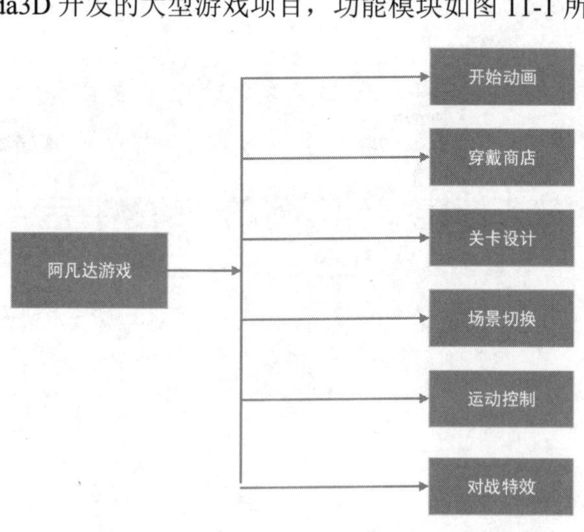


图11-1 功能模块

## 11.3 系统配置

扫码观看视频讲解

在本游戏项目中，经常用到的功能放在 base 模块中，主要包括游戏音效、启动游戏、地图纹理和 HUD 等信息。本节将详细讲解实现本项目系统配置模块的流程。

### 11.3.1 全局信息

编写文件 Globals.py，功能是设置本项目中多次用到的全局信息，包括 BitMask(在计算机应用中，BitMask 指的是一串二进制数字，通过与目标数字的按位操作达到屏蔽指定位的目的)、掩码、鼠标点击音效和场景滚动音效等信息。文件 Globals.py 的具体实现代码如下所示：

```
# 碰撞 BitMasks
WallBitmask = BitMask32.bit(1)
FloorBitmask = BitMask32.bit(2)
DoorBitmask = BitMask32.bit(3)
GagBitmask = BitMask32.bit(4)
DropBitmask = BitMask32.bit(5)
ProjBitmask = BitMask32.bit(6)

LARGE_BEAN_AMT = 20
THROW_COLOR = VBase3(255 / 255.0, 145 / 255.0, 66 / 255.0)

#声音
clickSound = None
rlvrSound = None

def getClickSound():
   clickSound = loader.loadSfx('phase_3/audio/sfx/GUI_click.ogg')
   return clickSound

def getRlvrSound():
   rlvrSound = loader.loadSfx('phase_3/audio/sfx/GUI_rollover.ogg')
   return rlvrSound

def getFont(name):
   return loader.loadFont("phase_3/models/fonts/%s" % (name))
```

### 11.3.2 初始信息

编写文件 Data.py，功能是加载系统数据文件 data.ini 中的初始配置信息，这些信息包括主角精灵的位置、头像动物、体重、身高、颜色和性别等，然后使用方法 set()将获取的初始信息添加到配置对象 config 中。文件 Data.py 的具体实现代码如下所示：

```
class Data(DirectObject):
```

```
    def saveGame(self):
        Toon = render.find('**/Toon')
        if(Toon != None):
            self.saveAvatar(Toon)

    def saveAvatar(self, toon):
        avatar = toon.getPythonTag("AvatarInstance")
        localAvatar = avatar.getAvatar()
        config = configparser.RawConfigParser()
        config.add_section('Avatar')
        config.set('Avatar', 'X', str(localAvatar.getX()))
        config.set('Avatar', 'Y', str(localAvatar.getY()))
        config.set('Avatar', 'Z', str(localAvatar.getZ()))
        config.set('Avatar', 'H', str(localAvatar.getH()))
        config.set('Avatar', 'P', str(localAvatar.getP()))
        config.set('Avatar', 'R', str(localAvatar.getR()))
        for key in avatar.getData().keys():
            print(key)
            if(avatar.getData()[key] != None):
                config.set('Avatar', key, str(avatar.getData()[key]))
        #config.add_section('Gag')
        with open('data.ini', 'w') as configFile:
            config.write(configFile)

    def loadAvatar(self):
        config = configparser.ConfigParser()
        config.read('data.ini')
        pos = VBase3(float(config.get('Avatar', 'X')), float(config.get
                        ('Avatar', 'Y')), float(config.get('Avatar', 'Z')))
        hpr = VBase3(float(config.get('Avatar', 'H')), float(config.get
                    ('Avatar', 'P')), float(config.get('Avatar', 'R')))
```

## 11.3.3 音效信息

编写文件 SoundBank.py，功能是首先设置系统音效素材文件的目录，然后在字典 sounds 中保存在游戏中用到各种各样的音效信息，每一种音效名字和素材资源文件夹中的一种声音文件相对应。最后编写方法 getSound()，根据实际场景使用 loadSfx()加载对应的音效。文件 SoundBank.py 的具体实现代码如下所示：

```
phase_35 = 'phase_3.5/audio/sfx/'
phase_4 = 'phase_4/audio/sfx/'
phase_5 = 'phase_5/audio/sfx/'

sounds = {
  'pie_throw' : phase_35 + 'AA_pie_throw_only',
  'tart_coll' : phase_35 + 'AA_tart_only',
  'pie_coll' : phase_4 + 'AA_wholepie_only',
  'propeller' : phase_4 + 'TB_propeller',
  'propeller_in' : phase_5 + 'ENC_propeller_in',
  'cog_laugh' : phase_35 + 'Cog_Death',
  'cog_explode' : phase_35 + 'ENC_cogfall_apart',
  'skelecog_grunt' : phase_5 + 'Skel_COG_VO_grunt',
  'skelecog_murmur' : phase_5 + 'Skel_COG_VO_murmur',
```

```
 'skelecog_question' : phase_5 + 'Skel_COG_VO_question',
 'skelecog_statement' : phase_5 + 'Skel_COG_VO_statement',
 'victory_dance' : phase_35 + 'ENC_Win',
 'door_open' : phase_35 + 'Door_Open_1',
 'door_close' : phase_35 + 'Door_Close_1',
 'toon_walk' : phase_35 + 'AV_footstep_walkloop',
 'toon_run' : phase_35 + 'AV_footstep_runloop',
 'make_a_toon' : 'phase_3/audio/bgm/create_a_toon',
 'drop_pickup' : 'phase_5.5/audio/sfx/mailbox_alert',
 'bgm_doom' : 'bgm/Doomsday Theme',
 'bgm_ceo' : 'bgm/BossBot_CEO_v2',
 'bgm_bigboss' : 'bgm/The Big Boss',
 'bgm_install' : 'bgm/Installer Theme',
 'bgm_hall' : 'bgm/Hall of Fame',
 'bgm_orchestra' : 'bgm/Installer Orchestrated.mp3',
 'bgm_bldg' : 'bgm/encntr_suit_winning_indoor',
 'cannon_fire' : phase_4 + 'MG_cannon_fire_alt',
 'cannon_adjust' : phase_4 + 'MG_cannon_adjust',
 'cannon_whizz' : phase_4 + 'MG_cannon_whizz',
 'toonhall_warning' : 'sfx/CHQ_GOON_tractor_beam_alarmed',
 'power_tie_throw' : phase_5 + 'SA_powertie_throw',
 'power_tie_impact' : phase_5 + 'SA_powertie_impact'
}

def getSound(soundName):
   if('mp3' not in sounds[soundName]):
      sound = loader.loadSfx(sounds[soundName] + '.ogg')
   else:
      sound = loader.loadSfx(sounds[soundName])
   return sound
```

### 11.3.4　地图纹理

编写文件 TextureBank.py，功能是在字典 textures 中保存在游戏中需要用到的地图纹理图片，代码如下：

```
textures = {
'flunky' : 'phase_3.5/maps/flunky.jpg',
'tightwad' : 'phase_3.5/maps/tightwad.jpg',
'blood_sucker' : 'phase_4/maps/blood-sucker.jpg',
'double_talker' : 'phase_4/maps/double-talker.jpg',
'mingler' : 'phase_4/maps/mingler.jpg',
'mover_shaker' : 'phase_4/maps/mover_shaker.jpg',
'name_dropper' : 'phase_4/maps/name-dropper.jpg',
'number_cruncher' : 'phase_4/maps/number_cruncher.jpg',
'pencil_pusher' : 'phase_4/maps/pencil_pusher.jpg',
'robber_baron' : 'phase_4/maps/robber-baron.jpg',
'spin_doctor' : 'phase_4/maps/spin-doctor.jpg',
'yes_man' : 'phase_4/maps/yes_man.jpg',
'bottom_feeder' : 'phase_3.5/maps/bottom-feeder.jpg',
'corporate_raider' : 'phase_3.5/maps/corporate-raider.jpg'
}

def getTexture(texName):
   return textures[texName]
```

## 11.3.5 实现 HUD 模块

在一款游戏中，能够用视觉效果向玩家传达信息的元素都可以称之为 HUD、常用的 HUD 元素是游戏子场景以及各种图标，它们负责向玩家传达各种关键信息，告诉玩家应该做什么，应该去哪里。编写实例文件 Hud.py，分别通过不同的方法实现 HUD 功能。实例文件 Hud.py 的具体实现流程如下所示。

(1) 创建 HUD 类 Hud，初始化时调用方法 setupHud()加载初始场景。代码如下：

```
class Hud(DirectObject):
    throwExpLbl = None
    def __init__(self, avatar):
        self.avatar = avatar
        self.setupHud()
        render.setPythonTag("Hud", self)
```

(2) 创建方法 showMessage()，功能是在玩游戏的过程中弹出提示信息“Successfully saved game!”。代码如下：

```
    def showMessage(self):
        ImpressBT = Globals.getFont('ImpressBT.ttf')
        saveTxt = OnscreenText(text = 'Successfully saved game!', pos = (0.025,
            0.5), font = ImpressBT, fg = (1, 0, 0, 1), scale = (0.08, 0.08, 0.08))
        base.taskMgr.doMethodLater(0.8, self.removeText, 'Remove Save TXT',
                                   extraArgs = [saveTxt], appendTask = True)
```

(3) 创建方法 removeText()，功能是删除子节点中的文本提示信息。代码如下：

```
    def removeText(self, text, task):
        text.removeNode()
        return Task.done
```

(4) 编写方法 showWaveInformation()，功能是显示当前 Wave 的信息。首先使用 getNextWave()获取接下来使用的 Wave，如果 icon 为 None，则加载使用 models 中的 cog 作为素材。如果 icon 不为 None，则使用 getCogName()遍历 degrees 深度的 cog。代码如下：

```
    def showWaveInformation(self):
        round = render.getPythonTag("Round")
        next_wave = round.getNextWave()
        icon = None
        if next_wave.getInvasion() == None:
        icon = loader.loadModel("phase_3/models/gui/cog_icons.bam").find('**/cog')
        else:
            degrees = [0, 90, 180, 270, 360, -90, -180, -270, -360]
            icon = Cogs().getCogHead(next_wave.getInvasion().getCogName())
            x = -1
            for degree in degrees:
                pass
```

(5) 编写方法 setupHud(self)，实现 HUD 主功能，加载 models 素材展示游戏界面的内容。代码如下：

```
    def setupHud(self):
        btnSrc = loader.loadModel('phase_3/models/gui/quit_button.bam')
        ImpressBT = Globals.getFont('ImpressBT.ttf')
        btnSrcUP = btnSrc.find('**/QuitBtn_UP')
        btnSrcDN = btnSrc.find('**/QuitBtn_DN')
        btnSrcRlvr = btnSrc.find('**/QuitBtn_RLVR')
        DirectButton(clickSound = Globals.getClickSound(),
            rolloverSound = Globals.getRlvrSound(), \
                geom = (btnSrcUP, btnSrcDN, btnSrcRlvr),\
            relief = None, pos = (1.12, 0, -0.95), command = self.saveGame)
        OnscreenText(text = 'Save', pos = (1.12325, -0.97), font = ImpressBT)
        DirectButton(clickSound = Globals.getClickSound(),
            rolloverSound = Globals.getRlvrSound(),
              geom = (btnSrcUP, btnSrcDN, btnSrcRlvr),
            relief = None, pos = (1.12, 0, -0.85), command = self.spawnCogs)
        OnscreenText(text = 'Spawn Cogs', pos = (1.12325, -0.87), font = ImpressBT)
        self.showWaveInformation()
```

(6) 编写方法 drawEXPBar()，功能是绘制游戏场景中表示经验值的进度条。代码如下：

```
    def drawEXPBar(self):
        self.throwExp = DirectWaitBar(relief=DGG.SUNKEN, frameSize=(-1,
            1,
            -0.15,
            0.15), borderWidth=(0.02, 0.02), scale=0.25,\
            frameColor=(Globals.THROW_COLOR[0] * 0.7, \
            Globals.THROW_COLOR[1] * 0.7, \
            Globals.THROW_COLOR[2] * 0.7, \
            1), barColor=(Globals.THROW_COLOR[0],
           Globals.THROW_COLOR[1],
            Globals.THROW_COLOR[2],
            1), pos=(0, 0, -0.8))
        self.throwExpLbl = OnscreenText(parent = self.throwExp,
                pos =(0, -0.05), scale = 0.18, fg = (0, 0, 0, 1), mayChange = 1,
align = TextNode.ACenter)
```

(7) 编写方法 updateEXPBar()，功能是及时更新经验值进度条。代码如下：

```
    def updateEXPBar(self, levelMgr):
        exp = levelMgr.getEXP()
        neededEXP = levelMgr.getEXPToNextLevel()
        level = levelMgr.getLevel()
        if(self.throwExpLbl == None):
            self.drawEXPBar()
        if not neededEXP == 0:
            value = ((float(exp) / float(neededEXP)) * 100)
        else:
            value = 0
        self.throwExp.update(value)
        if(level != levelMgr.getMaxLevel()):
            self.throwExpLbl.setText("%s/%s" % (exp, neededEXP))
        else:
            self.throwExpLbl.setText("%s/MAX" % (exp))
```

(8) 编写方法 saveGame()，功能是保存当前游戏的进度状态，这样在下次打开时知道上次玩到了哪一关。代码如下：

```
    def saveGame(self):
        self.showMessage()
        data = Data()
        data.saveGame()
```

### 11.3.6 游戏入口

本游戏的入口文件是 GameStart.py，功能是调用相关模块和方法运行游戏项目，代码如下：

```
loadPrcFile('../../config/Config.prc')
loadPrcFileData('', 'hardware-animated-vertices 1')

from direct.directnotify.DirectNotifyGlobal import directNotify

class GameStart(ShowBase):
    __module__ = __name__
    notify = directNotify.newCategory('GameStart')
    debug = False
    fps = False

    def __init__(self):
        ShowBase.__init__(self)

        if(self.debug):
            loadPrcFileData('', 'want-pstats 1')
            self.notify.warning('Debug Mode enabled.')
            base.startDirect()
        if(self.fps):
            base.setFrameRateMeter(True)
        base.camera.setPos(0, -15, 3)
        base.disableMouse()
        GameWorld()
launch = GameStart()
launch.run()
```

## 11.4 创建精灵

扫码观看视频讲解

在本游戏中，主角精灵是由类 Avatar 实现的，在此称为“阿凡达”。本节将详细讲解主角精灵类 Avatar 的具体实现过程。

### 11.4.1 主角精灵类 Avatar

编写文件 Avatar.py，创建主角精灵类 Avatar，具体实现流程如下所示。

(1) 首先创建字典 Anims，用于保存主角精灵的动作。代码如下：

```
Anims = {
"neutral", "walk", "run", "jump", "jump-idle", "running-jump",
```

```
"running-jump-idle", "pie-throw"
}
```

(2) 创建类 Avatar，在初始化时设置各种穿戴搭配的值为 None。代码如下：

```
class Avatar(DirectObject):

    def __init__(self):
        self.data = {
        'Animal' : None,
        'Gender' : None,
        'Weight' : None,
        'Height' : None,
        'Color' : None,
        'HeadType' : None,
        'HeadLength' : None,
        'Shirt' : None,
        'Bottoms' : None
        }
```

(3) 分别通过方法 setData()和 getData()设置、获取穿戴搭配的值。代码如下：

```
    def setData(self, key, value):
        self.data[key] = value

    def getData(self):
        return self.data
```

(4) 创建方法 generateAnims()，功能是加载 models 素材文件，生成不同 Anims 动作的动画。代码如下：

```
    def generateAnims(self, torso = False, legs = False):
        anims = {}
        if(torso):
            for anim in Anims:
                anims[anim] = loader.loadModel('phase_3/models/char\
                %s-torso-%s-%s.bam' % (self.data['Weight'],
                self.data['Gender'], anim))
        if(legs):
            for anim in Anims:
                anims[anim] = loader.loadModel('phase_3/models/char/%s-legs-%s.bam'\
                    % (self.data['Height'], anim))
        return anims
```

(5) 编写方法 removeOtherParts()，功能是删除其他不同节点的 data 数据。代码如下：

```
    def removeOtherParts(self, nodepath, impotent):
        if(impotent == None):
            matches = nodepath
            for part in range(0, matches.getNumPaths()):
                name = matches.getPath(part).getName()
                if(name):
                    if(name == 'muzzle-short-neutral' and self.data['Animal'] ==
'mouse'):
                        if(self.data['HeadType'] == 500 or self.data['HeadType'] ==
1000):
                            return
```

```
                matches.getPath(part).removeNode()
        else:
            matches = nodepath
            for part in range(0, matches.getNumPaths()):
                if(part != impotent):
                    if(matches.getPath(part).getName()):
                        matches.getPath(part).removeNode()
```

(6) 编写方法 setColor()，功能是设置穿戴着装的不同颜色。在本项目中预先设置了三种动物(dog、horse、monkey)的头像，在列表 body 中存储了身体的不同部位，然后遍历每一个身体部位并根据动物头像分别设置对应的颜色。代码如下：

```
    def setColor(self):
        colorEarAnimals = {'dog', 'horse', 'monkey'}
        body = ['**/head-*', '**/neck', '**/arms', '**/legs', '**/feet', '**/head']
        if(self.data['Animal'] not in colorEarAnimals):
            body.append('**/*ears*')
        for part in range(len(body)):
            self.Avatar.findAllMatches(body[part]).setColor(self.data['Color'])
        self.Avatar.findAllMatches("**/hands").setColor(AvatarAttributes().
            getColor('white'))
```

(7) 编写方法 generate()，功能是加载 models 素材并生成对应的穿戴搭配。需要注意的是，在每次更换穿着搭配时，会删除上次的穿着搭配。代码如下：

```
    def generate(self):
        head = loader.loadModel('phase_3/models/char/%s-heads-%s.bam' %
             (self.data['Animal'], self.data['HeadType']))
        torso = loader.loadModel('phase_3/models/char/%s-torso-%s.bam' %
              (self.data['Weight'], self.data['Gender']))
        legs = loader.loadModel('phase_3/models/char/%s-legs.bam' %
              (self.data['Height']))

        if(self.data['HeadLength'] == 'short'):
            self.removeOtherParts(head.findAllMatches('**/*long*'), None)
        else:
            self.removeOtherParts(head.findAllMatches('**/*short*'), None)
        muzzleParts = head.findAllMatches('**/*muzzle*')
        if(self.data['Animal'] != 'dog'):
            for partNum in range(0, muzzleParts.getNumPaths()):
                part = muzzleParts.getPath(partNum)
                if not 'neutral' in part.getName():
                    part.hide()
        self.removeOtherParts(legs.findAllMatches('**/boots*') +
             legs.findAllMatches('**/shoes'), None)
        torsoAnims = self.generateAnims(torso=True)
        legsAnims = self.generateAnims(legs=True)
        self.Avatar = Actor({'head' : head, 'torso' : torso, 'legs' : legs},
             {'torso' : torsoAnims, 'legs' : legsAnims})
        self.Avatar.attach('head', 'torso', 'def_head')
        self.Avatar.attach('torso', 'legs', 'joint_hips')
        if(self.data['Gender'] == 'girl'):
            femaleEyes = loader.loadTexture('phase_3/maps/eyesFemale.jpg',
                'phase_3/maps/eyesFemale_a.rgb')
            try:
```

```
            self.Avatar.find('**/eyes').setTexture(femaleEyes, 1)
        except: pass
        try:
            self.Avatar.find('**/eyes-%s' %
(self.data['HeadLength'])).setTexture(femaleEyes, 1)
        except: pass
    # Reseat pupils of old head models
    pupils = head.findAllMatches('**/*pupil*')
    for pupil in pupils:
        pupil.setY(0.02)
    self.setColor()
    shadow = ShadowCaster(self.Avatar)
    shadow.initializeDropShadow()
    self.Avatar.setPythonTag("AvatarInstance", self)
```

## 11.4.2 属性信息

编写文件 AvatarAttributes.py，设置主角精灵类 Avatar 的属性信息，这些属性信息包括体重信息、身高信息、动物头像类型、头部信息和颜色信息等。文件 AvatarAttributes.py 的具体实现代码如下所示：

```
animals = ['cat', 'dog', 'duck', 'mouse', 'pig', 'rabbit', 'bear', 'horse',
           'monkey']
colors = {
    'white' : Vec4(1.0, 1.0, 1.0, 1.0),
    'peach' : Vec4(0.96875, 0.691406, 0.699219, 1.0),
    'bright red' : Vec4(0.933594, 0.265625, 0.28125, 1.0),
    'red' : Vec4(0.863281, 0.40625, 0.417969, 1.0),
    'maroon' : Vec4(0.710938, 0.234375, 0.4375, 1.0),
    'sienna' : Vec4(0.570312, 0.449219, 0.164062, 1.0),
    'brown' : Vec4(0.640625, 0.355469, 0.269531, 1.0),
    'tan' : Vec4(0.996094, 0.695312, 0.511719, 1.0),
    'coral' : Vec4(0.832031, 0.5, 0.296875, 1.0),
    'orange' : Vec4(0.992188, 0.480469, 0.167969, 1.0),
    'yellow' : Vec4(0.996094, 0.898438, 0.320312, 1.0),
    'cream' : Vec4(0.996094, 0.957031, 0.597656, 1.0),
    'citrine' : Vec4(0.855469, 0.933594, 0.492188, 1.0),
    'lime green' : Vec4(0.550781, 0.824219, 0.324219, 1.0),
    'sea green' : Vec4(0.242188, 0.742188, 0.515625, 1.0),
    'green' : Vec4(0.304688, 0.96875, 0.402344, 1.0),
    'light blue' : Vec4(0.433594, 0.90625, 0.835938, 1.0),
    'aqua' : Vec4(0.347656, 0.820312, 0.953125, 1.0),
    'blue' : Vec4(0.191406, 0.5625, 0.773438, 1.0),
    'periwinkle' : Vec4(0.558594, 0.589844, 0.875, 1.0),
    'royal blue' : Vec4(0.285156, 0.328125, 0.726562, 1.0),
    'slate blue' : Vec4(0.460938, 0.378906, 0.824219, 1.0),
    'purple' : Vec4(0.546875, 0.28125, 0.75, 1.0),
    'lavender' : Vec4(0.726562, 0.472656, 0.859375, 1.0),
    'pink' : Vec4(0.898438, 0.617188, 0.90625, 1.0)
}

class AvatarAttributes(DirectObject):
    __module__ = __name__

    notify = DirectNotifyGlobal.directNotify.newCategory('AvatarAttributes')
```

```
def convertColorDictToTbl(self):
    colorsTbl = []
    for color in colors.values():
        colorsTbl.append(color)
    return colorsTbl

def getAnimals(self):
    return animals

def getHeadTypes(self, animal = None):
    return [250, 500, 1000]

def getWeights(self):
    return ['skinny', 'stubby', 'fat']

def getHeights(self):
    return ['short', 'medium', 'tall']

def getColors(self):
    return colors

def getColor(self, color):
    return colors[color]
```

## 11.4.3 选择穿戴着装

编写文件 MakeAToon.py，功能是为主角精灵选择穿戴着装，具体场景和很多儿童玩的穿衣搭配游戏类似。文件 MakeAToon.py 的具体实现流程如下。

(1) 创建类 ToonCreator，在字典 dataCache 中设置 9 种主角的装扮属性信息。代码如下：

```
class ToonCreator(DirectObject):
    __module__ = __name__

    notify = DirectNotifyGlobal.directNotify.newCategory("ToonCreator")

    dataCache = {
      'Animal' : None,
      'Gender' : None,
      'Weight' : None,
      'Height' : None,
      'Color' : None,
      'HeadType' : None,
      'HeadLength' : None,
      'Shirt' : None,
      'Bottoms' : None
    }
    avatar = Avatar()
    subject = None
```

(2) 实现初始化处理，创建装扮搭配场景，代码如下：

```
    def __init__(self):
        self.trans = Transitions(loader)
```

```
        self.scene = render.getPythonTag('SceneManager').\
        createScene('Make-A-Toon')
        self.setupEnvironment()
        self.randomizeData()
        self.setupHUD()
        self.bgm = SoundBank.getSound('make_a_toon')
        self.bgm.setLoop(True)
        self.bgm.play()
```

(3) 编写方法 setupHUD(self)，功能是加载显示 models 中对应的装扮信息，在 UI 界面中显示各种类型的元素。这些元素包括各个身体部位，能够将不同的装扮赋给不同的身体部位。代码如下：

```
    def setupHUD(self):
        gui1 = loader.loadModel('phase_3/models/gui/create_a_toon_gui.bam')
        gui2 = loader.loadModel('phase_3/models/gui/gui_toongen.bam')
        mickeyFont = loader.loadFont('MickeyFont.bam')
        self.guiElements = [
            OnscreenImage(image = gui1.find("**/CrtATn_TopBar"),
                          parent=render2d,
                         pos=(0, 0, 0.9), scale=(0.85, 0.85, 1)),
            OnscreenText(text = "Make Your Toon", pos = (0, 0.85),
                        font = mickeyFont, fg = (1, 0, 0, 1),
                       scale=(0.2, 0.2, 0.2)),
            DirectButton(clickSound=Globals.getClickSound(),
                        rolloverSound=Globals.getRlvrSound(),
                        geom = (gui1.find("**/CrtATn_R_Arrow_UP"),
                        gui1.find("**/CrtATn_R_Arrow_DN"),
                        gui1.find("**/CrtATn_R_Arrow_RLVR")),
                        pos=(0.66, 0, 0.4), relief=None,
                        command=self.setData, extraArgs=['Animal', 'prev']),
            DirectButton(clickSound=Globals.getClickSound(),
                        rolloverSound=Globals.getRlvrSound(),
                        geom = (gui1.find("**/CrtATn_R_Arrow_UP"),
                        gui1.find("**/CrtATn_R_Arrow_DN"),
                        gui1.find("**/CrtATn_R_Arrow_RLVR")),
                        pos=(0, 0, 0.41), hpr=(0, 0, 180),
                        command=self.setData, extraArgs=['Animal', 'next'],
                        relief=None),
            OnscreenText(text = "Animal", pos = (0.325, 0.38), font = mickeyFont,
                        fg = (1, 0, 0, 1),
                        scale=(0.08, 0.08, 0.08)),
            DirectButton(clickSound=Globals.getClickSound(),
                        rolloverSound=Globals.getRlvrSound(),
                        geom = (gui1.find("**/CrtATn_R_Arrow_UP"),
                        gui1.find("**/CrtATn_R_Arrow_DN"),
                        gui1.find("**/CrtATn_R_Arrow_RLVR")),
                        pos=(0.66, 0, 0.1), relief=None,
                        command=self.setData, extraArgs=['Weight', 'prev']),
            DirectButton(clickSound=Globals.getClickSound(),
                        rolloverSound=Globals.getRlvrSound(),
                        geom = (gui1.find("**/CrtATn_R_Arrow_UP"),
                        gui1.find("**/CrtATn_R_Arrow_DN"),
                        gui1.find("**/CrtATn_R_Arrow_RLVR")),
                        pos=(0, 0, 0.11), hpr=(0, 0, 180),
```

```
                command=self.setData, extraArgs=['Weight', 'next'],
                    relief=None),
            OnscreenText(text = "Body", pos = (0.325, 0.1),
                font = mickeyFont, fg = (1, 0, 0, 1),
                scale=(0.08, 0.08, 0.08)),
            DirectButton(clickSound=Globals.getClickSound(),
                rolloverSound=Globals.getRlvrSound(),
                geom = (gui1.find("**/CrtATn_R_Arrow_UP"),
                gui1.find("**/CrtATn_R_Arrow_DN"),
                gui1.find("**/CrtATn_R_Arrow_RLVR")), pos=(0.66, 0, -0.20),
                relief=None, command=self.setData,
                extraArgs=['Height', 'prev']),
            DirectButton(clickSound=Globals.getClickSound(),
                rolloverSound=Globals.getRlvrSound(),
                geom = (gui1.find("**/CrtATn_R_Arrow_UP"),
                gui1.find("**/CrtATn_R_Arrow_DN"),
                gui1.find("**/CrtATn_R_Arrow_RLVR")),
                pos=(0, 0, -0.19), hpr=(0, 0, 180),
                command=self.setData, extraArgs=['Height', 'next'],
                  relief=None),
            OnscreenText(text = "Height", pos = (0.325, -0.2),
                font = mickeyFont, fg = (1, 0, 0, 1),
                scale=(0.08, 0.08, 0.08)),
            DirectButton(clickSound=Globals.getClickSound(),
                rolloverSound=Globals.getRlvrSound(),
                geom = (gui1.find("**/CrtATn_R_Arrow_UP"),
                gui1.find("**/CrtATn_R_Arrow_DN"),
                gui1.find("**/CrtATn_R_Arrow_RLVR")),
                pos=(0.66, 0, -0.50), relief=None,
                command=self.setData, extraArgs=['HeadType', 'prev']),
            DirectButton(clickSound=Globals.getClickSound(),
                rolloverSound=Globals.getRlvrSound(),
                geom = (gui1.find("**/CrtATn_R_Arrow_UP"),
                gui1.find("**/CrtATn_R_Arrow_DN"),
                gui1.find("**/CrtATn_R_Arrow_RLVR")),
                pos=(0, 0, -0.49), hpr=(0, 0, 180),
                command=self.setData, extraArgs=['HeadType', 'next'],
                  relief=None),
            OnscreenText(text = "Head", pos = (0.325, -0.5), font = mickeyFont,
                  fg = (1, 0, 0, 1),
                scale=(0.08, 0.08, 0.08)),
            DirectButton(clickSound=Globals.getClickSound(),
                rolloverSound=Globals.getRlvrSound(),
                geom = (gui1.find("**/CrtATn_R_Arrow_UP"),
                gui1.find("**/CrtATn_R_Arrow_DN"),
                gui1.find("**/CrtATn_R_Arrow_RLVR")),
                pos=(0.66, 0, -0.8), relief=None,
                command=self.setData, extraArgs=['Color', 'prev']),
            DirectButton(clickSound=Globals.getClickSound(),
                rolloverSound=Globals.getRlvrSound(),
                geom = (gui1.find("**/CrtATn_R_Arrow_UP"),
                gui1.find("**/CrtATn_R_Arrow_DN"),
                gui1.find("**/CrtATn_R_Arrow_RLVR")),
                pos=(0, 0, -0.79), hpr=(0, 0, 180),
                command=self.setData, extraArgs=['Color', 'next'],
                   relief=None),
                OnscreenText(text = "Length", pos = (0.328, -0.67),
                font = mickeyFont, fg = (1, 0, 0, 1),
```

```
                    scale=(0.08, 0.08, 0.08)),
        DirectButton(clickSound=Globals.getClickSound(),
                    rolloverSound=Globals.getRlvrSound(),
                    geom = (gui1.find("**/CrtATn_R_Arrow_UP"),
                    gui1.find("**/CrtATn_R_Arrow_DN"),
                    gui1.find("**/CrtATn_R_Arrow_RLVR")),
                    pos=(0.57, 0, -0.65), scale = 0.4,
                    relief=None, command=self.setData,
                    extraArgs=['HeadLength', 'short']),
        DirectButton(clickSound=Globals.getClickSound(),
                    rolloverSound=Globals.getRlvrSound(),
                    geom = (gui1.find("**/CrtATn_R_Arrow_UP"),
                    gui1.find("**/CrtATn_R_Arrow_DN"),
                    gui1.find("**/CrtATn_R_Arrow_RLVR")),
                    pos=(0.11, 0, -0.64), scale = 0.4,
                    hpr=(0, 0, 180), command=self.setData,
                    extraArgs=['HeadLength', 'long'], relief=None),
        OnscreenText(text = "Color", pos = (0.325, -0.8),
                    font = mickeyFont, fg = (1, 0, 0, 1),
                    scale=(0.08, 0.08, 0.08)),
        DirectButton(clickSound=Globals.getClickSound(),
                    rolloverSound=Globals.getRlvrSound(),
                    geom = (gui2.find("**/tt_t_gui_mat_girlUp"),
                    gui2.find("**/tt_t_gui_mat_girlDown")),
                    pos=(0.66, 0, 0.68), relief=None,
                    command=self.randomizeData, extraArgs = ['girl'],
                    scale=(0.55, 0.55, 0.55)),
        DirectButton(clickSound=Globals.getClickSound(),
                    rolloverSound=Globals.getRlvrSound(),
                    geom = (gui2.find("**/tt_t_gui_mat_boyUp"),
                    gui2.find("**/tt_t_gui_mat_boyDown")),
                    pos=(0, 0, 0.68), command=self.randomizeData,
                    extraArgs = ['boy'],
                    relief=None, scale=(0.55, 0.55, 0.55)),
                    DirectButton(clickSound=Globals.getClickSound(),
                       rolloverSound=Globals.getRlvrSound(),
                    geom=(gui2.find("**/tt_t_gui_mat_okUp"),
                    gui2.find("**/tt_t_gui_mat_okDown")),
                    pos=(1, 0, -0.2), relief=None, command=self.done)
    ]
```

(4) 编写方法 done(),功能是完成主角的穿戴搭配功能,根据玩家选择的搭配装扮主角。这样在后面的游戏场景中，主角将使用当前选择的装扮。代码如下：

```
def done(self):
    self.trans.irisOut()
    self.bgm.stop()
    for element in self.guiElements:
        element.removeNode()
    base.taskMgr.doMethodLater(0.8, self.cleanUp, 'Cleanup Scene')
```

(5) 编写方法 setData()，功能是设置用户的装扮信息。在字典 tables 中保存了我们可选择的属性信息类型,并判断 value 的值是否是 prev 或 next,如果是则表示还在继续选择当中,如果不是则表示用户已经选择完毕。代码如下：

```
def setData(self, key, value):
   attr = AvatarAttributes()
   tables = {'Animal' : attr.getAnimals(),
           'Gender' : ['boy', 'girl'],
           'Color' : attr.convertColorDictToTbl(),
           'Weight' : attr.getWeights(),
           'Height' : attr.getHeights(),
           'HeadType' : attr.getHeadTypes(),
           'HeadLength' : ['short', 'long']}
   if(key not in self.dataCache):
      return
   if(value == 'prev' or value == 'next'):
      table = tables[key]
      index = 0
      for i in range(len(table)):
         if(table[i] == self.dataCache[key]):
            index = i
      if(value == 'prev'):
         if(index == 0):
            self.dataCache[key] = table[(len(table) - 1)]
         else:
            self.dataCache[key] = table[index - 1]
      else:
         if(index == (len(table) - 1)):
            self.dataCache[key] = table[0]
         else:
            self.dataCache[key] = table[index + 1]
   else:
      self.dataCache[key] = value
   self.generateAvatar()
```

(6) 编写方法 randomizeData()，功能是随机化处理选择数据，代码如下：

```
def randomizeData(self, gender = None):
   def setData(key, value):
      self.dataCache[key] = value

   attr = AvatarAttributes()
   setData('Animal', random.choice(attr.getAnimals()))
   if(gender == None):
      setData('Gender', random.choice(['boy', 'girl']))
   else:
      setData('Gender', gender)
   setData('Weight', random.choice(attr.getWeights()))
   setData('Height', random.choice(attr.getHeights()))
   setData('Color', random.choice(attr.convertColorDictToTbl()))
   setData('HeadType', random.choice(attr.getHeadTypes()))
   setData('HeadLength', random.choice(['short', 'long']))
   self.generateAvatar()
```

(7) 编写方法 generateAvatar()，功能是根据用户选择的装扮生成主角精灵，代码如下：

```
def generateAvatar(self):
   self.cleanupAvatar()
   for key in self.dataCache.keys():
      self.avatar.setData(key, self.dataCache[key])
   self.avatar.generate()
   self.avatar.getAvatar().setPosHprScale(-2, 0, 0, 200, 0, 0, 1, 1, 1)
   self.avatar.getAvatar().reparentTo(render)
```

```
        self.subject = self.avatar.getAvatar()
        self.subject.setName("Temp Toon")
        self.subject.loop('neutral')
        shirtColor = random.choice(AvatarAttributes().convertColorDictToTbl())
        bttColor = random.choice(AvatarAttributes().convertColorDictToTbl())
        self.subject.find("**/torso-top").setColor(shirtColor)
        self.subject.find("**/sleeves").setColor(shirtColor)
        self.subject.find('**/torso-bot').setColor(bttColor)
```

## 11.5　调试运行

扫码观看视频讲解

为了节省本书篇幅，在前面的内容中只讲解了第一个游戏场景的实现过程。运行本游戏的入口文件 GameStart.py 后会来到第一个场景，在此场景中可以使用鼠标为主角搭配不同的装扮，效果如图 11-2 所示。

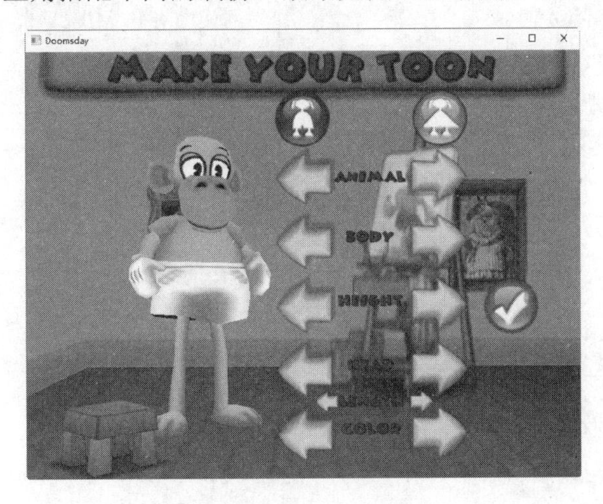

图 11-2　第一个游戏场景

使用鼠标选择好装扮后，单击右侧图标按钮☑后来到第二个游戏场景，效果如图 11-3 所示。

图 11-3　第二个游戏场景

# 第 12 章

# AI 人脸识别签到打卡系统

## (PyQt5+百度智能云+OpenCV-Python+SQLite3 实现)

为了提高上课签到的效率，实现无纸化办公的需求，某高校决定参考钉钉软件开发一个基于人脸识别的学生签到打卡系统。本章将详细讲解使用 PyQt5、百度智能云、Opencv-Python 和 SQLite3 实现一个学生签到打卡系统的过程。

扫码观看视频讲解

## 12.1 需求分析

本需求分析文档的目的是说明运行本签到系统的最终条件、性能要求及要实现的功能，为进一步设计与实现打下基础。本文档以文档形式将用户对软件的需求固定下来，是与用户沟通的成果，也供用户验收项目时参考。本文档预期读者为用户、项目管理人员、软件设计人员、编程人员、测试人员等项目相关人员。

### 12.1.1 背景介绍

为了保证现在大学生课堂出勤率以及对学生信息的管理，大部分学校纷纷采取不同的措施来对学生的出勤率进行管理和安排，故对合理、高利用率的学生签到系统有着迫切的需求。

- 系统用途：本系统利用相应的安卓平台，帮助学校各个部门更加电子化、智能化地管理学生的出勤，从而提高学校管理的效率。
- 系统开发人员：本系统由开发团队完成从可行性分析、概要设计、实现、代码调试等一系列过程。

### 12.1.2 任务目标

(1) 用户组：为了更加清晰明了地实现管理，对学生信息进行分类，将每个班级的学生信息作为一个分类。

- 添加用户组：添加代表某个班级的新的用户组。
- 删除用户组：删除一个已经存在的用户组。
- 查询用户组：根据输入的关键字查询某个用户组。

(2) 用户：每个用户对应一个学生信息。

- 添加用户：打开摄像头，向系统中添加新的学生信息，包括学生的人脸照片、学号、姓名和班级等。
- 删除用户：删除一个已经存在的用户信息。
- 更新用户：修改一个已经存在的用户信息。

(3) 签到：实现人脸识别签到功能。

- 启动签到：打开摄像头，识别当前人的信息，如果在系统数据库中存在此人的照片信息，则签到成功并显示此人的基本信息。
- 导出信息：导出此人的签到信息。
- 关闭签到：关闭当前的签到功能。

## 12.2　模块架构

在开发一个大型应用程序时，建立模块架构是一个非常重要的前期准备工作，是关系到整个项目的实现流程是否顺利完成的关键。本节将根据严格的市场需求分析，得出我们这个项目的模块架构。本系统的基本模块架构如图 12-1 所示。

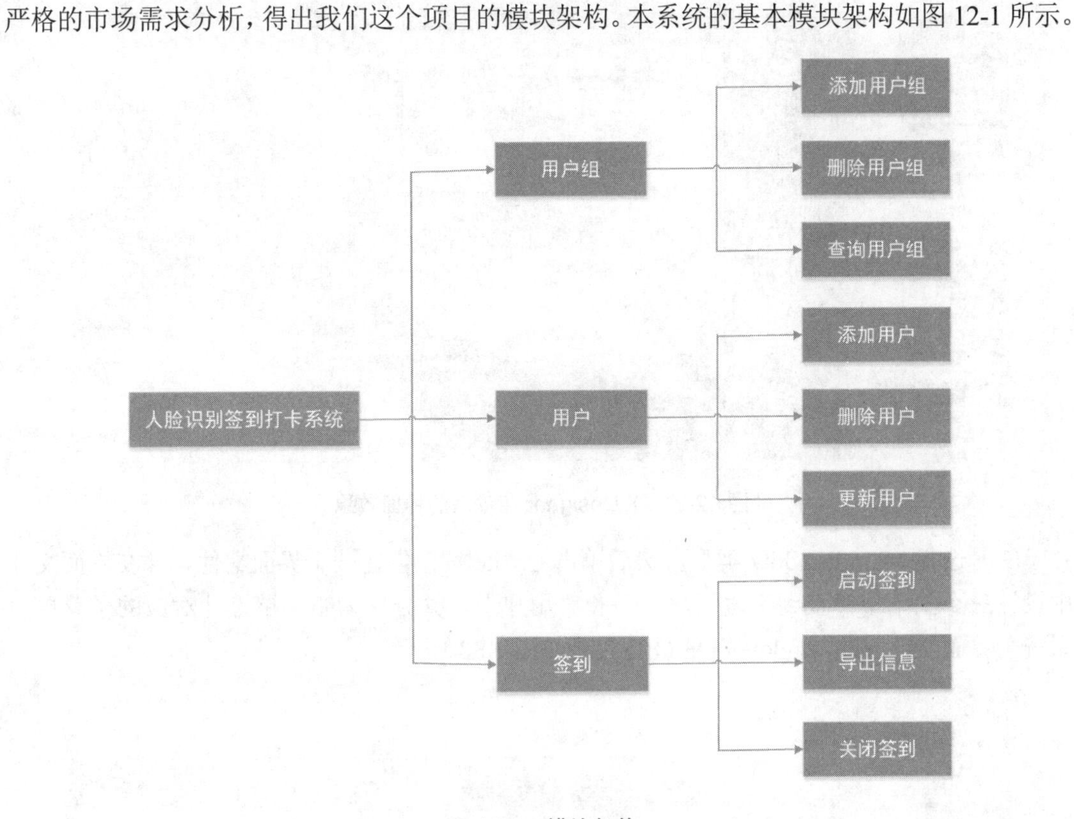


图 12-1　模块架构

## 12.3　使用 Qt Designer 实现主窗口界面

Qt Designer 是交互式可视化 GUI 设计工具，可以帮助我们提高开发 PyQt 程序的速度。Qt Designer 生成的 UI 界面是一个后缀为“.ui”的文件，可以通过 pyiuc 转换为“.py”文件。本节将详细讲解使用 Qt Designer 设计主窗口界面的过程。

### 12.3.1　设计系统 UI 主界面

(1) 在使用 Qt Designer 设计窗口界面之前，需要先试用如下命令安装 PyQt5-tools:

```
pip install PyQt5-tools
```

(2) 在 Pycharm 的命令行界面中输入下面的命令启动 Qt Designer，启动后的界面效果如图 12-2 所示。

```
pyqt5designer.exe
```

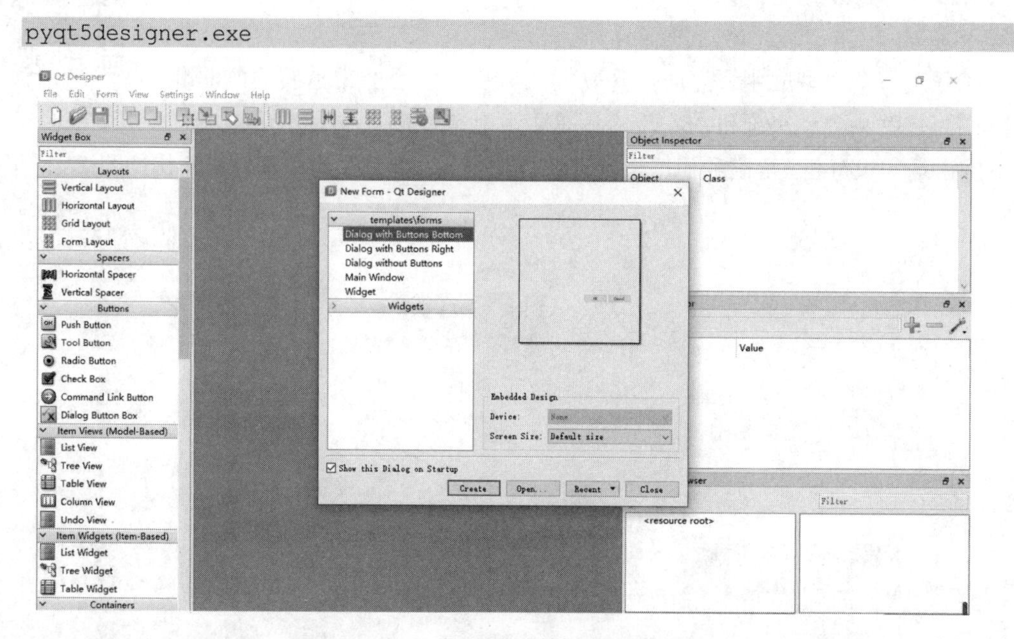


图 12-2 Qt Designer 启动后的界面效果

(3) 选择 Main Window 类型，然后单击 Create 按钮创建一个界面文件，并在界面文件中设计三个主菜单“签到”“用户组”和“用户”，并分别为主菜单设计对应的子菜单。系统主界面文件 mainwindow.ui 的最终效果如图 12-3 所示。

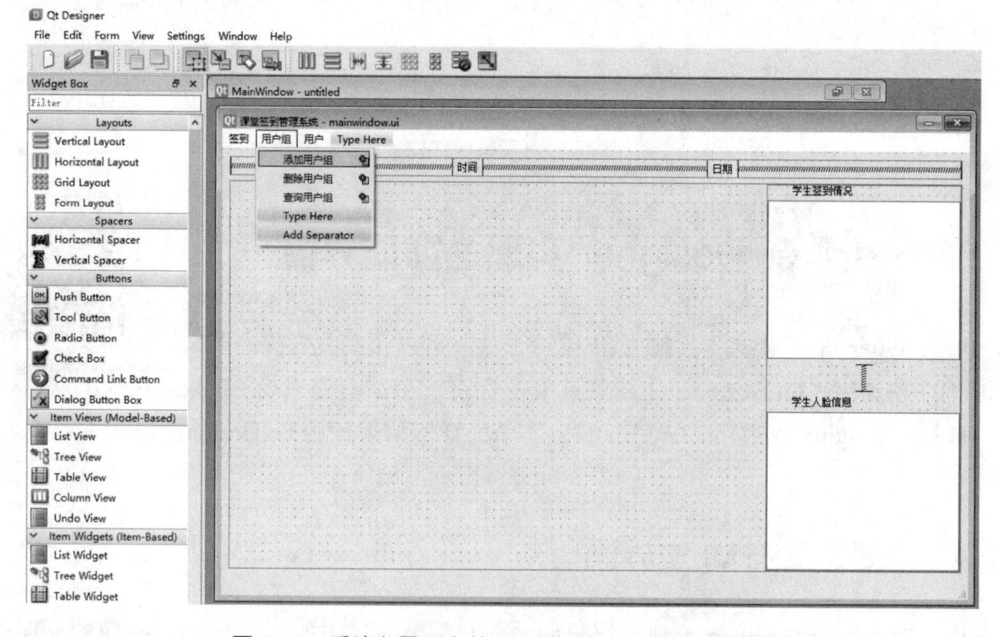


图 12-3 系统主界面文件 mainwindow.ui 的最终效果

## 12.3.2　将 Qt Designer 文件转换为 Python 文件

因为 Python 只会解释执行“.py”文件，不能识别“.ui”类型的文件，所以接下来需要把“.ui”类型的文件转换成“.py”文件。在 Pycharm 的命令行界面中输入下面的命令启动 Qt Designer，将 Qt Designer 设计文件 mainwindow.ui 转换成 Python 文件 mainwindow.py：

```
pyuic5 mainwindow.ui   -o  mainwindow.py
```

Python 文件 mainwindow.py 的具体实现代码如下所示：

```
from PyQt5 import QtCore, QtGui, QtWidgets

class Ui_MainWindow(object):
    def setupUi(self, MainWindow):
        MainWindow.setObjectName("MainWindow")
        MainWindow.resize(900, 545)
        self.centralwidget = QtWidgets.QWidget(MainWindow)
        self.centralwidget.setMinimumSize(QtCore.QSize(900, 500))
        self.centralwidget.setObjectName("centralwidget")
        self.gridLayout = QtWidgets.QGridLayout(self.centralwidget)
        self.gridLayout.setObjectName("gridLayout")
        self.horizontalLayout_2 = QtWidgets.QHBoxLayout()
        self.horizontalLayout_2.setObjectName("horizontalLayout_2")
        spacerItem = QtWidgets.QSpacerItem(40, 20,
           QtWidgets.QSizePolicy.Expanding, QtWidgets.QSizePolicy.Minimum)
        self.horizontalLayout_2.addItem(spacerItem)
        self.label_2 = QtWidgets.QLabel(self.centralwidget)
        self.label_2.setObjectName("label_2")
        self.horizontalLayout_2.addWidget(self.label_2)
        spacerItem1 = QtWidgets.QSpacerItem(40, 20,
           QtWidgets.QSizePolicy.Expanding, QtWidgets.QSizePolicy.Minimum)
        self.horizontalLayout_2.addItem(spacerItem1)
        self.label_3 = QtWidgets.QLabel(self.centralwidget)
        self.label_3.setObjectName("label_3")
        self.horizontalLayout_2.addWidget(self.label_3)
        spacerItem2 = QtWidgets.QSpacerItem(40, 20,
           QtWidgets.QSizePolicy.Expanding, QtWidgets.QSizePolicy.Minimum)
        self.horizontalLayout_2.addItem(spacerItem2)
        self.gridLayout.addLayout(self.horizontalLayout_2, 0, 0, 1, 1)
        self.horizontalLayout = QtWidgets.QHBoxLayout()
        self.horizontalLayout.setObjectName("horizontalLayout")
        self.label = QtWidgets.QLabel(self.centralwidget)
        self.label.setMinimumSize(QtCore.QSize(640, 480))
        self.label.setText("")
        self.label.setObjectName("label")
        self.horizontalLayout.addWidget(self.label)
        self.verticalLayout_2 = QtWidgets.QVBoxLayout()
        self.verticalLayout_2.setObjectName("verticalLayout_2")
        self.label_4 = QtWidgets.QLabel(self.centralwidget)
        self.label_4.setObjectName("label_4")
        self.verticalLayout_2.addWidget(self.label_4)
        self.plainTextEdit = QtWidgets.QPlainTextEdit(self.centralwidget)
        self.plainTextEdit.setMinimumSize(QtCore.QSize(200, 0))
```

```
        self.plainTextEdit.setLineWidth(1)
        self.plainTextEdit.setReadOnly(True)
        self.plainTextEdit.setTabStopWidth(80)
        self.plainTextEdit.setObjectName("plainTextEdit")
        self.verticalLayout_2.addWidget(self.plainTextEdit)
        spacerItem3 = QtWidgets.QSpacerItem(20, 40,
           QtWidgets.QSizePolicy.Minimum, QtWidgets.QSizePolicy.Expanding)
        self.verticalLayout_2.addItem(spacerItem3)
        self.label_5 = QtWidgets.QLabel(self.centralwidget)
        self.label_5.setObjectName("label_5")
        self.verticalLayout_2.addWidget(self.label_5)
        self.plainTextEdit_2 = QtWidgets.QPlainTextEdit(self.centralwidget)
        self.plainTextEdit_2.setMinimumSize(QtCore.QSize(200, 0))
        self.plainTextEdit_2.setObjectName("plainTextEdit_2")
        self.verticalLayout_2.addWidget(self.plainTextEdit_2)
        self.horizontalLayout.addLayout(self.verticalLayout_2)
        self.horizontalLayout.setStretch(1, 1)
        self.gridLayout.addLayout(self.horizontalLayout, 1, 0, 1, 1)
        MainWindow.setCentralWidget(self.centralwidget)
        self.menubar = QtWidgets.QMenuBar(MainWindow)
        self.menubar.setGeometry(QtCore.QRect(0, 0, 900, 23))
        self.menubar.setObjectName("menubar")
        self.menu = QtWidgets.QMenu(self.menubar)
        self.menu.setObjectName("menu")
        self.menu_2 = QtWidgets.QMenu(self.menubar)
        self.menu_2.setObjectName("menu_2")
        self.menu_3 = QtWidgets.QMenu(self.menubar)
        self.menu_3.setObjectName("menu_3")
        MainWindow.setMenuBar(self.menubar)
        self.statusbar = QtWidgets.QStatusBar(MainWindow)
        self.statusbar.setObjectName("statusbar")
        MainWindow.setStatusBar(self.statusbar)
        self.actionopen = QtWidgets.QAction(MainWindow)
        self.actionopen.setObjectName("actionopen")
        self.actionclose = QtWidgets.QAction(MainWindow)
        self.actionclose.setObjectName("actionclose")
        self.actionaddgroup = QtWidgets.QAction(MainWindow)
        self.actionaddgroup.setObjectName("actionaddgroup")
        self.actiondelgroup = QtWidgets.QAction(MainWindow)
        self.actiondelgroup.setObjectName("actiondelgroup")
        self.actionsave = QtWidgets.QAction(MainWindow)
        self.actionsave.setObjectName("actionsave")
        self.actiond = QtWidgets.QAction(MainWindow)
        self.actiond.setObjectName("actiond")
        self.actiongetlist = QtWidgets.QAction(MainWindow)
        self.actiongetlist.setObjectName("actiongetlist")
        self.actionadduser1 = QtWidgets.QAction(MainWindow)
        self.actionadduser1.setObjectName("actionadduser1")
        self.actionadduser = QtWidgets.QAction(MainWindow)
        self.actionadduser.setObjectName("actionadduser")
        self.actiondeluser = QtWidgets.QAction(MainWindow)
        self.actiondeluser.setObjectName("actiondeluser")
        self.actionupdateuser = QtWidgets.QAction(MainWindow)
        self.actionupdateuser.setObjectName("actionupdateuser")
        self.menu.addAction(self.actionopen)
```

```
        self.menu.addAction(self.actionclose)
        self.menu_2.addAction(self.actionaddgroup)
        self.menu_2.addAction(self.actiondelgroup)
        self.menu_2.addAction(self.actiongetlist)
        self.menu_3.addAction(self.actionadduser)
        self.menu_3.addAction(self.actiondeluser)
        self.menu_3.addAction(self.actionupdateuser)
        self.menubar.addAction(self.menu.menuAction())
        self.menubar.addAction(self.menu_2.menuAction())
        self.menubar.addAction(self.menu_3.menuAction())

        self.retranslateUi(MainWindow)
        QtCore.QMetaObject.connectSlotsByName(MainWindow)

    def retranslateUi(self, MainWindow):
        _translate = QtCore.QCoreApplication.translate
        MainWindow.setWindowTitle(_translate("MainWindow", "课堂签到管理系统"))
        self.label_2.setText(_translate("MainWindow", "时间"))
        self.label_3.setText(_translate("MainWindow", "日期"))
        self.label_4.setText(_translate("MainWindow", "     学生签到情况"))
        self.label_5.setText(_translate("MainWindow", "     学生人脸信息"))
        self.menu.setTitle(_translate("MainWindow", "签到"))
        self.menu_2.setTitle(_translate("MainWindow", "用户组"))
        self.menu_3.setTitle(_translate("MainWindow", "用户"))
        self.actionopen.setText(_translate("MainWindow", "启动签到"))
        self.actionclose.setText(_translate("MainWindow", "关闭签到"))
        self.actionaddgroup.setText(_translate("MainWindow", "添加用户组"))
        self.actiondelgroup.setText(_translate("MainWindow", "删除用户组"))
        self.actionsave.setText(_translate("MainWindow", "存储信息"))
        self.actiond.setText(_translate("MainWindow", "导出信息"))
        self.actiongetlist.setText(_translate("MainWindow", "查询用户组"))
        self.actionadduser1.setText(_translate("MainWindow", "添加用户"))
        self.actionadduser.setText(_translate("MainWindow", "添加用户"))
        self.actiondeluser.setText(_translate("MainWindow", "删除用户"))
        self.actionupdateuser.setText(_translate("MainWindow", "更新用户"))
```

编写文件 main.py，功能是调用上面的 UI 界面文件创建应用程序对象，然后分别创建并显示系统主界面的窗口，代码如下：

```
import sys
from PyQt5.QtWidgets import QApplication

from mywindow import mywindow

if __name__ == '__main__':
    # 创建应用程序对象
    app = QApplication(sys.argv)
    # 创建窗口
    ui = mywindow()
    # 显示窗口
    ui.show()
    # 应用执行
    app.exec_()
    sys.exit(0)
```

运行程序文件 main.py 会显示系统主界面，最终执行效果如图 12-4 所示。

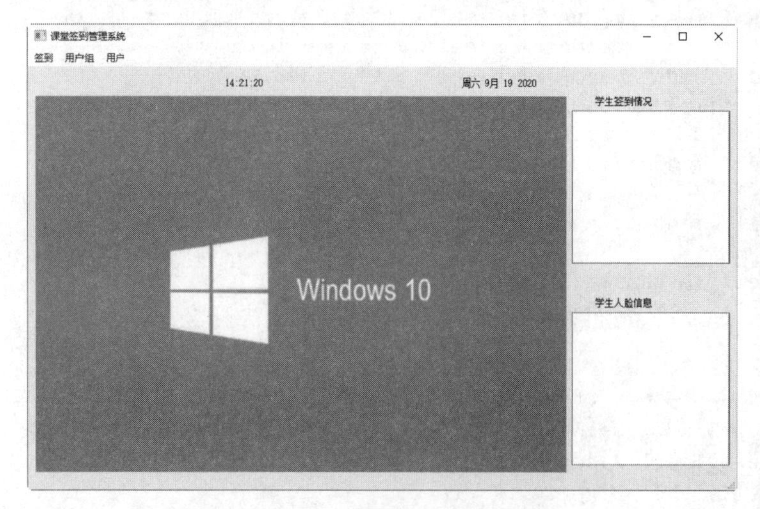


图 12-4 系统主界面的最终执行效果

## 12.4 签到打卡、用户操作和用户组操作

扫码观看视频讲解

签到打卡、用户操作和用户组操作是本项目的核心，首先使用 Qt Designer 分别设计打卡界面、用户操作界面和用户组操作界面，实现一个摄像头类打开摄像头，然后实现实时显示时间功能，并调用百度 AI 模块实现人脸识别功能。

### 12.4.1 使用百度 AI 之前的准备工作

为了提高开发效率，降低开发成本，我们使用百度公司提供的 AI 接口实现在线人脸识别功能，具体原理如下。

- 向百度 AI 发送人脸检测请求，让百度 AI 去完成人脸识别，百度 AI 返回识别结果。
- 发送请求时，不是任意的网络请求都能够接受，必须有百度提供的访问令牌(access_token)。

#### 1. 发送请求前的准备工作

在向百度 AI 发送人脸检测请求之前，必须先获取 access_token，并注册人脸识别 API 的如下参数。

- client_id：你当前应用程序的标识。
- client_secret：决定是否有访问权限。

也就是说，开发者需要注册百度 API，获取 id 与 secret，在注册时使用百度账号进行注册。注册后需要创建人脸识别应用，在创建应用后才会获得 id 与 secret。

- id：应用的 API Key，例如 kSD6zWfxpki2AKWtysCUe0nS。

- secret：应用的 Secret Key，例如 uNXjdRa7SbYwt0EgBdRwsmQYX6VADGx8。

在使用 requests.get(host)发送请求后，最终得到字典数据格式的数据，从字典中取出键为 access_token 的值即可。

### 2. 开始发送请求

发送请求，通过网络请求方式让百度 AI 进行人脸识别。百度 AI 会检测一张画面(图片)是否存在人脸以及人脸的一些属性。通过函数 requests.post()完成识别请求，返回检测到的结果。返回的结果数据是一个字典，在里面保存了多项数据内容，用键值对表示。

### 3. 完成人脸搜索

在百度 AI 库中搜索是否存在对应的人脸，有则实现签到功能。

综上所述，我们在使用百度 AI 实现人脸识别功能之前，需要先创建一个百度 AI 应用程序并获得 access_token，具体流程如下。

(1) 输入网址 https://ai.baidu.com/登录百度 AI 主页，如图 12-5 所示。

图 12-5　百度 AI 主页

(2) 单击顶部导航中的“开放能力”链接，然后在弹出的子链接中选择“人脸与人体识别”→“人脸识别”项，如图 12-6 所示。

图 12-6　选择“开放能力”→“人脸与人体识别”　→“人脸识别”项

(3) 在弹出的新界面中单击“立即使用”按钮，如图 12-7 所示。这一步需要输入百度账号登录百度智能云，如果没有百度账号则需要先申请一个。

图 12-7　单击“立即使用”按钮

(4) 在弹出的新界面中选择“人脸实名认证”→“创建应用”项，如图 12-8 所示。

图 12-8　选择“人脸实名认证”→“创建应用”项

(5) 在弹出的新界面中设置应用程序的名称和描述信息，例如都填写“人脸检测”，其他选项都是默认，最后单击下面的“立即创建”按钮，如图 12-9 所示。

(6) 创建成功后，在“应用列表”中会显示刚刚创建的应用，并可以查看这个应用的 API Key 和 Secret Key，如图 12-10 所示。通过使用 API Key 和 Secret Key 可以获取需要的 access_token。

图 12-9　点击下面的“立即创建”按钮

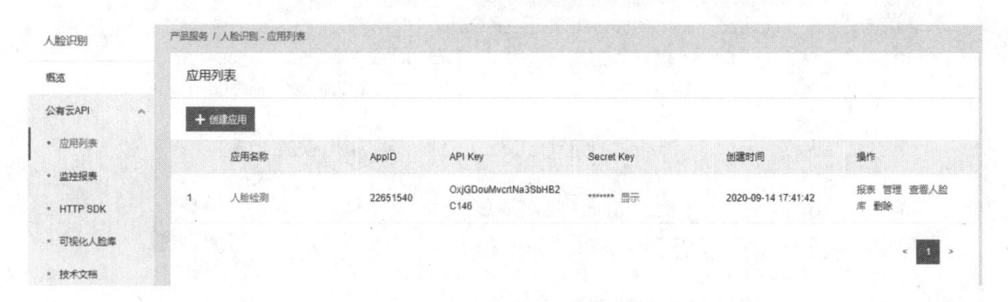


图 12-10　刚刚创建的应用

## 12.4.2　设计 UI 界面

使用 Qt Designer 设计签到打卡模块的 UI 界面文件 adduser.ui，在界面左侧显示摄像头界面，在右侧分别显示签到情况和识别的人脸信息，如图 12-11 所示。

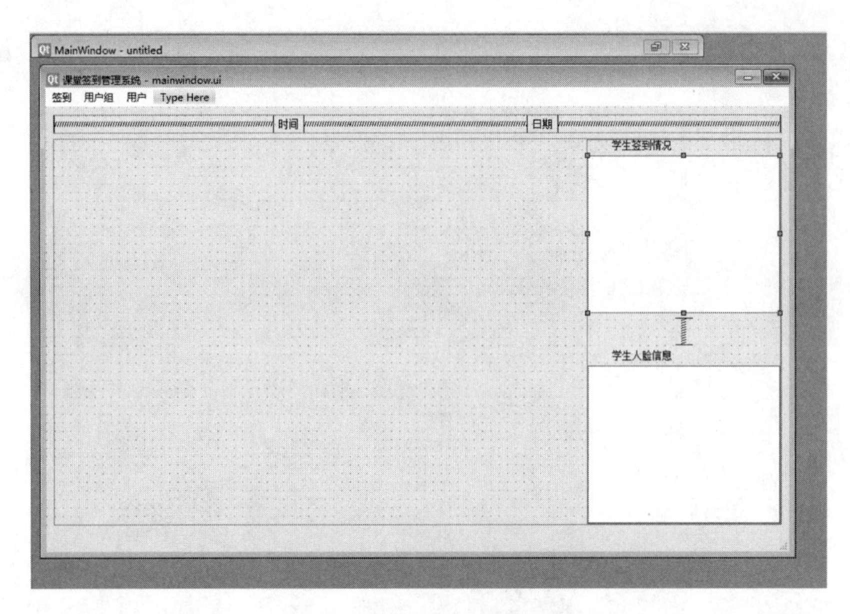


图 12-11　签到打卡模块的 UI 界面文件 adduser.ui

在使用 Qt Designer 实现签到打卡模块的 UI 界面文件 adduser.ui 后，接下来需要将此文件转换为 Python 文件 adduser.py，具体实现代码如下所示：

```
from PyQt5 import QtCore, QtGui, QtWidgets

class Ui_Dialog(object):
    def setupUi(self, Dialog):
        Dialog.setObjectName("Dialog")
        Dialog.resize(445, 370)
        self.gridLayout = QtWidgets.QGridLayout(Dialog)
        self.gridLayout.setObjectName("gridLayout")
        self.verticalLayout_4 = QtWidgets.QVBoxLayout()
        self.verticalLayout_4.setObjectName("verticalLayout_4")
        self.horizontalLayout = QtWidgets.QHBoxLayout()
        self.horizontalLayout.setObjectName("horizontalLayout")
        self.verticalLayout = QtWidgets.QVBoxLayout()
        self.verticalLayout.setObjectName("verticalLayout")
        self.label = QtWidgets.QLabel(Dialog)
        self.label.setMinimumSize(QtCore.QSize(305, 225))
        self.label.setText("")
        self.label.setObjectName("label")
        self.verticalLayout.addWidget(self.label)
        self.pushButton = QtWidgets.QPushButton(Dialog)
        self.pushButton.setObjectName("pushButton")
        self.verticalLayout.addWidget(self.pushButton)
        self.horizontalLayout.addLayout(self.verticalLayout)
        self.groupBox = QtWidgets.QGroupBox(Dialog)
        self.groupBox.setObjectName("groupBox")
        self.listWidget = QtWidgets.QListWidget(self.groupBox)
        self.listWidget.setGeometry(QtCore.QRect(10, 20, 81, 231))
        self.listWidget.setObjectName("listWidget")
        self.horizontalLayout.addWidget(self.groupBox)
        self.horizontalLayout.setStretch(0, 3)
```

```
        self.horizontalLayout.setStretch(1, 1)
        self.verticalLayout_4.addLayout(self.horizontalLayout)
        self.verticalLayout_3 = QtWidgets.QVBoxLayout()
        self.verticalLayout_3.setObjectName("verticalLayout_3")
        self.verticalLayout_2 = QtWidgets.QVBoxLayout()
        self.verticalLayout_2.setObjectName("verticalLayout_2")
        self.horizontalLayout_2 = QtWidgets.QHBoxLayout()
        self.horizontalLayout_2.setObjectName("horizontalLayout_2")
        self.label_2 = QtWidgets.QLabel(Dialog)
        self.label_2.setObjectName("label_2")
        self.horizontalLayout_2.addWidget(self.label_2)
        self.lineEdit = QtWidgets.QLineEdit(Dialog)
        self.lineEdit.setObjectName("lineEdit")
        self.horizontalLayout_2.addWidget(self.lineEdit)
        self.verticalLayout_2.addLayout(self.horizontalLayout_2)
        self.horizontalLayout_3 = QtWidgets.QHBoxLayout()
        self.horizontalLayout_3.setObjectName("horizontalLayout_3")
        self.label_3 = QtWidgets.QLabel(Dialog)
        self.label_3.setObjectName("label_3")
        self.horizontalLayout_3.addWidget(self.label_3)
        self.lineEdit_2 = QtWidgets.QLineEdit(Dialog)
        self.lineEdit_2.setObjectName("lineEdit_2")
        self.horizontalLayout_3.addWidget(self.lineEdit_2)
        self.label_4 = QtWidgets.QLabel(Dialog)
        self.label_4.setObjectName("label_4")
        self.horizontalLayout_3.addWidget(self.label_4)
        self.lineEdit_3 = QtWidgets.QLineEdit(Dialog)
        self.lineEdit_3.setObjectName("lineEdit_3")
        self.horizontalLayout_3.addWidget(self.lineEdit_3)
        self.verticalLayout_2.addLayout(self.horizontalLayout_3)
        self.verticalLayout_3.addLayout(self.verticalLayout_2)
        self.horizontalLayout_4 = QtWidgets.QHBoxLayout()
        self.horizontalLayout_4.setObjectName("horizontalLayout_4")
        spacerItem = QtWidgets.QSpacerItem(40, 20,
           QtWidgets.QSizePolicy.Expanding, QtWidgets.QSizePolicy.Minimum)
        self.horizontalLayout_4.addItem(spacerItem)
        self.pushButton_2 = QtWidgets.QPushButton(Dialog)
        self.pushButton_2.setObjectName("pushButton_2")
        self.horizontalLayout_4.addWidget(self.pushButton_2)
        spacerItem1 = QtWidgets.QSpacerItem(40, 20,
           QtWidgets.QSizePolicy.Expanding, QtWidgets.QSizePolicy.Minimum)
        self.horizontalLayout_4.addItem(spacerItem1)
        self.pushButton_3 = QtWidgets.QPushButton(Dialog)
        self.pushButton_3.setObjectName("pushButton_3")
        self.horizontalLayout_4.addWidget(self.pushButton_3)
        spacerItem2 = QtWidgets.QSpacerItem(40, 20,
           QtWidgets.QSizePolicy.Expanding, QtWidgets.QSizePolicy.Minimum)
        self.horizontalLayout_4.addItem(spacerItem2)
        self.verticalLayout_3.addLayout(self.horizontalLayout_4)
        self.verticalLayout_4.addLayout(self.verticalLayout_3)
        self.verticalLayout_4.setStretch(0, 3)
        self.verticalLayout_4.setStretch(1, 1)
        self.gridLayout.addLayout(self.verticalLayout_4, 0, 0, 1, 1)

        self.retranslateUi(Dialog)
```

```
        QtCore.QMetaObject.connectSlotsByName(Dialog)

    def retranslateUi(self, Dialog):
        _translate = QtCore.QCoreApplication.translate
        Dialog.setWindowTitle(_translate("Dialog", "Dialog"))
        self.pushButton.setText(_translate("Dialog", "选择人脸图片"))
        self.groupBox.setTitle(_translate("Dialog", "用户组选择"))
        self.label_2.setText(_translate("Dialog", "输入学号(字母数组下划线)："))
        self.label_3.setText(_translate("Dialog", "姓名："))
        self.label_4.setText(_translate("Dialog", "班级："))
        self.pushButton_2.setText(_translate("Dialog", "确定"))
        self.pushButton_3.setText(_translate("Dialog", "取消"))
```

## 12.4.3 创建摄像头类

编写文件cameravideo.py，创建一个摄像头类camera，分别实现如下功能：

- 打开摄像头。
- 获取摄像头的实时数据。
- 将摄像头的数据进行转换并提供给界面。

文件cameravideo.py的具体实现代码如下所示：

```
import cv2
import numpy as np
from  PyQt5.QtGui import QPixmap,QImage

class camera():
    def __init__(self):
        #VideoCapture 类对视频或调用摄像头进行读取操作
        #参数 filename; device
        #0 表示默认的摄像头进行打开
        #self.capture 表示打开的摄像头对象
        self.capture = cv2.VideoCapture(0, cv2.CAP_DSHOW)
        #isOpened 函数返回一个布尔值，来判断是否摄像头初始化成功
        # if self.capture.isOpened():
        #     print("isOpened")
        #定义一个多维数组，存取画面
        self.currentframe = np.array([])

    #读取摄像头数据
    def read_camera(self):
        #ret 是否成功，pic_data 数据
        ret,data = self.capture.read()
        if not ret:
            print("获取摄像头数据失败")
            return None
        return data

    #数据转换成界面能显示的数据格式
    def camera_to_pic(self):
        pic = self.read_camera()
        #摄像头是 BGR 方式存储，首先要转换成 RGB
```

```
        self.currentframe = cv2.cvtColor(pic,cv2.COLOR_BGR2RGB)
        #设置宽高
        #self.currentframe = cv2.cvtColor(self.currentframe,(640,480))

        #转换格式(界面能够显示的格式)
        #获取画面的宽度和高度
        height,width = self.currentframe.shape[:2]
        #先转换成 QImage 类型的图片(画面)，创建 QImage 类对象，使用摄像头的画面
        #QImage (data, width, height, format)创建:数据，宽度，高度,格式
        qimg = QImage(self.currentframe,width,height,QImage.Format_RGB888)
        qpixmap = QPixmap.fromImage(qimg)
        return qpixmap

    def close_camera(self):
        self.capture.release()
```

### 12.4.4　UI 界面的操作处理

在签到打卡模块的 UI 界面文件 adduser.py 中有许多元素，例如顶部菜单、顶部时间显示、素材背景图片和右侧的签到列表信息等，接下来编写文件 mywindow.py 为 UI 界面文件 adduser.py 中的元素实现操作处理。

(1) 创建界面类 mywindow。在 PyQt5 中存在一个机制：信号可以关联上另一个函数，另一个函数会在产生这个对应信号的时候调用“信号槽机制”。在组件中存在一些特定的信号，当组件执行某个操作的时候，这些对应的信号就会被激活(信号产生)。由于信号(如点击)执行的功能可能是不一样的，我们需要指定信号与槽函数的关联，设置信号要去执行的功能是什么信号对象。在类 mywindow 的初始化代码中，首先创建一个时间定时器，并设置在创建窗口的伊始就调用方法 get_accesstoken()申请访问百度 AI 的令牌，并设置了不同信号槽需要执行的对应功能函数。代码如下：

```
class mywindow(Ui_MainWindow,QMainWindow):
    detect_data_signal = pyqtSignal(bytes)
    group_id = pyqtSignal(str)
    camera_status = False

    def __init__(self):
        super(mywindow,self).__init__()
        self.setupUi(self)
        self.label.setScaledContents(True)
        self.label.setPixmap(QPixmap("1.jpg"))
        # 创建一个时间定时器
        self.datetime = QTimer()
        #启动获取系统时间/日期定时器,定时时间为 10ms，每 10ms 产生一次信号
        self.datetime.start(10)
        #创建窗口就应该完成进行访问令牌的申请(获取)
        self.get_accesstoken()
        #信号与槽的关联
        #self.actionopen：指定对象
        #triggered：信号
        #connect：关联(槽函数)
```

```
        #self.on_actionopen():关联的函数
        self.actionopen.triggered.connect(self.on_actionopen)
        self.actionclose.triggered.connect(self.on_actionclose)


        #添加用户组信号槽
        self.actionaddgroup.triggered.connect(self.add_group)
        #删除用户组信号槽
        self.actiondelgroup.triggered.connect(self.del_group)
        #查询用户组信号槽
        self.actiongetlist.triggered.connect(self.getgrouplist)
        #添加用户信号槽
        self.actionadduser.triggered.connect(self.add_user)
        #删除用户信号槽
        self.actiondeluser.triggered.connect(self.del_user)
        # 更新用户信号槽
        self.actionupdateuser.triggered.connect(self.update_user)
        #关联时间/日期的定时器信号与槽函数
        self.datetime.timeout.connect(self.data_time)
```

(2) 创建定时器函数 data_time(self)，获取当前的日期与时间，并添加到对应的定时器中。在类 mywindow 的初始化代码中，设置本项目的定时时间为 10ms，即每 10ms 产生一次信号。代码如下：

```
    def data_time(self):
        # 获取日期
        date = QDate.currentDate()
        #print(date)
        #self.dateEdit.setDate(date)
        self.label_3.setText(date.toString())
        # 获取时间
        time = QTime.currentTime()
        #print(time)
        #self.timeEdit.setTime(time)
        self.label_2.setText(time.toString())
        # 获取日期时间
        datetime = QDateTime.currentDateTime()
        #print(datetime)
```

(3) 创建启动打卡签到函数 on_actionopen(self)，首先弹出一个对话框，让用户输入所在的用户组，并不断调用摄像头识别当前打卡人员的脸部信息。代码如下：

```
    def on_actionopen(self):
        list = self.getlist()
        self.group_id, ret = QInputDialog.getText(self, "请输入所在用户组", "用户
            组信息\n" + str(list['result']['group_id_list']))
        if self.group_id == '':
            QMessageBox.about(self, "签到失败", "用户组不能为空\n")
            return
        group_status = 0
        for i in list['result']['group_id_list']:
            if i == self.group_id:
                group_status =1
                break
```

```
    if group_status == 0:
        QMessageBox.about(self, "签到失败", "该用户组不存在\n")
        return
    #启动摄像头
    self.cameravideo = camera()
    self.camera_status = True
    #启动定时器进行定时，每隔多长时间进行一次获取摄像头数据，用于流畅显示画面
    self.timeshow = QTimer(self)
    self.timeshow.start(10)
    #10ms 的定时器启动，每到 10ms 就会产生一个信号 timeout,信号没有()
    self.timeshow.timeout.connect(self.show_cameradata)
    #self.timeshow.timeout().connect(self.show_cameradata)
    # self.show_cameradata()
    #创建检测线程
    self.create_thread()
    # self.group_id.emit(str(group_id))
    # self.group_id.connect(self.detectThread.get_group_id)
    #当开启启动签到时，创建定时器 500ms，用作获取要检测的画面
    # facedetecttime 定时器设置检测画面获取
    self.facedetecttime = QTimer(self)
    self.facedetecttime.start(500)
    self.facedetecttime.timeout.connect(self.get_cameradata)

    self.detect_data_signal.connect(self.detectThread.get_base64)
    self.detectThread.transmit_data.connect(self.get_detectdata)
    self.detectThread.search_data.connect(self.get_search_data)
```

(4) 创建关闭打卡签到函数 on_actionclose(self)，功能是关闭定时器和摄像头，并显示本次签到情况是否成功。代码如下：

```
def on_actionclose(self):
    # 清除学生人脸信息(False)
    # self.plainTextEdit_2.setPlainText("  ")
    #关闭定时器，不再设置检测画面获取
    self.facedetecttime.stop()
    #self.facedetecttime.timeout.disconnect(self.get_cameradata)
    #self.detect_data_signal.disconnect(self.detectThread.get_base64)
    #self.detectThread.transmit_data.connect(self.get_detectdata)
    #关闭检测线程
    self.detectThread.OK = False
    self.detectThread.quit()
    self.detectThread.wait()
    print(self.detectThread.isRunning())
    # 关闭定时器，不再去获取摄像头的数据
    self.timeshow.stop()
    self.timeshow.timeout.disconnect(self.show_cameradata)
    # 关闭摄像头
    self.cameravideo.close_camera()
    self.camera_status = False
    print("1")
    #显示本次签到情况
    self.signdata = sign_data(self.detectThread.sign_list,self)
    self.signdata.exec_()
```

```
    if self.timeshow.isActive() ==\
        False and self.facedetecttime.isActive() == False:
        # 画面设置为初始状态
        self.label.setPixmap(QPixmap("1.jpg"))
        self.plainTextEdit.clear()
        self.plainTextEdit_2.clear()
    else:
        QMessageBox.about(self, "错误", "关闭签到失败\n")
```

(5) 创建函数 create_thread()，功能是调用函数 detect_thread()使用多线程技术实现人脸识别功能。函数函数 detect_thread()在文件 detect.py 中定义实现。函数 create_thread()的代码如下：

```
#创建线程完成检测
def create_thread(self):
    self.detectThread = detect_thread(self.access_token,self.group_id)
    self.detectThread.start()
```

(6) 编写函数 get_cameradata(self)，获取摄像头中的画面信息，并把摄像头画面转换成图片，然后将图片编码为 base64 编码格式的数据进行信号传递。代码如下：

```
def get_cameradata(self):
    # 摄像头获取画面
    camera_data = self.cameravideo.read_camera()
    # 把摄像头画面转换成图片，然后设置编码 base64 编码格式数据
    _, enc = cv2.imencode('.jpg', camera_data)
    base64_image = base64.b64encode(enc.tobytes())
    #产生信号，传递数据
    self.detect_data_signal.emit(bytes(base64_image))
```

(7) 编写函数 show_cameradata()，功能是获取摄像头数据并显示出画面。代码如下：

```
def show_cameradata(self):
    #获取摄像头数据，转换数据
    pic = self.cameravideo.camera_to_pic()
    #显示数据，显示画面
    self.label.setPixmap(pic)
```

(8) 编写函数 get_accesstoken(self)，功能是获取访问百度 AI 网络请求的访问令牌，代码如下：

```
def get_accesstoken(self):
    #host 对象是字符串对象存储是授权的服务地址-----获取 accesstoken 的地址
    host = 'https://aip.baidubce.com/oauth/2.0/\
    token? grant_type=client_credentials&client_id=\
    OxjGDouMvcrtNa3SbHB2C146&client_secret=xxxxxxxxxx'
    #发送网络请求   requests 网络库
    #使用 get 函数发送网络请求，参数为网络请求的地址，执行时会产生返回结果，
    response = requests.get(host)
    if response:
        # print(response.json())
        data = response.json()
        self.access_token = data.get('access_token')
```

(9) 编写槽函数 get_detectdata()，功能是获取检测数据，调用百度 AI 实现检测并返回学生信息和签到信息，然后将识别结果打印到主窗口的右侧界面中。代码如下：

```
def get_detectdata(self,data):
    if data['error_code'] != 0:
        self.plainTextEdit_2.setPlainText(data['error_msg'])
        return
    elif data['error_msg'] == 'SUCCESS':
        # 在 data 字典中，键为'result'对应的值才是返回的检查结果
        # data['result']才是检测结果
        # 人脸数目
        self.plainTextEdit_2.clear()
        face_num = data['result']['face_num']
        if face_num == 0:
            self.plainTextEdit_2.appendPlainText("未测到检人脸")
            return
        else:
            self.plainTextEdit_2.appendPlainText("测到检人脸")
        # 人脸信息 data['result']['face_list']是列表，每个数据就是一个人脸信息，
        # 需要取出每个列表数据
        # 每个人脸信息：data['result']['face_list'][0~i]人脸信息字典
        for i in range(face_num):
            # 通过 for 循环，分别取出列表的每一个数据
            # data['result']['face_list'][i]就是一个人脸信息的字典

            age = data['result']['face_list'][i]['age']
            beauty = data['result']['face_list'][i]['beauty']
            gender = data['result']['face_list'][i]['gender']['type']
            expression = data['result']['face_list'][i]['expression']['type']
            face_shape = data['result']['face_list'][i]['face_shape']['type']
            glasses = data['result']['face_list'][i]['glasses']['type']
            emotion = data['result']['face_list'][i]['emotion']['type']
            mask = data['result']['face_list'][i]['mask']['type']
            # 往窗口中添加文本，参数就是需要的文本信息
            self.plainTextEdit_2.appendPlainText("-----------------")
            self.plainTextEdit_2.appendPlainText("第" + str(i + 1) + "个学生信息：")
            self.plainTextEdit_2.appendPlainText("-----------------")
            self.plainTextEdit_2.appendPlainText("年龄：" + str(age))
            self.plainTextEdit_2.appendPlainText("颜值分数：" + str(beauty))
            self.plainTextEdit_2.appendPlainText("性别：" + str(gender))
            self.plainTextEdit_2.appendPlainText("表情：" + str(expression))
            self.plainTextEdit_2.appendPlainText("脸型：" + str(face_shape))
            self.plainTextEdit_2.appendPlainText("是否佩戴眼镜：" + str(glasses))
            self.plainTextEdit_2.appendPlainText("情绪：" + str(emotion))
            if mask == 0:
                mask = "否"
            else:
                mask = "是"
            self.plainTextEdit_2.appendPlainText("是否佩戴口罩：" + str(mask))
            self.plainTextEdit_2.appendPlainText("-----------------")
```

(10) 编写函数 add_group(self)，功能是向系统中添加新的用户组信息，添加的信息会同步到百度 AI 的应用程序中。代码如下：

```
    def add_group(self):
        #打开对话框，输入用户组
        group,ret = QInputDialog.getText(self,"添加用户组","请输入用户组(由数字、字
            母、下划线组成)")

        request_url =
"https://aip.baidubce.com/rest/2.0/face/v3/faceset/group/add"
        params = {
            "group_id":group
        }
        access_token = self.access_token
        request_url = request_url + "?access_token=" + access_token
        headers = {'content-type': 'application/json'}
        response = requests.post(request_url, data=params, headers=headers)
        if response:
            message = response.json()
            if message['error_msg'] == 'SUCCESS':
                QMessageBox.about(self,"用户组创建结果","用户组创建成功")
            else:
                QMessageBox.about(self,"用户组创建结果","用户组创建失败
                    \n"+message['error_msg'])
```

(11) 编写函数 del_group(self)，功能是删除系统中已经存在的某用户组信息，会同步在百度 AI 的应用程序中删除这个用户组。代码如下：

```
    def del_group(self):
        request_url = "https://aip.baidubce.com/rest/2.0/face/v3/faceset/
            group/delete"
        list = self.getlist()
        group, ret = QInputDialog.getText(self, "用户组列表", "用户组信息\n" +
str(list['result']['group_id_list']))
        # 删除，需要知道哪些组
        params = {
            "group_id": group  # 要删除的用户组的 id
        }
        access_token = self.access_token
        request_url = request_url + "?access_token=" + access_token
        headers = {'content-type': 'application/json'}
        response = requests.post(request_url, data=params, headers=headers)
        if response:
            data = response.json()
            if data['error_msg'] == 'SUCCESS':
                QMessageBox.about(self, "用户组删除结果", "用户组删除成功")
            else:
                QMessageBox.about(self, "用户组删除结果", "用户组删除失败\n" +
                    data['error_msg'])
```

(12) 编写函数 getlist(self)，功能是获取系统中已经存在的用户组信息，并将这些用户组信息列表显示出来。代码如下：

```
    #获取用户组
    def getlist(self):
        request_url =
"https://aip.baidubce.com/rest/2.0/face/v3/faceset/group/getlist"
```

```
    params = {
        "start":0,"length":100
    }
    access_token = self.access_token
    request_url = request_url + "?access_token=" + access_token
    headers = {'content-type': 'application/json'}
    response = requests.post(request_url, data=params, headers=headers)
    if response:
        data = response.json()
        if data['error_msg'] == 'SUCCESS':
            return data
        else:
            QMessageBox.about(self, "获取用户组结果", "获取用户组失败\n" +
                data['error_msg'])
```

(13) 编写函数 add_user(self)，功能是向系统中添加新的用户信息，在添加之前需要确认已经打开了摄像头，添加的信息会同步到百度 AI 的应用程序中。代码如下：

```
def add_user(self):
    request_url =
"https://aip.baidubce.com/rest/2.0/face/v3/faceset/user/add"
    if self.camera_status:
        QMessageBox.about(self,"摄像头状态","摄像头已打开，正在进行人脸签到\n 请关
            闭签到，再添加用户")
        return
    list = self.getlist()
    #创建一个窗口来选择这些内容
    window = adduserwindow(list['result']['group_id_list'],self)
    #新创建窗口，通过 exec()函数一直执行，阻塞执行，窗口不进行关闭
    #exec()函数不会退出，关闭窗口才会结束
    window_status = window.exec_()
    #进行判断，判断是否点击确定进行关闭
    if window_status != 1:
        return
    #请求参数，需要获取人脸：转换人脸编码，添加的组 id，添加的用户 id，新用户的 id 信息
    params = {
        "image":window.base64_image,#人脸图片
        "image_type":"BASE64",#人脸图片编码
        "group_id":window.group_id,#组 id
        "user_id":window.user_id,#新用户 id
        "user_info":'姓名：'+window.msg_name+'\n'
                +'班级：'+window.msg_class#用户信息
    }
    access_token = self.access_token
    request_url = request_url + "?access_token=" + access_token
    headers = {'content-type': 'application/json'}
    response = requests.post(request_url, data=params, headers=headers)
    if response:
        data = response.json()
        if data['error_msg'] == 'SUCCESS':
            QMessageBox.about(self, "人脸创建结果", "人脸创建成功")
        else:
```

```
            QMessageBox.about(self, "人脸创建结果", "用户组创建失败\n"+
data['error_msg'])
```

(14) 编写函数 update_user(self)，功能是修改系统中已经存在的某用户信息，会同步在百度 AI 的应用程序中更新这个用户的信息。代码如下：

```
#更新用户人脸
def update_user(self):
    request_url = "https://aip.baidubce.com/rest/2.0/face/v3/faceset/
        user/update"
    if self.camera_status:
        QMessageBox.about(self,"摄像头状态","摄像头已打开，正在进行人脸签到\n 请关
            闭签到，再添加用户")
        return
    list = self.getlist()
    #创建一个窗口来选择这些内容
    window = adduserwindow(list['result']['group_id_list'],self)
    #新创建窗口，通过 exec()函数一直执行，阻塞执行，窗口不进行关闭
    #exec()函数不会退出，关闭窗口才会结束
    window_status = window.exec_()
    #进行判断，判断是否点击确定进行关闭
    if window_status != 1:
        return
    #请求参数，需要获取人脸：转换人脸编码，添加的组 id，添加的用户 id，新用户的 id 信息
    params = {
        "image":window.base64_image,#人脸图片
        "image_type":"BASE64",#人脸图片编码
        "group_id":window.group_id,#组 id
        "user_id":window.user_id,#新用户 id
        "user_info":'姓名：'+window.msg_name+'\n'
                  +'班级：'+window.msg_class#用户信息
    }
    access_token = self.access_token
    request_url = request_url + "?access_token=" + access_token
    headers = {'content-type': 'application/json'}
    response = requests.post(request_url, data=params, headers=headers)
    if response:
        data = response.json()
        if data['error_msg'] == 'SUCCESS':
            QMessageBox.about(self, "人脸更新结果", "人脸更新成功")
        else:
            QMessageBox.about(self, "人脸更新结果", "用户组更新失败\n" +
                data['error_msg'])
```

(15) 编写函数 get_userlist()，获取系统中已经存在的用户列表信息，代码如下：

```
def get_userlist(self,group):
    request_url = "https://aip.baidubce.com/rest/2.0/face/v3/faceset/
        group/getusers"

    params = {
        "group_id":group
    }
    access_token = self.access_token
    request_url = request_url + "?access_token=" + access_token
```

```
    headers = {'content-type': 'application/json'}
    response = requests.post(request_url, data=params, headers=headers)
    if response:
        data = response.json()
        if data['error_msg'] == 'SUCCESS':
            return data
        else:
            QMessageBox.about(self, "获取用户列表结果", "获取用户列表失败\n" +
                              data['error_msg'])
```

(16) 编写函数 user_face_list()，功能是获取系统中已经存在的用户的人脸信息，代码如下：

```
#获取用户人脸列表
def user_face_list(self,group,user):
    request_url = "https://aip.baidubce.com/rest/2.0/face/v3/faceset/face/
      getlist"

    params = {
        "user_id":user,
        "group_id":group
    }
    access_token = self.access_token
    request_url = request_url + "?access_token=" + access_token
    headers = {'content-type': 'application/json'}
    response = requests.post(request_url, data=params, headers=headers)
    if response:
        data = response.json()
        if data['error_msg'] == 'SUCCESS':
            return data
        else:
            QMessageBox.about(self, "获取用户人脸列表结果", "获取用户人脸列表失败
                \n" + data['error_msg'])
```

(17) 编写函数 del_face_token()，功能是删除系统中已经存在的某个人脸信息，会同步在百度 AI 的应用程序中删除这个人脸信息。代码如下：

```
def del_face_token(self,group,user,face_token):
    request_url = "https://aip.baidubce.com/rest/2.0/face/v3/faceset/face/
        delete"

    params = {
        "user_id": user,
        "group_id": group,
        "face_token":face_token
    }
    access_token = self.access_token
    request_url = request_url + "?access_token=" + access_token
    headers = {'content-type': 'application/json'}
    response = requests.post(request_url, data=params, headers=headers)
    if response:
        data = response.json()
        if data['error_msg'] == 'SUCCESS':
            QMessageBox.about(self, "人脸删除结果", "人脸删除成功")
        else:
```

```
            QMessageBox.about(self, "人脸删除结果", "用户组删除失败\n" +
                data['error_msg'])
```

(18) 编写函数 del_user()，功能是删除系统中已经存在的某个用户信息，会同步在百度AI 的应用程序中删除这个用户信息。代码如下：

```
def del_user(self):
    #查询用户人脸信息(face_token)
    #获取用户组
    list = self.getlist()
    group,ret = QInputDialog.getText(self,"用户组获取","用户组信息
        \n"+str(list['result']['group_id_list']))
    group_status = 0
    if self.group_id == '':
        QMessageBox.about(self, "删除失败", "用户组不能为空\n")
        return
    for i in list['result']['group_id_list']:
        if i == group:
            group_status = 1
            break
    if group_status == 0:
        QMessageBox.about(self, "删除失败", "该用户组不存在\n")
        return
    #获取用户
    userlist = self.get_userlist(group)
    user,ret = QInputDialog.getText(self,"用户获取","用户信息
        \n"+str(userlist['result']['user_id_list']))
    user_status = 0
    if user == '':
        QMessageBox.about(self, "删除失败", "用户不能为空\n")
        return
    for i in userlist['result']['user_id_list']:
        if i == user:
            user_status = 1
            break
    if user_status == 0:
        QMessageBox.about(self, "删除失败", "该用户不存在\n")
        return
    #获取用户人脸列表
    face_list = self.user_face_list(group,user)
    for i in face_list['result']['face_list']:
        self.del_face_token(group,user,i['face_token'])
```

### 12.4.5 多线程操作和人脸识别

为了提高本项目的运行效率，使用多线程技术提高了人脸识别的能力。编写文件 detect.py，功能是使用多编程技术实现人脸识别；并创建 SQLite 数据库，将添加的用户信息和用户组信息添加到数据库中。文件 detect.py 的具体实现流程如下所示。

(1) 创建线程类 detect_thread，使用多线程技术向百度 AI 服务器传递识别信息，代码如下：

```
class detect_thread(QThread):
```

```
    transmit_data = pyqtSignal(dict)
    search_data = pyqtSignal(str)
    OK = True
    #字典用来存放签到数据
    sign_list = {}
    def __init__(self,token,group_id):
        super(detect_thread,self).__init__()#初始化操作
        self.access_token = token
        self.group_id = group_id
        self.condition = False
        self.add_status = 0
        # self.create_sqlite()

    #run 函数执行结束，代表线程结束
    def run(self):
        print("run")
        '''
        self.time = QTimer(self)
        self.time.start(500)
        self.time.timeout.connect(self.detect_face)
        '''
        while self.OK:
            if self.condition:
                self.detect_face(self.base64_image)
                self.condition = False
        print("while finish")
```

线程类只会执行线程类中的函数 run()，如果需要实现新的功能，则需要重新写一个 run() 函数。

(2) 编写函数 get_base64()，功能是获取传递的识别数据，将 base64 格式的图片发送给百度 AI，以便于获取人脸信息。代码如下：

```
    def get_base64(self,base64_image):
        #当窗口产生信号，调用槽函数，就把传递的数据，存放在线程的变量中
        self.base64_image = base64_image
        self.condition = True

        con = sqlite3.connect(r"stu_data.db")
        c = con.cursor()
```

(3) 编写函数 detect_face()，功能是使用多线程类向百度 AI 发送请求，并获取百度 AI 返回的人脸信息。向百度 AI 发送的是 base64 格式的图片，将人脸信息返回到文件 mywindow.py 中。代码如下：

```
    def detect_face(self,base64_image):
        '''
        #对话框获取图片
        #获取一张图片(一帧画面)
        #getOpenFileName 通过对话框的形式获取一个图片(.JPG)路径
        path,ret = QFileDialog.getOpenFileName(self,"open picture",".","图片格
            式(*.jpg)")
        #把图片转换成 base64 编码格式
        fp = open(path,'rb')
```

```
        base64_imag = base64.b64encode(fp.read())
        print(base64_imag)
        '''
        # 摄像头获取画面
        # camera_data = self.cameravideo.read_camera()
        # # 把摄像头画面转换成图片，然后设置编码 base64 编码格式数据
        # _, enc = cv2.imencode('.jpg', camera_data)
        # base64_image = base64.b64encode(enc.tobytes())
        # 发送请求的地址
        request_url = "https://aip.baidubce.com/rest/2.0/face/v3/detect"
        # 请求参数是一个字典，在字典中存储,百度 AI 要识别的图片信息，属性内容
        params = {
            "image": base64_image,  # 图片信息字符串
            "image_type": "BASE64",  # 图片信息格式
            "face_field": "gender,age,beauty,expression,face_shape,glasses,
                emotion,mask", # 请求识别人脸的属性， 各个属性在字符串中用,逗号隔开
            "max_face_num": 1
        }
        # 访问令牌
        access_token = self.access_token
        # 把请求地址和访问令牌组成可用的网络请求
        request_url = request_url + "?access_token=" + access_token
        # 参数：设置请求的格式体
        headers = {'content-type': 'application/json'}
        # 发送网络 post 请求,请求百度 AI 进行人脸检测,返回检测结果
        # 发送网络请求，就会存在一定的等待时间，程序就在这里阻塞执行，所以会存在卡顿现象
        response = requests.post(request_url, data=params, headers=headers)
        if response:
            data = response.json()
            if data['error_code'] != 0:
                self.transmit_data.emit(data)
                self.search_data.emit(data['error_msg'])
                return

            if data['result']['face_num'] > 0:
                #data 是请求数据的结果，需要进行解析，单独拿出所需的数据内容，分开
                self.transmit_data.emit(dict(data))
                self.face_search(self.group_id)
```

(4) 编写函数 ace_search()，功能是使用百度 AI 实现人脸识别功能。代码如下：

```
    def face_search(self,group_id):
        request_url = "https://aip.baidubce.com/rest/2.0/face/v3/search"
        params = {
            "image": self.base64_image,
            "image_type": "BASE64",
            "group_id_list": group_id,
        }
        access_token = self.access_token
        request_url = request_url + "?access_token=" + access_token
        headers = {'content-type': 'application/json'}
        response = requests.post(request_url, data=params, headers=headers)
        if response:
            data = response.json()
            if data['error_code'] == 0:
```

```
            if data['result']['user_list'][0]['score'] > 90:
                #存储要保存的签到数据，方便显示
                del(data['result']['user_list'][0]['score'])
                datetime = QDateTime.currentDateTime()
                datetime = datetime.toString()
                data['result']['user_list'][0]['datetime'] = datetime
                key = data['result']['user_list'][0]['group_id']+\
                    data['result']['user_list'][0]['user_id']
                if key not in self.sign_list.keys():
                    self.sign_list[key] = data['result']['user_list'][0]
                self.search_data.emit("学生签到成功\n 学生信息: "
                    +data['result']['user_list'][0]['user_info'])
                stu_data = data['result']['user_list'][0]['user_info']
                info = stu_data.split('\n')
                _, info_name = info[0].split(': ')
                _, info_class = info[1].split(': ')
                id = data['result']['user_list'][0]['user_id']
                # self.add_sqlite(id, info_name, info_class, datetime)
                self.search_sqlite(id)
                search_id=0
                for i in self.values:
                    search_id = i[0]
                if search_id == id:
                    self.update_sqlite(id,info_name,info_class,datetime)
                else:
                    self.add_sqlite(id,info_name,info_class,datetime)
            else:
                self.search_data.emit("学生签到失败，找不到对应学生")
```

(5)　编写函数 create_sqlite(self)，创建数据库 stu_data.db 和表 student，代码如下：

```
def create_sqlite(self):
    con = sqlite3.connect(r"stu_data.db")
    c = con.cursor()
    c.execute("create table student(id primary
        key ,name ,stu_class,datetime)")
    print("创建成功")
```

(6)　编写函数 add_sqlite()，功能是将添加的学生信息添加到数据库中，代码如下：

```
def add_sqlite(self,id,name,stu_class,datetime):
    con = sqlite3.connect(r"stu_data.db")
    c = con.cursor()
    value = (id,name,stu_class,datetime)
    sql = "insert into student(id,name,stu_class,datetime) values(?,?,?,?)"
    c.execute(sql,value)
    print("添加成功")
    # 提交
    con.commit()
```

(7) 编写函数 update_sqlite()，功能是更新数据库中某个学生的信息，代码如下：

```
def update_sqlite(self,id,name,stu_class,datetime):
    con = sqlite3.connect(r"stu_data.db")
    c = con.cursor()
    # value = (name,stu_class,datetime,id)
    sql = "update student set name=?,stu_class=?,datetime=? where id =?"
    c.execute(sql,(name,stu_class,datetime,id))
    con.commit()
    print("更新成功")
```

(8) 编写函数 search_sqlite(self,id)，功能是查询数据库中某个学生的信息，代码如下：

```
def search_sqlite(self,id):
    con = sqlite3.connect(r"stu_data.db")
    c = con.cursor()
    sql = "select * from student where id=?"
    self.values = c.execute(sql,(id,))
```

## 12.4.6 导出打卡签到信息

某个学生成功打卡签到后，可以导出本次签到信息，将信息保存到本地文件中。

(1) 首先使用 Qt Designer 设计 UI 界面文件 sign_indata.ui，效果如图 12-12 所示。

图 12-12 UI 界面文件 sign_indata.ui

(2) 将 Qt Designer 文件 sign_indata.ui 转换为 Python 文件 sign_indata.py，代码如下：

```
class Ui_Dialog(object):
    def setupUi(self, Dialog):
        Dialog.setObjectName("Dialog")
        Dialog.resize(440, 353)
        self.gridLayout = QtWidgets.QGridLayout(Dialog)
        self.gridLayout.setObjectName("gridLayout")
        self.verticalLayout = QtWidgets.QVBoxLayout()
```

```
        self.verticalLayout.setObjectName("verticalLayout")
        self.tableWidget = QtWidgets.QTableWidget(Dialog)
        self.tableWidget.setMinimumSize(QtCore.QSize(0, 0))
        self.tableWidget.setObjectName("tableWidget")
        self.tableWidget.setColumnCount(4)
        self.tableWidget.setRowCount(0)
        item = QtWidgets.QTableWidgetItem()
        self.tableWidget.setHorizontalHeaderItem(0, item)
        item = QtWidgets.QTableWidgetItem()
        self.tableWidget.setHorizontalHeaderItem(1, item)
        item = QtWidgets.QTableWidgetItem()
        self.tableWidget.setHorizontalHeaderItem(2, item)
        item = QtWidgets.QTableWidgetItem()
        self.tableWidget.setHorizontalHeaderItem(3, item)
        self.verticalLayout.addWidget(self.tableWidget)
        self.horizontalLayout = QtWidgets.QHBoxLayout()
        self.horizontalLayout.setObjectName("horizontalLayout")
        spacerItem = QtWidgets.QSpacerItem(40, 20,
           QtWidgets.QSizePolicy.Expanding, QtWidgets.QSizePolicy.Minimum)
        self.horizontalLayout.addItem(spacerItem)
        self.pushButton = QtWidgets.QPushButton(Dialog)
        self.pushButton.setObjectName("pushButton")
        self.horizontalLayout.addWidget(self.pushButton)
        spacerItem1 = QtWidgets.QSpacerItem(40, 20,
           QtWidgets.QSizePolicy.Expanding, QtWidgets.QSizePolicy.Minimum)
        self.horizontalLayout.addItem(spacerItem1)
        self.pushButton_2 = QtWidgets.QPushButton(Dialog)
        self.pushButton_2.setObjectName("pushButton_2")
        self.horizontalLayout.addWidget(self.pushButton_2)
        spacerItem2 = QtWidgets.QSpacerItem(40, 20,
           QtWidgets.QSizePolicy.Expanding, QtWidgets.QSizePolicy.Minimum)
        self.horizontalLayout.addItem(spacerItem2)
        self.verticalLayout.addLayout(self.horizontalLayout)
        self.gridLayout.addLayout(self.verticalLayout, 0, 0, 1, 1)

        self.retranslateUi(Dialog)
        QtCore.QMetaObject.connectSlotsByName(Dialog)

    def retranslateUi(self, Dialog):
        _translate = QtCore.QCoreApplication.translate
        Dialog.setWindowTitle(_translate("Dialog", "Dialog"))
        item = self.tableWidget.horizontalHeaderItem(0)
        item.setText(_translate("Dialog", "姓名"))
        item = self.tableWidget.horizontalHeaderItem(1)
        item.setText(_translate("Dialog", "班级"))
        item = self.tableWidget.horizontalHeaderItem(2)
        item.setText(_translate("Dialog", "学号"))
        item = self.tableWidget.horizontalHeaderItem(3)
        item.setText(_translate("Dialog", "签到时间"))
        self.pushButton.setText(_translate("Dialog", "导出"))
        self.pushButton_2.setText(_translate("Dialog", "取消"))
```

(3) 编写文件 sign_indata.py，功能是导出某次打卡签到的信息，代码如下：

```
class sign_data(Ui_Dialog,QDialog):
    def __init__(self, signdata,parent=None):
        super(sign_data, self).__init__(parent)
        self.setupUi(self)#创建界面内容
        #设置窗口内容不能被修改
        self.signdata = signdata
        self.tableWidget.setEditTriggers(QAbstractItemView.NoEditTriggers)
        for i in signdata.values():
            info = i['user_info'].split('\n')
            rowcount = self.tableWidget.rowCount()
            self.tableWidget.insertRow(rowcount)
            info_name = info[0].split(': ')
            self.tableWidget.setItem(rowcount, 0, QTableWidgetItem(info_name[1]))
            info_class = info[1].split(': ')
            self.tableWidget.setItem(rowcount, 1, QTableWidgetItem(info_class[1]))
            self.tableWidget.setItem(rowcount, 2, QTableWidgetItem(i['user_id']))
            self.tableWidget.setItem(rowcount, 3, QTableWidgetItem(i['datetime']))
        #导出按钮
        self.pushButton.clicked.connect(self.save_data)
        #取消按钮
        self.pushButton_2.clicked.connect(self.close_window)
    def close_window(self):
        self.reject()

    def save_data(self):
        #打开对话框，获取要导出的数据文件名
        filename,ret = QFileDialog.getSaveFileName(self,"导出数据",".","TXT(*.txt)")
        f = open(filename,"w")
        for i in self.signdata.values():
            info = i['user_info'].split('\n')
            _,info_name = info[0].split(': ')
            _,info_class = info[1].split(': ')
            f.write(str(info_name+"  "+info_class+"  "+i['user_id']+"
                "+i['datetime'] ))
        f.close()
        self.accept()
```

## 12.5 调试运行

扫码观看视频讲解

运行文件 main.py，系统主界面的执行效果如图 12-13 所示。

选择菜单栏中的“用户组”→“添加用户组”命令，弹出“添加用户组”对话框，输入合法的用户组名字，然后单击 OK 按钮，会向系统数据库中添加这个用户组的信息，并且在百度 AI 的应用中也会同步创建这个用户组，如图 12-14 所示。

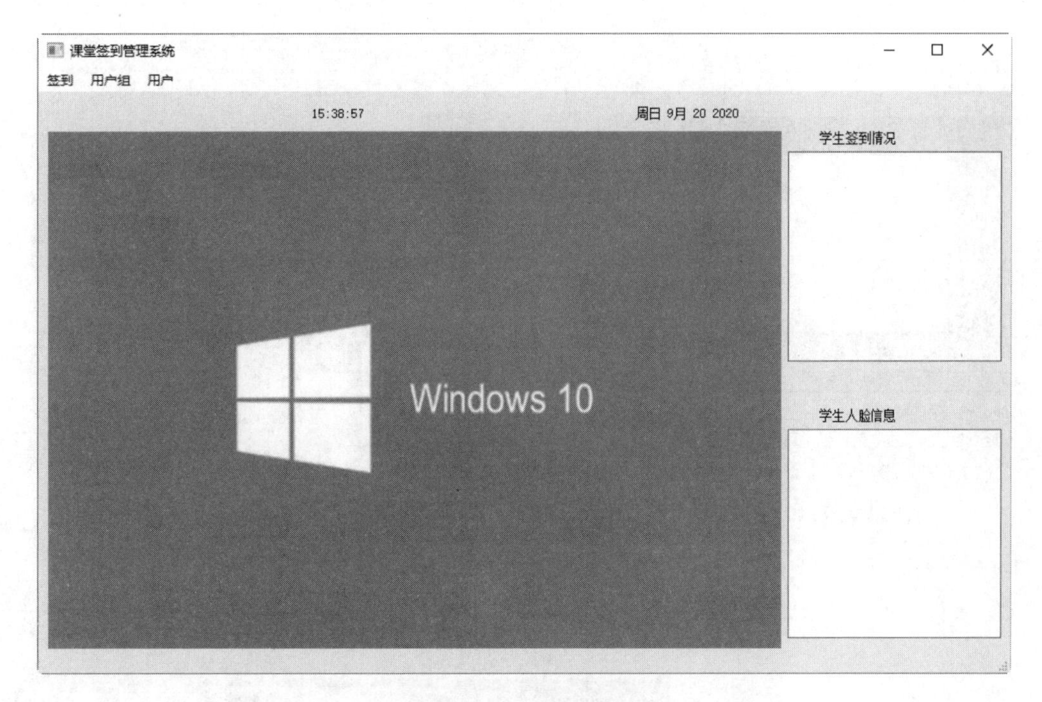


图 12-13　系统主界面

“添加用户组”对话框　　　　同步在百度 AI 应用中创建用户组

图 12-14　新建用户组操作

选择菜单栏中的“用户”→“添加用户”命令，弹出“添加用户”对话框，首先选择这个用户所属的用户组，然后输入这个用户的姓名、学号和班级信息，最后单击“确定”按钮会向系统数据库中添加这个用户的信息，并且在百度 AI 的应用中也会同步添加这个用户的信息，如图 12-15 所示。

选择菜单栏中的“签到”→“启动签到”命令，弹出“请输入所在用户组”对话框，输入用户组并单击 OK 按钮后，会打开摄像头实现人脸识别功能，如果识别成功则自动进行打卡签到，如图 12-16 所示。

选择菜单栏中的“签到”→“关闭签到”命令，在弹出的新对话框中单击“导出”按钮，会将本次打卡签到信息导出到文本文件中，如图 12-17 所示。

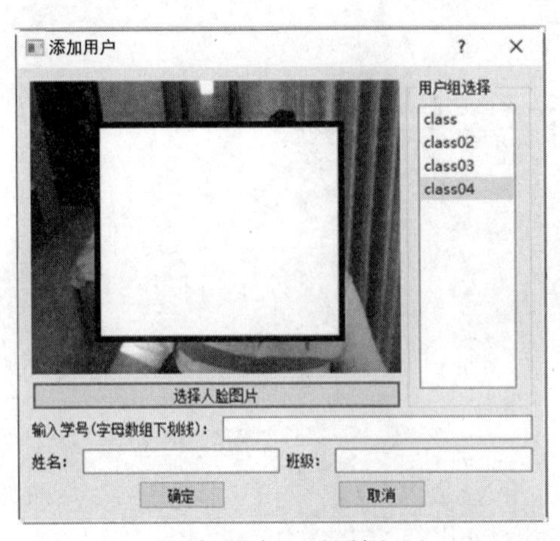


“添加用户”对话框

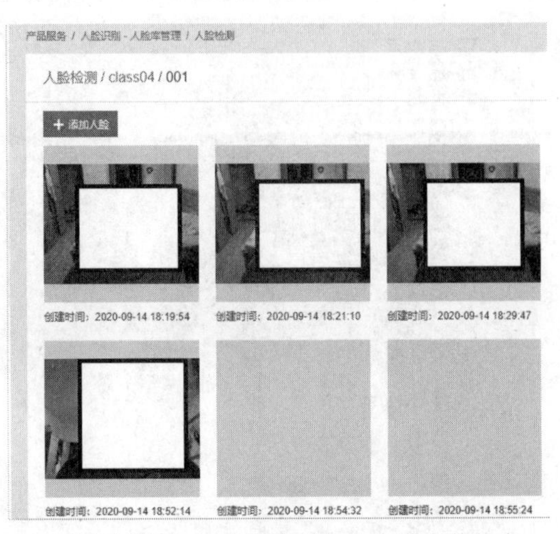


同步在百度 AI 应用中添加这个用户的信息

图 12-15 添加用户操作

请输入所在用户组

用户组信息

['class', 'class02', 'class03', 'class04']

OK Cancel

输入所在用户组

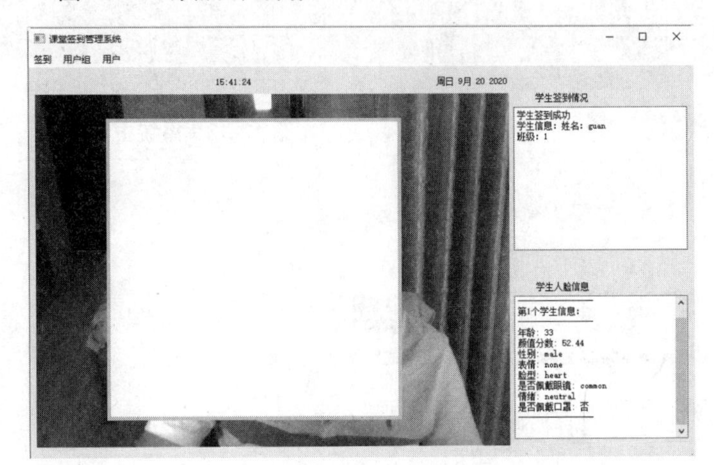


自动打卡签到

图 12-16 签到

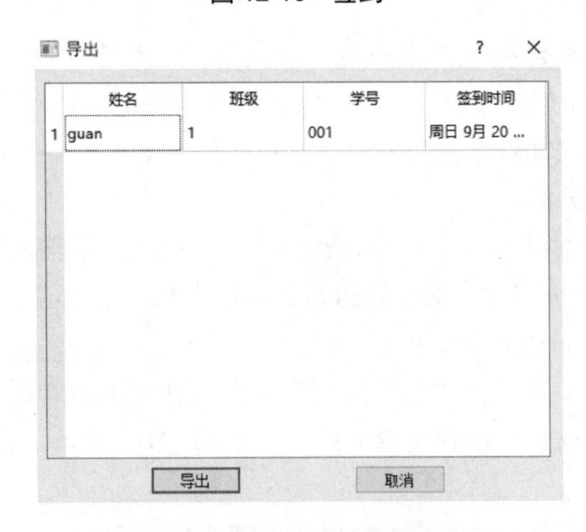


图 12-17 导出本次打卡签到信息